CONTENTS

Part I

The Evolution of Networking

Telecommunications:
A Beginner's Guide

HILL ASSOCIATES, INC.

McGraw-Hill/Osborne
New York Chicago San Francisco
Lisbon London Madrid Mexico City Milan
New Delhi San Juan Seoul Singapore Sydney Toronto

McGraw-Hill/Osborne
2600 Tenth Street
Berkeley, California 94710
U.S.A.

To arrange bulk purchase discounts for sales promotions, premiums, or fund-raisers, please contact McGraw-Hill/Osborne at the above address. For information on translations or book distributors outside the U.S.A., please see the International Contact Information page immediately following the index of this book.

Telecommunications: A Beginner's Guide

1234567890 CUS CUS 01987654321

ISBN 0-07-219356-5

Publisher
  Brandon A. Nordin
Vice President & Associate Publisher
  Scott Rogers
Editorial Director
  Tracy Dunkelberger
Acquisitions Editor
  Steve Elliot
Senior Project Editor
  Carolyn Welch
Acquisitions Coordinator
  Alex Corona
Technical Editor
  David A. Train

Copy Editor
  Lisa Theobald
Proofreader
  Stefany Otis
Indexer
  Claire Splan
Computer Designer
  Tabitha M. Cagan
  Mickey Galicia
Illustrators
  Jackie Sieben
Cover Series Design
  Amparo Del Rio
Series Design
  Peter F. Hancik

This book was composed with Corel VENTURA™ Publisher.

ABOUT THE AUTHORS

Mitchell S. Moore

Mitchell specializes in the strategic use of telecommunications technology—especially the Internet—to enhance the value proposition that companies bring to their customers. His particular interests and expertise include industry convergence and Internet-enabled collaborative communities. He is also an expert in optical networking technology. Prior to his tenure with Hill Associates, Mitchell was a Product Designer for AT&T Bell Laboratories, a Regional Sales Manager for Siemens Corporation, and is Founder and President of MSM Enterprises, Inc. He holds a B.S. in psychology from Georgetown University and an M.S. in pharmacology from Georgetown University School of Medicine. He has published articles in scientific journals and in the telecommunications trade press, and is the author of *The Telcommunication Manager's Plain English Guide to Packet Switching*. Mitchell is a member of IEEE.

N. Todd Pritsky

Todd specializes in computer networking and the Internet. Currently Director of E-Learning Courseware and a Senior Member of Technical Staff for Hill Associates, Todd has also managed a computer center, led workshops on computer applications, and written multimedia training materials. He holds a B.A. in philosophy and Russian/Soviet Studies from Colby College, and is a contributing author to the upcoming *Computer Security Handbook, 4th ed.*, soon to be published by John Wiley and Sons in 2002. Todd is a member of IEEE, ICANN, EFF, and Internet Society.

Cliff Riggs

Cliff's areas of technical expertise include TCP/IP, LANs, routing and switching, security, e-commerce, and VPNs. At the same time, he is a Cisco Certified Network Professional, Cisco Certified Design Professional, Certified Novell Administrator, Citrix Certified Administrator, Microsoft Certified Professional + Internet, and Microsoft Certified Trainer. Cliff is currently a Member of Technical Staff for Hill Associates. He holds an M.Ed. from Johnson State College and is a member of the IEEE and ICANN. Cliff has authored articles on IP addressing, BGPv4, IP Multicasting, IP QoS, and MPLS.

Peter V. Southwick

For 20 years, Peter has been in communications—designing, implementing, and training on voice and data systems. He specializes in wide area technologies, network design, Telephony, ISDN, SS7, LANs, and SNA. Before joining Hill Associates, he was a Senior Systems Engineer at GTE Government Systems. Peter is the co-author of *ISDN: Concepts, Facilities, and Services* (published by McGraw Hill), and a contributing author to *The Handbook of Local Area Networks* from CRC Press. His speaking engagements have included ComNet and ICA/SUPERCOM. Peter holds a B.S. in electrical engineering from Clarkson College. He is a member of IEEE and has Cisco Networking Certifications.

ABOUT THE TECHNICAL EDITOR

David A. Train

With 22 years of experience in network design, implementation, and instruction, Dave is recognized as a leader in technical education and training. He possesses a deep and broad knowledge of both technology trends and learning issues. Currently, Dave is CTO for Hill Associates. He holds a Ph.D. in computer science from the University of Manchester, UK. He has authored numerous articles and is the co-author of *Metropolitan Area Networks*, published by McGraw-Hill.

Part II

Network Foundations

Part III

The PSTN

Part V
The Internet

Part VI

Convergence

INTRODUCTION

Telecommunications technology—perhaps more than any other technology—continually shapes the very fabric of our global society. In 1991, how many of us watched real-time video accounts of military strikes in the Persian Gulf War? Such access to significant events halfway around the world attests to the power of telecommunications to provide instant and uncensored information to a massive, worldwide audience. More recently, the coverage of the tragic events of September 11th, as they were unfolding before our very eyes, served to galvanize the United States and its allies against an unspeakable evil. Undoubtedly, telecommunications technology will play an indispensable role in resolving the conflict and, ultimately, stabilizing this century's postwar world. Moreover, the past decade has ushered in the Internet revolution, and with it, radical and sometimes unexpected changes in the way we conduct business on a global scale.

The purpose of this book is to introduce the reader to the fascinating world of telecommunications technology. Our goal was not to delve into any one facet of this technology too deeply, but rather to use broad strokes to paint a picture of an entire industry. To that end, we have tried to cover every important aspect of the telecommunications domain. From telephone networks to the Internet, from wired networks to wireless networks, and from mature technologies to those on the very leading edge—you will find them all in this book. We have tried to describe telecommunications technology in a way that is both understandable to the novice and engaging to the previously initiated.

Hill Associates has been providing training in the telecommunications industry since 1981, specializing in service providers and equipment manufacturers. In that time, our audiences have ranged from corporate executives to field technicians, and every level in between. This has provided our staff with unique insight into the issues surrounding the selection, deployment, and sale of telecommunications networks and equipment. The wealth of experience of our technical staff—as both content developers and instructors—provides a foundation for providing a clear and in-depth presentation of complex issues for the reader. We this book will be a first step in demystifying the complex world of telecommunications.

PART I

The Evolution of Networking

CHAPTER 1

Telecommunications Basics

This chapter provides an introduction to telephony and telephone systems basics, including sound propagation, telephone system management and structure, networking, regulatory controls, associations, and a little history. First we'll look at the nature of sound and its propagation over electronic media.

SOUND PROPAGATION AND THE TELEPHONE SET

Telephony involves the transmission of sound over distances. This sound is most often voice, although it can also be music or data. The public telephone network constructed during the last century was built primarily to carry voice. We know, however, that individuals and businesses today can transmit voice, data, images, video, and other types of information over this network. Before we discuss how transmission occurs electronically, let's consider the nature of sound, human or otherwise, and how it propagates.

Sound is produced by vocal cords, musical instruments, airplanes, the wind, and millions of other sources. Sound is created by regions of high and low pressure in the surrounding air that stimulate the inner ear to generate impulses that the brain recognizes as sound. Air is the transmission medium for sound. Transmission occurs mechanically as the regions of high and low pressure rapidly move through the air away from a source, in the same way that ripples move across a pond.

As in all mechanical transmission, sound incurs losses as it moves away from the source—that is, it becomes softer. This is because its energy, which at first is concentrated, spreads out over a larger area as the pressure wave moves away from the source—again, think of ripples. It also becomes softer due to the inelastic way that air molecules collide with one another. These losses limit the distance over which intelligible speech can be sent through the air. We can, however, amplify the sound as it leaves the source, but even this limits the sound transmission to distances of 3000 feet or less. Transmitting sound over longer distances requires a mechanism other than mechanical transmission through air.

The invention of the telephone set in 1876 heralded the beginning of our ability to send voice conversations over long distances. It provided a way to convert mechanical energy to electrical energy and back again. This conversion technique meant that a signal that modeled the pressure wave of voice could be sent over copper wires and periodically amplified to overcome electrical losses. This enabled transmission across hundreds or thousands of miles, rather than just the few thousand feet allowed by pure mechanical transmission over air.

One type of telephone handset contains a microphone powered by a constant voltage from the network. This microphone is filled with carbon granules and is in series with the battery potential. Its resistance varies as the voice pressure wave alternately compresses and releases the granules. The circuit obeys Ohm's Law relating voltage, current, and resistance (voltage = current × resistance). Therefore, the voice pressure wave produces a varying current signal that models the pressure wave.

This electrical signal can be transmitted over the network to another telephone set, where it encounters the speaker in the receiver. In some of these devices, the varying current signal alters the strength of an electromagnet that sets up vibrations in a thin metal disc. These vibrations cause a varying pressure wave to occur in the air between the receiver and a person's ear. This pressure wave is then converted to sound by the ear and brain.

The telephone set thus gives us the transducer needed to convert mechanical energy to electrical energy and back again. While the signal is in an electrical form, we can transmit it over long distances.

TELEPHONE NETWORK STRUCTURE

If you wished, you could create a simple telephone network by running a line between each person's telephone and the telephone of every other subscriber to whom that person might wish to talk. However, the amount of wire required for such a network would be overwhelming. Interestingly enough, the first telephone installations followed exactly this method; with only a few telephones in existence, the number of wires were manageable.

As the telephone caught on, this approach proved to be uneconomical. Therefore, the telephone industry of today uses a *switched network*, in which a single telephone line connects each telephone to a centralized switch. This switch provides connections that are enabled only for the period during which two parties are connected. Once the conversation/transmission is concluded, the connection is broken. This switched network allows all users to share equipment, thereby reducing network costs. The amount of equipment that is shared by the users is determined by the traffic engineers and is often a cost trade-off. Indeed, a guiding principle of network design is to provide a reasonable grade of service in the most cost-effective manner. The switched network takes advantage of the fact that not everyone wants to talk at the same time.

Figure 1-1 identifies the elements of the switched network. The direct connection from each telephone to a local switch is called the *local loop* (or line) that, in the simplest case, is a pair of wires. Typically, each subscriber has a dedicated wire pair that serves as the connection to the network. In party-line service, this local loop is shared by multiple subscribers (in early rural networks, eight-party service was common).

Most telephone networks require that each switch provide connections between the lines of any two subscribers that connect to that switch. Because there is a community of interest among the customers served by a switch, most calls are just line-to-line connections within one switch. However, any two subscribers should be able to connect, and this requires connections between switches so customers served by two different switches can complete calls. These switch-to-switch connections are called *trunks*.

If 10 trunks connect offices A and B, only 10 simultaneous conversations are possible between subscribers on A talking to B. But, as soon as one call is concluded, the trunk becomes free to serve another call. Therefore, many subscribers can share trunks sequentially. Traffic engineers are responsible for calculating the proper number of trunks to provide between switches.

Local Loop

Local loop is the connection between a subscriber and a central office (CO) switch (i.e., the network). As the customer's access line to the network, the loop is typically dedicated to

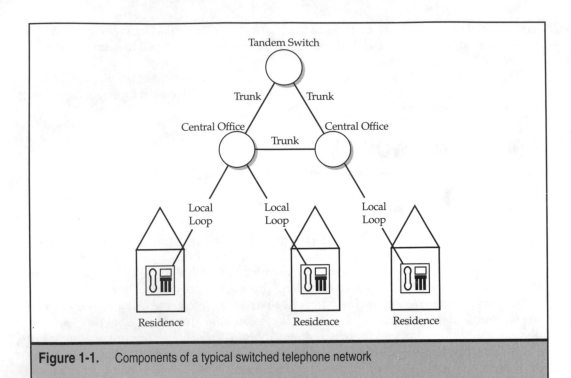

Figure 1-1.   Components of a typical switched telephone network

one subscriber (party lines are scarce). As such, when a customer wants to make a call, the local loop is always available.

The loop is typically a two-wire circuit. This permits the network service provider, or *local exchange carrier* (LEC), to provide reasonable transmission quality while keeping costs low. Over a two-wire local loop, transmissions must use the same electrical path in both directions (both wires are used for transmitting voice from the telephone to the switch and from the switch to the telephone). This is generally not a problem in voice environments, because usually only one person talks at a time (i.e., voice is *half duplex*), and anyway, the talker is seldom listening at the same time. However, data applications typically employ simultaneous two-way transmission (i.e., data is *full duplex*). This characteristic requires special techniques to make full-duplex data transmission possible over the two local loops. This issue is only one of the challenges we encounter as we try to carry data over a network that was built to carry voice traffic.

Central Office Switch

The CO switch is the core network element that establishes temporary connections between two subscribers. To accomplish this, the CO switch must terminate subscriber lines and interoffice trunks. For a subscriber to place a call and be provided indication of the progress of the call, service circuits are also terminated at the switch. These service circuits support providing tones and announcements to the customer, ringing the telephone, and collecting digits. Figure 1-2 illustrates the major components of a modern CO telephone switch.

The switching network, or switch *fabric*, is the structure that connects terminations, either lines or trunks. All lines, trunks, and service circuits must terminate on the switching network. The switching network may be one of the following types:

▼ *Electromechanical*, in which relay closures connect a separate wire transmission path between two customers

■ *All electronic*, using separate solid-state switching elements to connect terminations

▲ *Fully digital*, performing the connections by rearranging the time slots when samples of conversations are passed from one termination to another (more on digital voice encoding in Chapter 10).

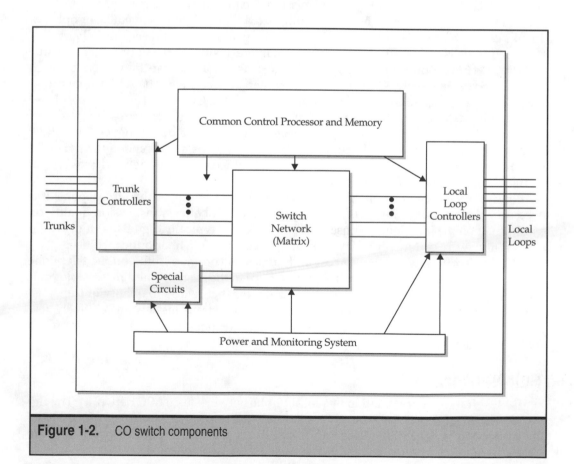

Figure 1-2. CO switch components

The type of switching network typically indicates the age of the CO switch. CO switches from the 1950s and 1960s are electromechanical, those from the '70s and '80s are electronic, and those from the '90s and beyond are digital.

The control element is necessary for controlling and monitoring call setup and calls in progress. Its function can be distributed throughout the switching elements or collected in centralized equipment. The control function can be implemented in electromechanical relay logic or via digital computer. Another function often performed by the control element is billing and usage monitoring.

The last major set of CO switch components is not actually part of the switch. The CO switch is typically found in a building called a *wire center*. This facility houses not only the telephone switch, but power and battery equipment, carrier equipment, frame terminations for private lines, and test boards as well. These other components are all mandatory for proper CO switch operation.

Trunks

As stated earlier, trunks are the transmission facilities between switches. Typically, trunks are provisioned on high-quality multiplexed digital facilities (metallic or optical). Trunks can be hundreds or thousands of miles long and are shared sequentially by customers. If there is an insufficient community of interest between a pair of switches to justify direct-trunk connections (e.g., high-usage trunks), a call can be routed through several intervening switches over trunks that interconnect them. The AT&T network of the 1970s was designed as a five-layer pyramid, with the CO switch as the lowest layer of the pyramid. If a direct trunk was not found between CO switches, the call was handed to the next layer. The next layer of switches again looked for a direct connection. If they failed to find one, they would hand the call up again. As the call ascended the pyramid, the amount of shared facilities used grew. Traffic engineering in this network was critical to minimize the cost of routing a call.

Because trunks can be quite long and are far fewer in number than local loops, it is necessary and economically justifiable to provide much better transmission quality on trunks than on the local loops. Consequently, trunks are typically digital and four wire if copper, and more commonly today, they are provisioned via fiber-optic facilities.

This trade-off between service cost and quality is common in the public telephone network. The objective is to yield affordable service to the customer and reasonable profits to the provider. We have illustrated one trade-off that relates to quality of transmission. It makes sense to spend money for high-quality transmission on relatively few trunks. On the user side, the sheer number of local loops makes such an approach cost-prohibitive.

The Full Network

Figure 1-3 shows a representation of a local public network (e.g., LEC network). The figure shows two CO switches that serve customer lines, with trunks to other offices. Note the distinction between high-usage and final-trunk groups—the difference lies in the

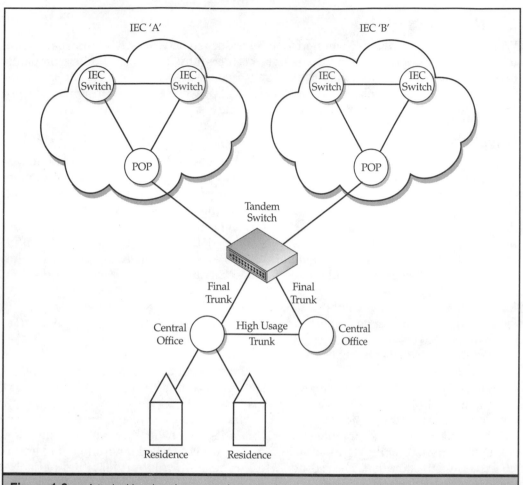

Figure 1-3.   A typical local exchange carrier network

way they are accessed. A high-usage group has insufficient trunks to handle all traffic. Calls between two switches connected by a high-usage group will first be routed to the high-usage group. If no idle trunk is found, the call is routed to a final group, which may go to another CO switch to be through switched. Switches that are positioned and provisioned as through switches are called *tandem* switches. (In old network parlance, tandem switches were class 4 switches, where CO switches were class 5 switches. These classes were relative to the pyramid structure used by the pre-1983 AT&T network.)

The tandem switch can be configured to terminate only trunks or trunks and customer lines. In the first configuration (which is most common), it performs only through switching.

Private Networks

While our discussion has been limited to public switched networks, it is important to note that private networks also exist. Some have been built by utilities and other companies that have the right of way to build their own facilities. Others may be built by leasing facilities from the telephone company. Elements of private networks are similar to those of the public network, except that the names of the components are changed.

Private Branch Exchange

A *private branch exchange* (PBX) is a telephone switch owned by a customer and located on its premises. The PBX establishes connections between its stations in the same way that a CO switch does. It also provides a connection to the public network or to a larger private network. PBXs typically employ a private numbering plan that is not part of the North American Numbering Plan (discussed further later in this chapter). To access a public number, a second dial-tone scenario is typically used. By dialing "9," for example, a customer is given access to the public network; this creates a line appearance on a CO switch where dial tone is obtained and passed to the subscriber. The subscriber then dials the public telephone number.

PBX lines and trunks can be installed and maintained by the private company or leased from a local network provider. They can also go through a wire center (i.e., nonswitched) to another PBX.

Another option exists for companies wishing for a private telephone network. Centrex is the generic name for PBX-like service furnished by a public telephone service provider, from a CO switch. This service might include a private numbering plan, class of service screening, call hold, call transfer, and other features.

Companies should consider two factors when choosing between Centrex service or a PBX: its ability to install, operate, and maintain its own switching infrastructure, and more important, its ability to monitor and maintain a controlled environment to ensure effective switch operation. The second factor can be a difficult hurdle for inexperienced communications customers.

Traditional vs. IP Telephony

So far, we have covered the components and operation of a typical circuit-switched telephone network. This section describes a typical scenario for a long-distance telephone call over this typical circuit-switched network. When you pick up your phone, you connect to your LEC's CO switch, which provides a dial tone to indicate that you are connected and that it is ready for you to enter the telephone number. Your telephone has a dedicated path to and from the LEC switch for which you pay a monthly fee.

When you make a long-distance call, the originating switch seizes a trunk to your preferred interexchange carrier and passes the number dialed to the *interexchange carrier* (IEC). The IEC routes the call to the appropriate LEC for termination (i.e., to the LEC that "owns" the dialed number). The IEC then routes the call to the appropriate destination *point of presence* (POP) and seizes a trunk to the destination LEC switch. The destination switch attempts to complete the call by ringing the telephone associated with the dialed number. When the called party picks up, the setup is complete. This call is actually a con-catenation of circuit-mode connections: calling party to LEC switch, LEC switch to IEC POP, IEC POP to IEC POP, IEC POP to LEC switch, and LEC switch to the called party. Billing begins when the called party answers (i.e., when the circuit is complete). When either party hangs up, the circuit is shut down, and all shared equipment (trunks) is returned to the pool.

An emerging trend in telecommunications is the use of packet-switching technology for transporting voice traffic. The packet-switching protocol of choice today is the Internet Protocol (IP). The advantage of packet switching over circuit switching is cost. More voice calls can share trunks in a packet mode than in circuit-switched mode. The reason for this is the statistical nature of multiplexing in a packet-switching network. As users have some-thing to say, bandwidth is allocated to these users. When a user is idle (e.g., not talking) no bandwidth is being used. Conversely, in circuit switching, the same bandwidth is allocated for the call regardless of the amount of talking going on. A simple example might help make this clear. You place a call to a friend and converse for 10 minutes. In a cir-cuit-switched environment you have used up two times the bandwidth (*from* you and *to* you) for 10 minutes for a total of 20 minutes of bandwidth. In a packet-switched environ-ment, 50 percent of the time you are listening, so no bandwidth is being used, and 30 per-cent of the time when you are talking is silence (this is a regional number; your mileage may vary). So of 20 minutes of bandwidth used for the circuit-switched example, the packet-switched examples uses 7 minutes of bandwidth. A three-to-one savings of band-width is incurred.

Here's a scenario describing an IP voice call using standard telephones. The call starts out in the same manner as the other example, with user A picking up the phone and being connected to his LEC. However, user A calls a voice over IP (VoIP) gateway, which is a local call. The gateway answers and requests a user ID, which is entered via the telephone keypad. If authorized, a second dial tone is returned and the called party's number can now be entered. Buffering the phone number, the originating gateway must now find the appropriate terminating gateway. In many circumstances, gateway operators inform each other of their presence and keep a database of gateway IP addresses associated with telephone numbers. Alternatively, the originating gateway could issue a special (multicast) IP packet to locate a terminating gateway. Once located, the originating gate-way sends its counterpart a series of IP packets to request and set up a connection at the destination. If accepted, the terminating gateway dials the called party just as if it were the originating telephone.

Once user B answers the phone, active circuit switched connections occur between user A and the originating gateway, and between the terminating gateway and user B. The digitized voice signal is sent through the network as a stream of IP packets routed to the destination gateway. The voice transmission between the line card on the LEC's switch and the respective gateway uses traditional PCM digitization at 64,000 bits per second (bps); however, the IP packet transport will likely employ one of the new low bandwidth (e.g., 8000bps) compression schemes, so the gateways will perform any neces-sary conversions.

IP Telephony and Internet Telephony Service Providers (ITSPs) is a growing portion of the telephone industry, with more mainstream and alternative providers offering IP Telephony. As a final note to this section, a number of long-distance service providers are carrying voice traffic as IP packets (VoIP). This is being transparently performed and at high quality.

NORTH AMERICAN NUMBERING PLAN

The international numbering plan defined by the International Telecommunication Union-Telecommunication Standardization Sector (ITU-T) is defined in ITU-T Recommendation E.164. It divides the world into nine geographic zones. For example, World Zone 1 includes all of North America except Mexico. Within a zone, the ITU-T assigns country codes of one, two, or three digits. The country code for the United States is 1, and the remaining digits of a telephone number are assigned by that county's numbering plan administrators and considered to be a national significant number (NSN). For the United States, the North American Numbering Plan (NANP) defines the national significant numbers. AT&T administered the NANP beginning in 1947 and Telcordia Technologies began administering the plan after divestiture. The current administrator is Lockheed Martin CIS (Communications Industry Services).

The NANP is a hierarchical scheme that uses ten digits, in the form of a NPA-NXX-XXXX. The first three digits define the numbering plan area (NPA) or area code. Prior to 1995, the form of the NPA was N 0/1 X, where N was a digit in the range 2–9 and X was a digit in the range 0–9. The middle digit was 0 or 1. This limited the number of NPAs to 8×2×10, or 160. Some codes (such as 500, 800, and 900) are reserved for special services.

The next part of the number is the CO code. This takes the form NXX, where N and X are as just defined. This allows 8×10×10, or 800, CO codes, of which only 792 are useable: the N11 codes (e.g., 411) are reserved for services such as directory assistance. Just a note: the NXX code was initially NNX, but was expanded to allow more exchanges per NPA.

The last part of the number is the line or subscriber number. It is four digits long and in the range 0000–9999. This means that 10,000 numbers are available within a CO code. Some CO switches serve 60,000 to 80,000 lines and therefore will have several assigned CO codes. Others might serve only a few hundred subscribers. In the latter case, an entire CO code is usually allocated to the switch, even though only a few line numbers will be used. Historically, it was rare to split a CO code over two switches. By avoiding such splits, routing of calls to the proper switch can be performed using only the NPA or NPA-NXX codes. Another reason for keeping a NPA-NXX to one CO switch is billing; local exchange carriers use a construct called a *rate center* to determine the billing for a call. The originating NPA-NXX and the destination NPA-NXX are entered into a table and the cost of the call is extracted from the table. If a NPA-NXX should span multiple CO switches, both of the switches must be in the exact same rate centers, otherwise billing becomes a major problem.

Explosive growths in telephones for cellular and fax use, and for other business purposes, caused the last available NPA codes to be assigned well before the year 2000, as originally anticipated. Consequently, as of January 1, 1995, the form of the NPA code changed to NXX; thus, instead of only 160 area codes, there are now 800.

Special Numbers

You are all familiar with special service codes, such as 411 for local directory assistance, 611 for repair service, and 911 as the universal emergency number. In addition to these are a number of N00 codes called *service access codes*. The most familiar of these is the 800 code used for toll-free calling, where the called party, rather than the calling party, is charged for the call. This code is an important marketing tool for businesses that allows customers to call toll-free to place orders, obtain customer assistance, and the like. Within an NPA there can be only 800 NXX codes, and within an NXX only 10,000 line numbers can exist. The explosive growth of 800 services has brought about the imminent depletion of 800 numbers. Consequently, Telcordia Technologies identified the additional codes for use with toll-free services. The 888 codes for toll-free service were introduced in March 1996, and the 877 codes were introduced in January 1997.

The introduction of Personal communications services (PCS) opens up the opportunity for people to use a single number for telephone, fax, paging, and other services. This number will follow the person everywhere, just as a cellular customer can be reached while roaming. Telcordia Technologies has already begun assigning blocks of numbers to carriers for PCS use from the 500 service access code.

Running Out of Numbers?

While it may be hard to believe, we are running out of available numbers in the NANP. The conversion to interchangeable area codes and CO codes has increased the supply of both; however, the rate of consumption of codes has increased significantly and threatens the scheme with exhaustion in the foreseeable future.

By the end of 1998, 248 area codes were in use. With that many assigned from a pool of 800, one would think they would last indefinitely. Indeed, examination of recent history shows that only 9 new area codes were assigned between the end of 1984 and 1994. However, since that time, 22 codes were assigned in 1996, another 43 in 1997, and 26 in 1998. Consider just the state of California: at the end of 1992, the state had 13 area codes; now they project that by the end of 2002, 41 will be needed. This dramatic increase in the rate of consumption sets off alarms.

The obvious question is this: "What causes such an increase in the need for telephone numbers?" Part of the answer lies in the increased use of fax machines, cellular telephones, and second lines at home for Internet access. However, these account for only part of the problem. An additional factor is the need for numbers by competitive local exchange carriers (CLECs).

Historically, we have assigned numbers in blocks of 10,000. That is, we have assigned a CO code for a new switch, and that code is capable of supporting 10,000 line numbers. The effect is considerable waste in the numbering plan. However, recall that we did not want to split a CO code over more than one switch because we use the code for routing to a particular switch. Add to that the fact that a CLEC that wishes to offer service throughout an NPA with 50 rate centers typically wants at least one CO code for each rate center, even though all those codes will not be used initially. With a maximum of 792 CO codes in an NPA, as few as 16 CLECs could exhaust the supply of codes if each was to access each rate center. When this happens, the NPA must be split or overlaid with a new area code.

Clearly, this situation forces us to seek a solution that delays the exhaustion of the supply of area codes. One such solution that is being trailed is *number pooling*. This term is applied to the assignment of numbers in blocks of 1000 rather than 10,000. This means that the same CO code could be assigned to multiple switches owned by multiple carriers. What makes this possible is the implementation of local number portability (LNP). LNP forces the implementation of an infrastructure that can look up individual line numbers and route to a specific switch based on the carrier serving that customer. We will cover LNP later in this chapter.

U.S. REGULATORY ISSUES IN TELEPHONY

In this section, details of the modified final judgment (MFJ) are discussed along with how it changed the structure of telecommunications in the United States. As a result of the MFJ, telecommunications in the United States supports a myriad of service providers and a large selection of neat acronyms to include: RBOCs, LECs, CSAs, LATAs, IECs, POPs, CAB, CAPs, CLECs, ILECs, and TA96. Each of these will be expanded and explained the in following paragraphs. Also covered in this section are the effects on the industry and consumer as a result of competition in the telecommunications arena.

The Modification of Final Judgment

The MFJ agreement between AT&T and US District Judge Harold H. Greene that took effect in 1984 transformed the telecommunications industry. It also set the stage for the Telecommunications Act of 1996 (TA96), 12 years later. TA96 completed the task begun by the MFJ.

The MFJ got its name because it modified the "final judgment" document of the 1956 consent decree, which was itself a settlement of a huge antitrust proceeding. With the MFJ, the court (specifically Judge Greene's court) became an "extra" regulator for both the new AT&T system and the local Bell Operating Companies (BOCs) that spun off (divested) from the Bell System. The BOCs became known as the local exchange carriers (LECs) and in later years the *incumbent* local exchange carriers (ILECs). The breakup of AT&T (Bell System) was called *divestiture*.

The 22 BOCs were reorganized into seven Regional Bell Operating Companies (RBOCs) of roughly equal asset and subscriber size, but of vastly different geographic scope. These RBOCs (NYNEX, Bell Atlantic, BellSouth, Ameritech, Southwestern Bell, US WEST, and Pacific Telesis) offered local service and used only AT&T Long Lines for long-distance service, unless customers specified otherwise through a series of increasingly easier procedures (called *equal access*).

AT&T retained Western Electric (WECO) and most of Bell Labs. The core long-distance business, AT&T Long Lines, was the service arm of this operation. As an interesting note, AT&T spun off the manufacturing arm of its operations (WECO and Bell Labs) into a new corporation called Lucent Technologies.

Under the MFJ, the RBOCs were required to provide equal access for their customer's long-distance service to AT&T's competitors, such as MCI, Sprint, and a whole raft of smaller companies. There were requirements both on the part of the RBOCs (called *feature groups*) and the long distance companies (called *points of presence*) to make this happen. A feature group is a signaling plan in which a subscriber designates a long-distance provider by dialing a specific sequence of digits, and the RBOC hands off the call to the specified provider. The points of presence (POPs) are the connection points between a local-exchange provider and the long-distance provider. If a long-distance provider wishes to accept service from a local-exchange provider, it must provision a POP for that LEC.

The RBOCs retained a monopoly for calls within specially established geographic zones called *local access and transport areas* (LATAs), whose size and shape was a topic of fierce debate. All calls originating in one LATA and terminating in the same LATA had to be handled by the RBOC. All calls originating in one LATA and terminating in another had to be handled by one of the long-distance providers, a.k.a. *interexchange carriers* (IECs), with a POP in the LATA. The monopoly on local service and the restriction from long-distance service are areas that have been revisited in TA96.

To distribute the wealth of universal service, the MFJ established a system of access charges whereby the LECs billed the relevant IEC for the local portion of the long-distance call. Access charges have become the center of a great debate in recent years. The issue is whether an Internet service provider (ISP) is a local service provider or a long-distance provider (the Internet, of course, spans the globe and can transport voice calls). At this point in time, ISPs are considered a local service provider and do not have to pay access charges for terminating calls.

Local Access and Transport Areas

For the purposes of the MFJ, the entire United States was divided into 245 LATAs. The LECs could provide only what were known as intra-LATA services—that is, only calls where the origination point and the termination point were both within the same LATA could be totally handled on the LEC facilities. AT&T, or one of AT&T's competitors, had to provide inter-LATA service, under equal access rules. This meant that the LEC had to allow any subscriber to access any IEC, on an equal basis, as long as the IEC had a switching office access point (a POP) in the LATA.

Each LATA was designed to have about 1 million subscribers and similar usage patterns in terms of local and long-distance calling. Independent local companies, giants like GTE and SNET, and the small providers like the Champlain Telephone Company, had some choices to make. They could establish their own LATAs in their service areas (which some large independents did), join an established LATA (which most small companies did), or do nothing at all. The LATA structure and restriction was binding only on the seven RBOCs.

Points of Presence

In 1984, the MFJ established a definite demarcation point between local services and long-distance services, although it became much more accurate to call these intra-LATA and inter-LATA. It didn't address calls that were within a LATA but not in your calling area. These calls were covered under a special category known as intra-LATA toll calls—calls that originated and terminated within a single LATA but were billed at a higher rate, called a *toll*. Today long-distance calls are often cheaper than toll calls.

The MFJ structure left customers with three basic types of telephone calls. *Local* calls were almost universally billed at flat rates regardless of connection time. *Toll* calls were billed by the LEC on a time and distance basis. And *long-distance* calls (technically inter-LATA toll calls) were billed on a separate section of the customer's bill and reflected a distance and time pricing structure.

The place where local service ended and long distance began was at the IEC's POP within the LATA. The POP was the IEC's equipment within a LATA that handled local traffic into and out of the LATA. But what would happen if I placed a call to my buddy in California using an IEC that had POPs only on the East Coast? To handle this type of situation, if an IEC did not have a POP in a particular LATA, it handed the call off to a competing IEC for termination. In this case, the billing IEC (originating end) paid out a portion of the billed amount to the other IEC (terminating end). This happened whenever an IEC with a POP in the originating LATA did not have a POP in the terminating LATA, which happened, and still happens, frequently. In fact, as recently as 1995, only AT&T had a POP in every LATA in the United States.

Under the MFJ, all local subscribers had to have "equal access" to any and all IECs maintaining POPs in the LATA. Often this meant dialing a prefix to the area code and number, but IEC preferences could be presubscribed. Therefore, by a customer expressing a preference for MCI (for example), unless special dialing rules were followed, all "normal" calls outside the LATA were carried by and billed by MCI. Several categories of special dialing rules to override the presubscription arrangements were known as *feature groups*. The ultimate feature group, Feature Group D, allowed dialing a "10288" prefix to reach AT&T (for example). The *288* (or *A-T-T*) was AT&T's carrier identification code (CIC). Today we are bombarded by commercials that extol the virtues of using the Feature Group D dialing plan, and we are often told to dial 10-10-321 to get cheap long-distance service.

NOTE: The unauthorized transfer of customers from one presubscribed IEC to another is common, annoying, and costly to users. This technique is known as *slamming* and came under the scrutiny of the FCC, but continues even today. In some jurisdictions you can put a nonslamming tag on your phone that prohibits any change to your long-distance service that is not authorized specifically by you.

Competition Comes to Local Service: Bypass

The MFJ of 1984, which opened inter-LATA toll calling to competition, preserved the local monopoly on local service for the LECs. However, even before the MFJ and divestiture, this monopoly was not absolute, secure, or universal.

Companies like MCI had been offering competitive long-distance service since the 1970s. Many organizations were attracted to these competitive IECs. Of course, to access the competitive IEC's switching office, an organization had to establish local access to it. And since the LEC retained a monopoly on local service, the LEC had to provide this access to the IEC.

In most cases, the LEC providing access was part of the "old" AT&T Bell System. Even though a company such as MCI was being used for long distance, the LECs at each end were still under the AT&T umbrella. This dissatisfied many organizations that felt that long-distance and local services were overpriced. These organizations sought not only to cut AT&T out of their long distance plans, but cut "AT&T" (the LEC) out of the local-access portion as well. Naturally, the LEC still has to be used for local calling, but for accessing IECs, why pay the LEC?

In the mid 1970s, the first viable alternative became available. This was known as *local service bypass,* or just bypass. Bypass did not require that an access link to the IEC switching office be purchased from the LEC. Several small access providers began to do business in metropolitan areas. These providers used their own fiber and wire systems within an area and allowed access to IEC facilities directly from customer locations. Since their services and pricing competed with the LECs for IEC access, they came to be called *competitive access providers* (CAPs).

Ironically, in many cases the CAPs did not own their facilities. They purchased facilities from the LECs at bulk-rate discounts and then resold them to their own customers, the organizations that had the most to gain from bypass. Some of these resellers prospered enough to build and operate their own facilities in metropolitan areas, although this did not become common until the late 1980s.

Many bypass arrangements were possible. Under the most common arrangement, instead of using a LEC access line to access an IEC POP, the organization could gain access to a CAP switching office, which then ran to the IEC POP. This cut the LEC out of the IEC access arrangement.

Competitive Access Providers

By the early 1990s, divestiture was deemed a rousing success. Competition had two lasting effects on long-distance calling. First, it became common; an awareness of alternatives to AT&T Long Lines stimulated the public to call further and for longer times than ever before. Second, the price of long distance fell, and then fell again. MCI's promise of 50 percent reductions turned out to be closer to 67 percent, an amazing fact that led some to consider that long distance had been not only overpriced, but vastly overpriced.

Years of regulation had led to a mentality of maximizing revenues, not closely watching (or even knowing) costs. When AT&T invented area codes and encouraged direct-distance dialing (DDD) in 1961, the cost of a long-distance call coast-to-coast was pegged at $12 a minute. Today's prices of 5 cents a minute and, in some advertisements, 2 cents a minute show how much it really costs to place a long-distance call.

But local rates, still regulated monopolies, rose about 13 percent nationally. This was perceived as the difference between a competitive, deregulated, long-distance environment and a monopolistic, regulated, local-service environment. Faced with this perception, the states tried to bring local rates down by encouraging some local competition. After all, it had worked nationally and brought glory to the FCC. Why not in our state?

Many states decided in the 1990s to encourage, but at the same time control, the CAPs. The fostered competition encouraged the LECs to lower their prices. The LECs had to lower rates voluntarily to compete with the CAPs, so the state regulators would not have to order rate cuts and face court challenges and bad publicity.

Most states instituted a policy of certification for CAPs and other companies (like major corporate telecommunications departments). Certification involved hearings and voluminous paperwork, and most states eventually certified one or more CAPs as competitive LECs (CLECs, pronounced either "see-lecks" or "clecks").

The former sole LEC in an area became the *incumbent* LEC, or ILEC ("eye-leck"). Because the ILECs were still regulated by the state, they stood by while new CLEC competitors wrote contracts at prices based on undercutting the ILEC's own tariff by 10 or even 20 percent! To enhance competition, the regulatory bodies required the ILECs to sell services (access lines and trunks) to the CLECs at reduced prices. The CLECs then resold these same services to the ILEC's customers at a reduced rate (lower than the tariffed rate that the ILECs could not reduce). As one colleague put it, "Imagine that you had a lemonade stand and were doing good business and making a little money. The local authorities came by and said that you had to sell your lemons, water, and sugar to a competitor at the price that you had paid for them. Then the authorities said that you had to allow the competitor to use your stand to sell lemonade. By the way, the competitor sold a glass of lemonade for less than your price, but because you were established, you had to keep your prices as is." This is the competitive environment in the local exchange environment today.

The Telecommunications Act of 1996

The Telecommunications Act of 1996 (TA96) came together quickly due to the widespread perception that local competition was not only here already, but also a good idea. Although TA96 is sometimes positioned as a replacement of the Communications Act of 1934, this is not the case. The 1934 act had far-reaching provisions untouched by TA96, which was much more focused and limited in scope. The overall goals Congress had in mind for TA96 seem clear from many public statements key players made during and after the TA96 framing process:

▼ First and foremost, TA96 eliminates the increasingly artificial distinctions between local and long-distance calling.

■ A close second is the elimination of distinctions between telcos, cable TV companies, wireless service providers, ISPs, and other enhanced service providers (ESPs).

▲ TA96 allows the LECs to offer long-distance service, and this was probably the major incentive for LEC support of TA96. But TA96 also allows IECs to offer local services as well.

At its core, TA96 is an acknowledgment of the fact of convergence. Converging of networks and services breaks down the barriers between telco and ISP, between local and long distance, and even among voice, video, and data.

TA96 Simplified

The effects and provisions of TA96 can be boiled down to a few main ideas and goals.

In the long-distance service arena, TA96 allows a LEC, specifically an ILEC, to handle long distance (inter-LATA toll calling) on their own facilities if they can prove to their respective states that their local markets are open to competition and meet a few guidelines. Although this is only one aspect of TA96, it has been seized on as the major characteristic of TA96.

As far as local service is concerned, local markets are open to "all," which translates to state-certified and qualified entrants. Access charges for calls completed by ILECs are still to be determined, with good arguments to be made for both continuing the current structure and abolishing access charges.

TA96 addresses telephone companies and other telecommunications companies. For broadcast TV, TA96 allows a single owner to control up to 35 percent of a given market. This means that if a local market is served by four broadcast stations (e.g., CBS, NBC, ABC, and Fox), the same company or individual can own more than one station, as long as the market share does not exceed 35 percent. Also, TV sets must include some means for blocking programming with sexual or violent content.

For cable TV companies, upper-tier programming services rate regulation expired in March 1999, but local basic-tier regulation is still in effect. This means that the cable TV companies will be able to raise and lower rates more or less at will. Rate regulation is immediately lifted for cable TV companies with less than 1 percent of the national market, according to TA96.

Wholesaling

One of the most controversial aspects of TA96 regards the issue of *wholesaling*. TA96 insists that service providers must allow other providers to resell their services as if they were their own. If operator services are included, a Company X LEC operator is even supposed to say "Thank you for using Company Z," which strikes some as silly.

Reselling must be done on a "nondiscriminatory" basis, which translates to any certified common carrier, not necessarily anyone who knocks on the door. A key section of TA96, Section 252, states that the same terms of interconnection be available to all carriers. Obviously, since the pricing has to be the same for all, a key factor is just what the price should be in the first place.

According to Section 251(c)(4)(A) of TA96, pricing must be done on a wholesale (discount from retail) basis. But TA96 lets the FCC figure out what this discount should be. Whatever the discount, the packaging offered to local competitors can be end-to-end or unbundled. End-to-end packaging sells the loop, the operator, the switch port, and so on, up to the door of the competitor's switch. Unbundling establishes a number of points where local competitors can access the incumbent LEC's network, usually at the local loop itself.

Theoretically, end users would be able to "mix and match" both incumbent and reseller services. Users would choose the best in each area, thus stimulating competition even more. For instance, a user/customer/subscriber could get the local loop from a reseller (obtained wholesale from the ILEC), switching services from a CLEC (perhaps an IEC), and operator services from the ILEC (who now says "Thank you for using ???").

LEC Long Distance

One of the key provisions of TA96 is the promise that the LECs can get into the lucrative long-distance service market (that is, the *inter-LATA toll market*, since LECs were always able to provide intra-LATA toll—long distance—service on their own facilities) if some basic competitive local guidelines are met. Before offering long-distance service in their markets, the LECs must demonstrate that viable local competition exists. Individual state regulators get to interpret the FCC guidelines on local competition, but there seems to be little argument over the local competition principle.

Instead, the controversy seems to revolve around the exact nature of this local competitive environment. The TA96 concept of "real competition" is tied up with the concept of offering unbundled interconnection at "reasonable rates" to resellers. Not surprisingly, there is much discussion about what rate is reasonable.

Is the leasing of unbundled ILEC facilities at rates 5 to 10 percent below retail reasonable? The ILECs certainly think so, and they have been busily securing their own arrangements along these lines with several potential competitors in many states. Or, is the leasing of unbundled ILEC facilities at rates 25 to 35 percent below retail reasonable? The CLECs, IECs, and cellular companies find this more to their liking.

On one side, the FCC fully "expects" (as they put it, hinting at further action if results along these lines are not forthcoming) the rate of 25 to 35 percent below retail to prevail. The FCC is reluctant to "approve" local competition at only 10 percent or so.

On the other side, the United States Telephone Association (USTA), a strong group of ILECs, points out that if the 10 percent rates were unreasonable, no one would ever agree to them. Yet clearly many have, and therefore, local competition exists at reasonable prices.

In the middle, state utility commissions have approved interconnection agreements allowing discounts of about 20 percent. As a result, ILECs across the country are entering the long-distance business. The benefit is not only in the additional revenue, but the business positioning of being the end-to-end provider of services regardless of the geographic distance. This, for once, allows the ILEC account representative to be on equal footing with the IEC and CAPs for national services and accounts.

TA96 Competitive Guidelines

TA96 is explicit about what constitutes local competition. These guidelines are usually called the "14 points" because of their number and the importance they have in allowing ILEC entry into the long distance market.

First and foremost is the concept of *comparably efficient interconnection* (CEI). To be used in a long-distance entry argument, an ILEC must show that interconnection of competitors is allowed on a nondiscriminatory basis (i.e., same price for any and all) and at a reasonable rate (which no one yet knows). The term *comparably efficient* means that inter-connections must be comparable for all, regardless of exact physical arrangement, and efficient in the sense that competitors cannot be given older or unreliable facilities.

Moreover, the ILEC must provide access to the unbundled local loop from switching office to customer premises. The ILEC must also allow trunk access to the local switching office from the competitor's switching office. Access to local "network elements" such as operator services must be given as well.

Competitors must be able to access local loop poles, conduits, and rights-of-way to perhaps run or repair their own cables and other equipment. Competitors can even access a local switch at the loop and trunk side with their own facilities, buying only unbundled switching from the ILEC.

The ILEC must list the competitors' customers in their own white pages directory, and in fact all subscribers will continue to receive one directory, not several. Also, access to 911, 411, and other "core" services must be available.

Since each competitor with its own switching office must have telephone numbers, the ILECs must supply these in a nondiscriminatory fashion. And because 800/888 numbers must be translated to "normal" telephone numbers at many points, and many other services require signaling system processing, nondiscriminatory access to the signaling databases must be granted.

Easy and open competitor access to telephone numbers and signaling databases may take some time. Until then, the ILECs must show some form of viable local number portability, which would allow a local subscriber to keep his current ILEC number regardless of the local service provider. After all, any service performed for the number by the ILEC can now be performed by the CLEC, even if there is no open access to numbers and signaling databases; all this will be transparent to the subscriber.

Competitors must have access to the resources needed for local dialing parity. This means that the ILECs cannot make subscribers dial 20 extra digits to reach a CLEC subscriber. If ILEC customers dial no extra digits, neither may the CLEC customers. This was a common ploy in the old MCI/AT&T long-distance days, when MCI customers had to dial seven extra digits.

There must be some reciprocal billing compensation in effect for local calls from the ILEC to the CLEC and vice versa. The mechanism is not spelled out and can be a simple "bill and keep" arrangement, which solves the problem by ignoring it (i.e., just bill your own customers and keep it all). Bill and keep would favor the CLECs, while an elaborate software/hardware billing system network fund transfer would tend to favor the ILECs.

Finally, there must be resale of local telecommunication services, not just interconnection. These must be at nondiscriminatory and reasonable rates, of course.

Impact of the Bill

TA96 has far-reaching impact on the telecommunications business in the United States. Some of the effects appear obvious; others might not become apparent for some time. What is clear is that the opening of widespread competition has resulted in the availability of a variety of new equipment and services.

Almost as soon as the bill was passed, several companies, including BellSouth, Bell Atlantic/NYNEX Mobile, and SBC Communications announced they were offering inter-LATA service to their cellular subscribers. Several mergers were also announced soon after TA96 was signed. All the merger companies were loudly criticized for trying to eliminate competitors by merging, but they were able to give regulators assurances that such was not the case. In some cases, the FCC imposed conditions on the parties that had to be met before it approved the merger.

Other impacts of TA96 include LEC attempts to offer inter-LATA toll service, the FCC's mandate of local number portability, and increasing local competition for both corporate accounts and residential subscribers.

LEC Inter-LATA Toll

Since passage of TA96, LECs have relished the prospect of getting into the $100-billion interexchange toll market in their serving areas. They have every right to expect the pickings will be good, if the experience of SNET (now part of SBC), a Connecticut company, is an example. SNET offers long-distance service and has already captured some 30 to 40 percent of the market from AT&T.

The LECs have most of the network facilities in place to offer long-distance service. However, before beginning they must satisfy three requirements under Section 271 of TA96. The first two involve demonstrating that they have opened up their local markets, by showing the presence of a facilities-based competitor in their markets and by satisfying a 14-point checklist. Then they must convince the FCC that their entry into the market is in the public interest.

Several companies have tried unsuccessfully to do this. In the aftermath of the failed petitions, recent RBOC attempts to offer long-distance service have focused on bypassing Section 271. Bell Atlantic, US WEST, and Ameritech have filed TA96 Section 706 petitions to create broadband data networks. Intended to foster such infrastructure deployment, the "Advanced Telecommunications Incentives" Section 706 allows the FCC to forbear (i.e., ignore) existing inter-LATA restrictions, as the petitioners' desire.

Another tactic to provide single-source local and long-distance communications is RBOC promotion of IEC services. US WEST (prior to their merger with Qwest) and Ameritech (prior to SBC) both also filed petitions with the FCC to review their marketing agreements to offer long-distance service provided by interstate carrier Qwest. The FCC denied these requests.

In November 1999, the FCC approved Bell Atlantic's application to offer toll service in New York only. Following Bell Atlantic's success, the quest for entry into long distance has increased pace.

Local Number Portability

The TA96 did much more than allow LECs and IECs into each others' markets. By requiring ILECs to show that local competition exists before offering long-distance services in a given market, TA96 has indirectly required local number portability (LNP), which allows local subscribers to change from an ILEC to a CLEC and keep their telephone numbers. Moreover, the FCC mandated LNP in the top 100 metropolitan statistical areas (MSA) by the end of 1998.

This has several benefits to the customer and network alike. The customer does not have to worry about dealing with more paperwork. The network directories keep the same information as before. Unfortunately, a subscriber's telephone number does more than just identify the customer; it locates the subscriber within the Public Switched Telephony Network (PSTN) for call routing purposes. The entire NANP system operates on the numbering plan area (NPA), NXX (end office code), XXXX (access line code) structure. The combination of NPA and NXX codes locates any subscriber in the country within a few miles. As such, for example, we can be certain that 914–555 is in Westchester County, New York, and 802–555 is in Vermont.

Without LNP, all the software in switching offices, ILECs, CLECs, and IECs would have to be changed to route the call properly. The FCC has specified that the implementation of LNP will be the local routing number (LRN) technique. The RBOCs favored a method called Query on Release, in which calls would be routed to the switch first, assigned an NXX code, and then a database would be interrogated for routing instructions only if the call could not be completed.

LRN, however, requires that once a switch has been LNP-enabled and at least one telephone number in an NXX has been ported to another carrier, a call to any number in that NXX causes a "database dip" to determine how the call should be routed. Figure 1-4 illustrates the method. A caller dials the directory number of the customer they wish to reach. If the network determines that the call is to an NXX that is LNP-enabled, a packet is sent to a database in the signaling network to determine proper routing. The database returns the address of the CO that now hosts the number, and a call setup sequence is initiated to complete the call. In this example, the signaling network is the Signaling System Seven (SS7) network and the database is housed in a computer system called a service control point (SCP). More information on each of these subjects is provided in Chapter 11.

Most ILECs have implemented LNP such that once an office is enabled, calls to all numbers in the office undergo the database dip regardless of whether any numbers have actually been ported. This does not slow down call processing but does put a heavy reliance on the SS7 network for all call processing.

Effect on Competition

Competition was intended as the major outcome of the TA96. Many complain this has not happened. Both the LECs and IECs stand to lose shares as their markets are opened to competition, so neither wants to act first. The ILECs know they must allow interconnection with competitors under TA96, but some are accused of allowing too few carriers into their markets. The ILECs complain that competitors want into only lucrative markets, leaving the ILECs with a dominant share in smaller, more expensive markets.

More slowly than hoped, as we write this, the local access market is opening up to more providers, allowing a choice of carriers to a growing number of customers. At the

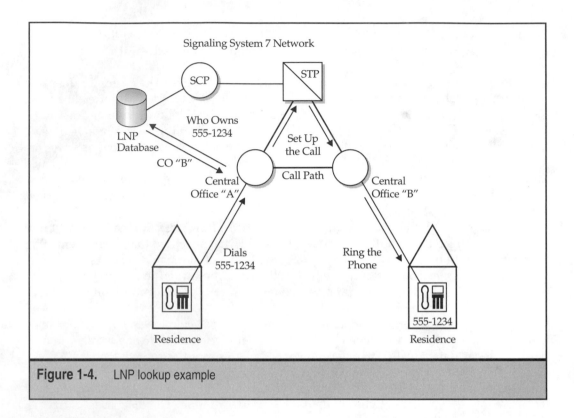

Figure 1-4. LNP lookup example

end of 1998, the FCC classified 355 carriers as CLECs or CAPs, based on revenues earned predominantly in the local access market. A more inclusive use of CLEC includes another 59 local competitors such as local resellers and private carriers.

Although ILECs still provide dial tone to nearly 99 percent of residential subscribers, competitors are making inroads to lucrative business customers. Omitting smaller carriers, CLECs large enough to contribute to the Universal Service Fund captured 8.1 percent of the private line and special access revenues in 1998. Of the 1.3 percent local exchange market held by the same reporting CLECs, about 60 percent was due to switched services provided to businesses rather than residential end users. Additionally, ILECs still held about 96 percent of the $102 billion total local service revenues, but CLECs accounted for 2.4 percent; another 1.1 percent went to other competitors, primarily IECs. This competition has resulted in lower prices. For example, the basic price of a T-1 line in Maryland has dropped from more than $1000 to less than $500 per month.

A big obstacle to local competition remains—the cost of building local facilities. On the other hand, logistic difficulties and financial disincentives encountered by competitors attempting resale strategies have encouraged large companies to pursue facilities-based approaches. Rather than build, however, the dominant IECs are entering the local market largely through acquisition.

Taken together, facilities-based CLECs were established in every state, and in 175 of 245 LATAs by the end of 1998. The local service competitors are expanding in numbers and reach, while the mergers and acquisitions continue.

Hot Issues and TA96

TA96 has changed the U.S. telecommunications industry significantly. And like all major changes, TA96 is accompanied by its fair share of controversies and hot issues that will occupy the time and challenge the wisdom of carriers, regulators, legislators, judges, companies, and even ordinary citizens for many years to come.

Some of the TA96 controversies continue to this day. Some of the current issues are discussed next.

Access Charges

The ILECs have tried unsuccessfully to have access charges levied against ISPs, arguing that ISPs are carrying telephone calls over the Internet. In addition, access charges are still one of the largest expense items for IECs and one of the largest revenue sources for ILECs. In August 1999 the Coalition for Affordable Local and Long Distance Services (Bell Atlantic, BellSouth, AT&T, Sprint, SBC, and GTE) submitted an access charge reform proposal to the FCC. The proposal calls for reductions in access charges from 1.2 cents per minute on each end of a call down to about .55 cents per minute for the RBOCs and GTE, and to .65 cents per minute for other price-cap ILECs. In exchange, the ILECs would be allowed to raise the subscriber line charge (SLC) from a current average of $5 per month to $7 per month by July 1, 2003. The proposal should be revenue neutral to the ILECs but would make the ILECs less vulnerable to loss of revenue from CLECs taking large business customers, because a larger fraction of their revenue would come from residential customers.

Internet Telephony

The increasing popularity of VoIP networks including the Internet causes an interesting dilemma for the ILECs. ILECs not only cannot assess an access charge against an ISP, but they are forced by TA96 to pay reciprocal compensation to an IEC (that is also an ISP) for local calls they deliver to the IEC. The FCC ruled in February 1999 that calls delivered to an ISP could be considered long-distance calls (because they go out over the Internet) and thus paved the way for ILECs to cease paying reciprocal compensation.

ADSL and the Local Loop

A more recent FCC ruling determined that ILECs must unbundle the high frequencies on the local loop for use by nonvoice asymmetric digital subscriber line (ADSL) providers (such as Covad).

Unbundled Network Elements

UNEs have been a source of controversy since TA96 was passed. UNEs are the network elements the ILECs must sell wholesale to a CLEC if access to the element is necessary for the CLEC to compete and failure to obtain access would impair the CLEC's ability to provide service. The Supreme Court ruled in January 1999 that the method the FCC used to construct the UNE list was inconsistent with the language in TA96. The FCC published a new list of UNEs in September 1999.

TELEPHONE ASSOCIATIONS

This section provides a list of major standards bodies that affect telecommunications, including the ITU-T, ISO, ANSI, ATIS, EIA, TIA, and IEEE. Also included is a look at the regulatory bodies that control the telecommunications industry in the United States.

Standards Bodies

A public telephone network in which equipment from various vendors can be attached requires interfaces and standards that specify rules for obtaining service. Similarly, standards are also required for data networking. The following list identifies several important standards bodies and their Internet locations.

▼ **International Telecommunication Union-Telecommunication Standardization Sector (ITU-T)**   A United Nations–sponsored agency for defining telecommunications standards. **http://www.itu.int/ITUTELECOM/index.html**

■ **International Organization for Standardization (ISO)**   An international organization that produces standards for industry and trade (e.g., film speed, screw threads, and telecommunications). **http://www.iso.ch/**

■ **American National Standards Institute (ANSI)**   Publishes standards produced by accredited committees, such as T1 for telecommunications. **http://www.ansi.org/**

■ **Alliance for Telecommunications Industry Solutions (ATIS)**   Sponsored by the telephone carriers. Sponsors the T1 committee for development of telecommunications standards. **http://www.atis.org/**

■ **Electronics Industry Association (EIA)**   Standards body representing manufacturers that produce standards such as the EIA-232 interface. **http://www.eia.org/**

■ **Institute of Electrical and Electronics Engineers (IEEE)**   Sponsors the 802 committee that has developed many of the local area networking standards. **http://www.ieee.org/**

▲ **The European Telecommunications Standards Institute (ETSI)**   Standards body that represents the telecommunications companies from 52 countries inside and outside of Europe. **http://www.etsi.org/**

Of this list, the two bodies that have the greatest impact on telephone standards in the United States are ITU-T and the ATIS T1 committees. The types of standards with which each are associated are listed here:

▼ **ITU-T standards**   Transmission performance, signaling, numbering plans, transmission and call setup delay, ISDN

▲ **ATIS T1 committee**   Performance and signal processing; interfaces, power and protection from networks; internetwork operations, administration, maintenance, and provisioning (IOAM&P); wireless/mobile services and systems; services, architectures, and signaling; Digital hierarchy and synchronization

Regulatory Bodies

The regulatory landscape in the United States is divided into multiple strata. At the national level are the regulatory bodies that control interstate communications, international communications, and internetwork issues such as numbering plans and interconnectivity. The state level deals with local connectivity, competition, and services offered by the local carriers. Local principalities deal with the local licensing of the carriers. This hodgepodge of regulators and jurisdictions create a maze for service providers and consumers alike. Which regulations apply and which regulatory body has oversight depend on the size of the carrier. Two carriers, for example Verizon and Champlain Telephone Company, requesting the same approval for the same service for similar locations might have two totally different regulatory paths. Listed here are a few of the national level regulatory bodies and the national level forums for telephone companies.

▼ **Federal Communications Commission (FCC)**   This national level commission has the overall responsibility for communications in the Unites States. It regulates radio, television, and telecommunications activities. **http://www.fcc.gov/**

■ **National Exchange Carriers Association (NECA)**   This nonprofit corporation was formed by the FCC to administer the access charge plan for telephone companies. **http://www.neca.org/**

■ **Industry Numbering Committee (INC)**   This standing committee of the Carrier Liaison Committee (CLC) provides a forum for issues related to the NANP. **http://www.atis.org/atis/inc**

▲ **United States Telecom Association (USTA)**   This industry forum represents the telephone companies in the United States. It serves some 1200 telecommunications companies across all carrier classes. **http://www.usta.org/**

CHAPTER 2

Introduction to Data Networks

This chapter sets the stage for computer networking. Computing environments have evolved from a few huge and expensive computers to our current world of compact PCs, laptops, and palm devices in use in most businesses and homes. Today's computers are most effective when they're networked; this chapter provides an introduction to how computer networks are built using the Open Systems Interconnection (OSI) Model as our guide. We examine two other models for networking: a proprietary model called Systems Network Architecture (SNA) and an open model called the Internet Protocol (IP) suite. We'll also discuss some common computer network scenarios to illustrate how computers and communications make a powerful combination.

THE EVOLUTION OF COMPUTING

It is hard to imagine life without computers. Computers are everywhere—from small microprocessors in watches, microwave ovens, cars, calculators, and PCs, to mainframes and highly specialized supercomputers. A series of hardware and software developments, such as the development of the microchip, made this revolution possible. Moreover, computers today are rarely stand-alone devices. They are connected into networks that span the globe to provide us with a wealth of information. Thus, computers and communications have become increasingly interdependent.

The nature and structure of computer networks have changed in conjunction with hardware and software technology. Computers and networks have evolved from the highly centralized mainframe systems of the 1950s and 1960s to the distributed systems of the 1990s and into the new millennium. Today's enterprise networks include a variety of computing devices, such as terminals, PCs, workstations, minicomputers, mainframes, and supercomputers. These devices can be connected via a number of networking technologies: data is transmitted over local area networks (LANs) within a small geographic area, such as within a building; metropolitan area networks (MANs) provide communication over an area roughly the size of a city; and wide area networks (WANs) connect computers throughout the world.

Mainframes

The parent of all computers is the *mainframe*. The first mainframe computers were developed in the 1940s, but they were largely confined to research and development uses. By the late 1950s, however, government and some corporate environments were using computers. These machines were huge—in size and in price. Together with connected input/output (I/O) devices, they occupied entire rooms. The systems were also highly specialized; they were designed for specific tasks, such as military applications, and required specialized environments. Not surprisingly, few organizations could afford to acquire and maintain these costly devices.

Any computer is essentially a device to accept data (i.e., *input*), process it, and return the results (i.e., *output*). The early mainframe computers in the 1950s were primarily large systems placed in a central area (the computer center), where users physically brought programs and data on punched cards or paper tapes. Devices, such as card or paper-tape readers, read jobs into the computer's memory; the central processing unit (CPU) would then process the jobs sequentially to completion. The user and computer did not interact during processing.

The systems of the 1950s were stand-alone devices—they were not connected to other mainframes. The processor communicated only with peripheral I/O devices such as card readers and printers, over short distances, and at relatively low speeds. In those days, one large computer usually performed the entire company's processing. Because of the long execution times associated with I/O-bound jobs, turnaround times were typically quite long. People often had to wait 24 hours or more for the printed results of their calculations. For example, by the time inventory data had been decremented to indicate that a refrigerator had been sold, a day or two might have passed with additional sales to further reduce inventory.

In such a world, the concept of *transaction processing*, in which transactions are executed immediately after they are received by the system, was unheard of. Instead, these early computing systems processed a collection, or *batch*, of transactions that took place over an extended time period. This gave rise to the term *batch processing*.

In batch jobs, a substantial number of records must be processed, with no particular time criticality. Several processing tasks of today still fit the batch-processing model perfectly (such as payroll and accounts payable).

Although the mainframe industry has lost market share to vendors of smaller systems, the large and expensive mainframe system, as a single component in the corporate computing structure, is still with us today and is not likely to disappear in the near future. IBM is still the leading vendor of mainframes, with its System/390 computers, and SNA is still the predominant mainframe-oriented networking architecture.

Although IBM has been developing noncentralized networking alternatives, the model for mainframe communications remains centralized, which is perfectly adequate for several business applications in which users need to access a few shared applications. In an airline reservation database application, for example, a users' primary goal is not to communicate with each other, but to get up-to-date flight information. It makes sense to maintain this application in a location that is centrally controlled and can be accessed by everybody. Moreover, this application requires a large disk storage capacity and fast processing—features a mainframe provides. Banks and retail businesses also use mainframes and centralized networking approaches for tasks such as inventory control and maintaining customer information.

Minicomputers

In the 1950s and early 1960s, individual departments within a company did not typically buy their own mainframes. Why not? Because computers were prohibitively expensive. Centralized systems led to high communications costs; however, those costs practically vanished when compared to the price of the computers themselves. Purchasing dedicated lines to connect users to a large mainframe was much cheaper than buying several computers and distributing applications over several sites.

All of this changed when Digital Equipment Corporation (DEC) introduced the first minicomputer (the PDP5) in 1963. The minicomputer, or midrange system, has the same basic components as the mainframe; however, it's smaller, less powerful, and less expensive. The construction of minicomputers was made possible by advances in integrated circuit technology—the electronic components of a computer became smaller, faster, and cheaper to produce. The minicomputer cannot handle the vast array of I/O devices associated with centralized mainframes, but it is perfect for handling the computing needs of smaller companies that cannot afford a mainframe, and it is suited for department-oriented computing.

Minicomputers started the evolution (or revolution) from centralized to *distributed* computing, in which no central authority, or computer, controls all the others. Companies such as banks and retail chains purchased multiple minicomputers and placed them in their regional offices. Each system could run applications relevant to the individual branch office. For example, the PriceMart store in Denver can maintain its own inventory database; it does not have to compete for computing time with the Dallas PriceMart store that maintains a separate inventory database. Consequently, users can expect short response times to database queries.

Typically, users access a minicomputer via terminals over low-speed links. Terminal servers might be employed to concentrate traffic collected from several terminals.

The minicomputer changed the cost structure of computing, so that the price of the computer is equal to the cost of networking. It costs less to distribute the computing power across all minicomputers than to gather it in one central mainframe.

Minicomputers see wide use today, although the industry has lost market share due to the proliferation of even smaller computer systems. Minis often coexist with mainframes and PCs in business environments and, like other types of computers, are just one component in the corporate network.

Personal Computers

In 1981, IBM released its first personal computer (PC), and since the minute the PC appeared in the corporate environment, life hasn't been the same. PCs are small, low-cost desktop systems, designed to support an individual's computing needs. Even more advances in circuit technology made it possible to shrink the circuitry more and to mass-produce circuitry *chips*. PCs are sometimes called *microcomputers* for that reason. A microprocessor, the CPU in a PC, fits on the tip of your finger.

With the PC, the power of the computer is brought directly to the desktop, and users are no longer at the mercy of the current load on the mainframe or minicomputer. The user has sole possession of the processor. The PC gives users access to a new set of applications that execute locally with no variation in the quality of execution from one day to another. Plenty of software is available for the PC—from word processing, spreadsheet, database, and graphics packages, to games and music software. PC hardware and software are inexpensive (compared to mainframe and minicomputer platforms), making it easy for individuals or small businesses to invest in the technology and take the concept of distributed computing one step further.

Client/Server Computing

The rapid proliferation of PCs in the workplace quickly exposed a number of their weaknesses. A stand-alone PC can be extremely inefficient.

Any computing device requires some form of I/O system. The basic keyboard and monitor system is dedicated to one user, as it is hardly expected that two or more users will share the same PC at the same time. The same is not true for more specialized I/O devices, with which, for example, two or three printers attached to a mainframe or minicomputer environment can be accessed by any system user who can download a file to the printer. It might mean a walk to the I/O window to pick up a printout, but a few printers can meet many users' needs.

In the stand-alone PC world, the printer is accessible only from the computer to which it is attached. Because the computer is a single-user system accessible only from the local keyboard, the printer cannot be shared, and therefore, must be purchased for each PC that needs to print; otherwise, PCs with dedicated printers must be available for anyone's use, in which case a user would take a file or data (usually on a floppy disk) to the printer station to print it. This is affectionately referred to as *sneakernet*. It doesn't take a rocket scientist to note the waste in time, resources, and flexibility of this approach.

We use printers here as just one example of the inefficiency of stand-alone PCs. Any specialized I/O device faces the same problems (i.e., plotters, scanners, and so on), along with such necessities as hard disk space (secondary storage) and even the processing power itself.

Software is another resource that cannot be shared in stand-alone systems. Separate programs must be purchased and loaded on each station. If a department or company maintains database information, the database needs to be copied to any station that needs it. This is a sure-fire formula for inconsistent information or for creating an information bottleneck at some central administrative site.

Finally, the stand-alone PC is isolated from the resources of the mainframe or minicomputer environment. Important information assets are not available, usually leading to two or more separate devices on each desk (such as a PC and a terminal into the corporate network). A vast array of computing power develops that is completely out of the control of the Information Technology (IT) group. The result can be (and often is) chaotic.

It rapidly became evident that a scheme was necessary to provide some level of interconnection. Enter the local area network (LAN). The LAN became the medium to connect these PCs to shared resources. However, the simple connection of resources and computers was not all that was required. Sharing these resources effectively requires a *server*. As an example of server function, consider again the problem of sharing a printer among a collection of end users. A printer is inherently a *serial* device (it prints one thing at a time). A printer cannot print a few characters submitted by user A, then a few from user B, and so on; it must print user A's complete document before printing user B's job. Simply connecting a printer to the network will not accomplish the serialization of printing, since users A and B are not synchronized with respect to job submission. A simple solution to this problem is to attach the printer to the network via a specialized processor, called a *printer server*. This processor accumulates inputs from users A and B, assembles each collection of inputs into a print job, and then sends the jobs to the printer in serial fashion. The printer server can also perform such tasks as initializing the printer and downloading fonts. The server must have substantial memory capability to assemble the various jobs, and it must contain the logic required to build a number of print queues (to prioritize the stream of printer jobs).

A second example of a server's function involves a shared database connected to the network. In most systems, different users have different privileges associated with database access. Some might not be allowed access to certain files, others might be allowed to access these files to read information but not write to the files, while still others might have full read/write access. When multiple users can update files, a gate-keeping task must be performed, so that when user A has accessed a given file, user B is denied access to the file until user A is finished. Otherwise, user B could update a file at the central location while user A is still working on it, causing file overwrites. Some authority must perform *file locking* to assure that databases are correctly updated. In sophisticated systems, locking could be performed on a record (rather than a file) basis—user B can access any record that user A has not downloaded, but B cannot obtain access to a record currently being updated.

The job of the file (or database) server is to enforce security measures and guarantee consistency of the shared database. The file server must have substantial resources to store all the requisite databases and enough processing power to respond quickly to the many requests submitted via the network.

Many other server types are available. For example, a *communications* server might manage requests for access to remote resources (offsite resources that are not maintained locally). This server would allow all requests for remote communication to be funneled through a common processor, and it would provide an attachment point for a link to a WAN. *Application* servers might perform specialized computational tasks (graphics is a good example), minimizing the requirement for sophisticated hardware deployed at every network location.

Servers are sometimes simply PCs, but they are often specialized high-speed machines called *workstations*. In some environments, the servers might even be minicomputers or mainframes. Those computers that do not provide a server function are typically called *clients*, and most PC networks are client/server oriented.

THE OSI-ISO REFERENCE MODEL

OSI stands for *Open Systems Interconnection*. OSI is a concept for applying open standards to ensure that networking compatibility exists between different systems.

Let's discuss what is meant by the term "open systems." Companies sometimes build "closed" systems that are proprietary and vendor-specific. When several closed systems need to be tied together, a popular approach is to use proprietary solutions. This works, but it severely limits options for end users. Proprietary solutions tie customers to a specific manufacturer's application and/or hardware. User dissatisfaction with these proprietary solutions prompted the need for an "open" approach to interconnection.

The OSI Reference Model (the Model), created by the International Organization for Standardization (ISO), is the defining document for the concept of developing open standards for interconnection. It provides a reference for network and protocol designers to use when designing open systems and allows the exchange of information among systems that meet all standards adopted for interconnection.

Models are formulated as a step to developing standards. A *model* is a broad description of the components of some system. If there is prevailing agreement about the structure of the model, it is then possible to take the next step—defining a set of standards (a *specification*) defining the nature of the components of the system.

Consider the task of providing an international set of standards for building a house. We can agree that a house has such features as a roof, to keep out precipitation and sunlight; walls, to provide a barrier to the exchange of air with the outside; doors, to provide a means of entrance and egress; and so on. Our model describes the component in a general way and specifies its function. It does not specify that doors must be steel, of a certain width and height, four panel, and so on. This type of detail is provided by standards, which follow agreements on the components of the model. If there is agreement on the model, standards can be developed. If there is no agreement on the model, standards are an impossibility. For example, if the Hawaiian representative insists that walls are not necessary, a house standard is impossible.

For any communication to take place, a sequence of agreements must facilitate or allow the communication process to proceed. Such agreements are called *protocols*. The study of communication in general is the study of protocols. All communication requires protocols, not just communication associated with computer networks. For example, for two people to communicate successfully, they must agree on a language and an alternation scheme, or protocol.

Layers in the Model

The Model was initially developed by the ISO in 1978. Today ISO and the International Telecommunication Union-Telecommunication Standardization Sector (ITU-T) perform major development of the Model and OSI protocols cooperatively. The goal of the Model is to allow the open interconnection of computers, while requiring a minimum of technical agreement outside of the interconnection itself. This includes connecting computers built by different manufacturers, owned by different organizations, having different levels of complexity, and/or using different technologies. The Model consists of seven layers (described in the following sections), in which services and interfaces are specified for each layer. OSI standards specify the rules, but not how the rules must be implemented.

The main principles guiding the layering of the Model are listed here:

▼ *An appropriate amount of modularization*   There should not be so many layers as to make network design and implementation overly complex, nor so few layers as to make the operation of each layer overly complex.

■ *Transparency*   The operation of one layer should remain transparent to other layers.

- ■ *A minimal amount of information carried across the layer-to-layer interfaces*
  To preserve transparency.

- ■ *A well-defined task or set of tasks*   Each layer should perform these tasks.

- ▲ *A new layer should be created whenever a different level of abstraction is
  required*   For example, if one function acts on individual bits and another
  function acts on blocks of bits, these functions should be in different layers.
  Layering is a mechanism used for breaking a large task into smaller, more
  manageable pieces.

The total set of protocols required for complete interoperability of two applications is
enormous. Consider the problem of exchanging a document between two different word
processing systems. We must not only successfully transmit the bits from processor to
processor, but we must also take steps to ensure that the transmitted information is in a
form understandable by both word processing systems. Rather than attempting to study
all the necessary protocols as a single group, we can break the total set of protocols into
subsets, each addressing a specific set of closely related issues.

A set of agreements that addresses a specific class of communication issues is called a
protocol layer. Generally, a protocol layer is responsible for a limited set of services. A protocol layer expects certain services from the layers beneath it and provides an enhanced set of
services to the layers above. For example, the protocol layer responsible for assuring that
the format of word processing documents is interpretable by both systems depends on the
services of other protocol layers to ensure that the bits are transmitted successfully.

Protocol layers expect *services* from other layers and are not concerned with the
specific method by which those services are provided. Continuing with our example,
the protocol layer responsible for formatting word processing documents is oblivious
to the specific method by which the bits are transmitted, as long as they are successfully
transmitted. Figure 2-1 illustrates the relationship between protocols and services.

In the OSI terminology, peer layers communicate via protocols. That is, layer N in one
processor communicates to layer N in another processor using OSI protocols. On the
other hand, layer N in a processor communicates with layer N ± 1 in the same processor
using an *interface*. Since interfaces might depend on the architecture of the processor, they
are not part of the OSI standard.

Consider a communication between two executives located at different sites. The
transmitting executive expects a set of services from his subordinates. In particular,
the executive understands that if he/she provides the name of a recipient of a document
to the administrative assistant layer (the next lower layer), the document in question will
be safely delivered to the named recipient. The administrative assistant layer understands that if he/she places the document in an envelope, appends a full address to the
envelope, and clips on a sum of money, the next lower layer (the courier layer) will provide the delivery service. The courier layer delivers the envelope in a two-day airfreight
package to the next lower layer. Finally, some layer is responsible for the physical delivery of the document to the receiving executive.

In a perfect system, the sending executive has no concept of the nature of the delivery
system. The executive expects certain services from the lower layers but does not care how

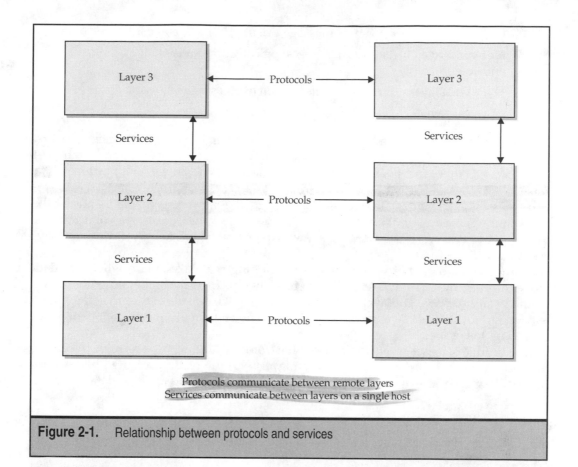

Figure 2-1. Relationship between protocols and services

they are provided. This is a critical issue that suggests that in a correctly layered system, a layer can change the mechanism by which it provides a service without affecting any other layer. So, for example, if all data communications systems used fully layered protocol structures, we could change from copper to fiber facilities with absolutely no impact on any software or hardware at higher layers of the protocol stack. Or the courier could ride a bicycle instead of driving a truck, and still the package would arrive intact and on time.

Following are the main functions of the seven layers of the OSI Model. By stringing together the services of the layers, we can provide interoperability, assuming the various protocol implementations adhere to standards for the specific layer.

▼ **Application layer**   Data transformations specific to a given application

■ **Presentation layer**   Data transformations of a general nature

■ **Session layer**   Adds user friendliness to the transport function; dialog control during session

- **Transport layer**   Error-free, message-oriented, end-to-end communication
- **Network layer**   Routing and congestion control; provision of a "raw" end-to-end path
- **Data Link layer**   Error-free transmission over a single data link
- ▲ **Physical layer**   Actual physical transmission of data bits

The layers are divided into two groups: the *end-to-end* layers and the *chained* layers. The end-to-end layers consist of the Application, Presentation, Session, and Transport layers. We call them end-to-end because they typically reside only in end machines. The lower three layers (Network, Data Link, and Physical) are the chained layers, which typically appear in all end machines (hosts) and intermediary switches (nodes). Thus, as the name implies, they chain the end systems and nodes together, creating a connection or path between end systems. Figure 2-2 shows a depiction of end-to-end and chained layers at work.

Figure 2-3 shows the concept of OSI enveloping that takes place whenever data moves through the OSI stack. As each layer receives data from the higher layer, the data is placed in an envelope. The layer adds some writing to the outside of the envelope, which is intended for its peer layer in the next device. The envelope is passed unopened and unread by lower layers.

In Figure 2-3, a user wants to transfer a file from the sending process to the receiving process. The user supplies the Application layer with the file, represented by the data block at the top of the figure. The Application layer protocol accepts the data and adds

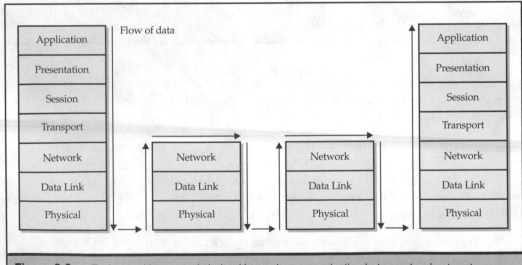

Figure 2-2.   End-to-end layers and chained layers in communication between two host systems

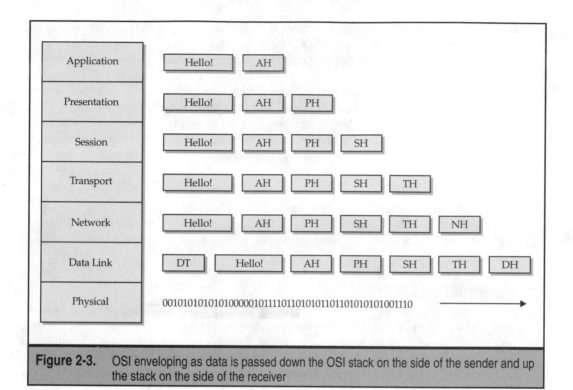

Application	Hello!	AH					
Presentation	Hello!	AH	PH				
Session	Hello!	AH	PH	SH			
Transport	Hello!	AH	PH	SH	TH		
Network	Hello!	AH	PH	SH	TH	NH	
Data Link	DT	Hello!	AH	PH	SH	TH	DH
Physical	0010101010101000001011110110101011011010101001110 →						

Figure 2-3.   OSI enveloping as data is passed down the OSI stack on the side of the sender and up the stack on the side of the receiver

some information of its own. This protocol-specific information is placed in the Application layer header (AH) and is intended for use only by the receiving Application layer protocol. The header and data together form the Application protocol data unit (APDU).

The APDU is passed down to the Presentation layer, which treats the entire APDU as data. The Presentation layer adds a Presentation layer header (PH) to form a Presentation PDU (PPDU). The PPDU is then passed down to the Session layer and the pattern continues. The rest of the diagram is straightforward, except for a couple of points. First, the Data Link layer typically adds both a header (DH) and trailer (DT). This is because error detection and other information is placed at the end of the transmission.

This diagram assumes the original data did not have to be segmented for transport across the subnet. If the file is too large for transmission through the subnet, it must be segmented into packets. Then, the Session protocol data unit becomes several Transport protocol data units, and each TPDU becomes a Network protocol data unit (or *packet*). Each protocol layer at the receiving host strips off the appropriate header (and trailer) information, just as layers in the sending host added that information.

To further explain the function of the OSI Model with regard to data transmission, let's examine the process of a fictional data transfer between two companies. Big Cars Inc.

and Round Tires Corp. have agreed to enter into a partnership where Big Cars Inc. uses Round Tires Corp. as the primary supplier of tires for the vehicles that they manufacture. To streamline the order process, the two companies agree to share purchasing information by linking their ordering systems.

Prior to the creation of this *extranet*, the companies had each used proprietary order-tracking software. To change either version of software would be expensive and would likely introduce bugs into the ordering process. Instead of changing their internal order processing procedures, Big Cars Inc. and Round Tires Corp. both agree on a common format to be used when sending the information from one network to another.

The Application Layer

With this agreement in place, when an order is placed from a computer at Big Cars Inc. for a new batch of tires, the order is passed to the Application layer, which uses the Big Cars Inc. order format. The Application layer then passes the information to the Presentation layer.

The Presentation Layer

The Presentation layer is responsible for generalized data transformation. This means accepting the information from the Application layer interface and reformatting it in the common format agreed upon by both companies. Because sensitive information such as costs and inventory information is being passed between the two companies, the information contained in the order is also encrypted at the Presentation layer.

The Session Layer

After processing by the Presentation layer, the encrypted order is passed through to the Session layer that initiates a request from the order processing software at Big Cars Inc. to the order processing software at Round Tires Corp. The Session layer will take responsibility for the initial establishment of the connection, monitor its progress, and finally close the connection after the order has been completed. From the point of view of the Session layer, the connection that it is initiating is established directly to the Session layer on the remote end. However, for this to be physically possible, the Session layer must pass down a connection request to the Transport layer.

The Transport Layer

The Transport layer is responsible for end-to-end correctness of data transfers. In the message from Big Cars Inc., it is likely that the order's data is larger than the underlying layers can transmit at one time. As such, the order placed by Big Cars Inc. must be broken into smaller sections. Because of this segmentation, it is possible that the smaller sections could be reordered as they pass through the network on their way to Round Tires Corp. To allow these segments to be reassembled in the correct order, the Transport layer will number each segment. Each message section can then be reassembled by the Transport layer on the computer at Round Tires Corp.

This numbering of segments allows the receiving Transport layer at Round Tires Corp. not only to reassemble the segments but to also communicate back to Big Cars Inc. that "I have received segment numbers 1, 2, and 4 from you. Could you please resend 3?" In this way, the Transport layer can reassemble the message and pass it up intact and correct to the Session and higher layers.

Once the Transport layer has segmented and numbered each section of the Big Cars Inc. order by adding this identifying data to the original data passed to it by the Session layer, the information is ready to be passed to the Network layer.

The Network Layer

While Application, Presentation, Session, and Transport layers are generally found on each end system, Network, Data Link, and Physical layers are found along every interconnection device between the communicating hosts.

These layers are generic in the sense that they can support virtually any type of data passed to them by a higher layer. Indeed, this is often a necessity, since modern networks support many different types of data at once. These three layers generally do not care what the data they are carrying is. They are simply occupied with the matter of getting that data from one point to another.

The Network layer is responsible for routing and addressing functions. Unlike Data Link layer addresses, which are described next, Network layer addresses are responsible for getting data from one end of the network to another over many possible Data Link and Physical layers. Thus, an addressing scheme that is independent of any hardware or Physical layer identification is desired. Routing is the process of a device reading a Network layer address and deciding which way to send it from one piece of hardware to another.

In the case of the Big Cars Inc. order, the Network layer receives the segmented and numbered pieces of the original order from the Transport layer and determines that the ultimate destination of this sequence of datagrams is located on a remote machine not on the same physical section of the network as itself. To fulfill the request of the Transport layer that these segments be sent remotely, the Network layer adds another header to the extra information (sequence numbers) added by the Transport layer. The information on the front of the Transport layer data gives each datagram the information it needs to be routed through multiple different networks.

With this addressing information added to the Big Cars Inc. order, the order is ready to be transferred to the Data Link layer for insertion to the local network segment.

The Data Link Layer

Most communication networks in the world are composed of a wide variety of network types. One part of the network may run over shared copper cables, like the cable that connects to your television. Another part of the network may use heavy-duty phone wire, while other sections are using fiber optics or parts of the radio spectrum to transmit data. Furthermore, because of variation in standards and disagreement as to what is the best way, each physical transmission medium has a different and often incompatible definition of how data should be formatted. While this disorganization may seem like a detriment to the efficiency of communication networks, the ability of the Data Link layer to adapt information to the local link on a case-by-case basis enables the upper layers to have a global reach and encourages innovation regarding data transmission.

Each of these network types may require its own modifications to the data passed from the upper layers to work properly. To accommodate each section of the network, every computer attached to the network makes use of the Data Link layer to format the information in a manner appropriate for each link.

In our example, the Data Link layer will add another header to the passed-down data, with information specific to each link of the network. This information is added to the header of each datagram of the Big Cars Inc. order passed to the Data Link layer. This prepares the data for transmission on the Physical layer.

The Physical Layer

At the bottom of the OSI Model, the Physical layer is responsible for finally placing the bits—1s and 0s—onto the physical medium. How this occurs on copper, fiber, and wireless technologies varies.

For the data from Big Cars Inc. that has passed through the layers from entering the order, to formatting, to breaking up the message by the Transport layer, addressing by the Network layer, and formatting the local link by the Data Link layer, the Physical layer receives the bits directly from the Data Link layer and inserts the data into the communication medium and onto the network.

Since Big Cars Inc. and Round Tires Corp. are separated by a great deal of physical distance, it is likely that more than one Physical layer exists for the data being sent from Big Cars Inc. At the termination of each Physical layer is some sort of intermediate system (in modern networks, this device is often known as a *router*) that contains another Physical layer to accept the bits from the physical medium. These bits are presented to the Data Link layer on the intermediate device, which checks the frame to ensure that nothing was corrupted as the data passed from one system to the next. If the check is successful, the intermediate node removes the Data Link layer information and passes up the datagram with the Network layer address to the Network layer for its inspection.

When presented with the datagram from Big Cars Inc. to Round Tires Corp., the intermediate system examines the Network layer address of the datagram and based on this information makes a decision on where to send the datagram next. If each packet were received on an interface called "interface 1" and an examination of the Network layer address reveals that Round Tires Corp. is found in the direction of "interface 2," the intermediate system routes the packet to interface 2 and passes it down to the Data Link layer for reinsertion on a different physical medium.

This process will occur as many times as needed until the stream of datagrams from Big Cars Inc. reaches its ultimate destination of the order processing host at Round Tires Corp.

From there, the data will be received by the remote Physical layer interface, assembled and error corrected by the Data Link layer and passed up to the Network layer, which will remove its addressing header. The Transport layer will then receive the packet and place it in the correct sequence with the other packets received. The data will then be passed to the Session layer and then the Presentation layer for decryption. Finally, the complete order information will be presented to the Application layer for interpretation as an order for new tires!

The OSI Today

In reality, the OSI Model isn't widely deployed as a network architecture foundation. So why use the Model?

Examining the Model is an excellent way to learn about network architectures, without being product-specific. The basics of networking could just as easily be taught using TCP/IP or IBM's SNA, but this would certainly slant learning toward a particular methodology. It would also add to the complexity of the materials. Proprietary architectures, while efficient, are often much more complicated than the OSI Model. The bottom line: While the OSI Model is not as popular as its creators envisioned, it is an excellent model for teaching network architecture basics. It also provides a common vocabulary for discourse. Since the industry is familiar with OSI, there is no confusion if someone observes, for example, that a specific failure is "a Transport layer problem." It is perhaps in this sense of common language that the Model has its greatest impact.

OTHER MODELS

Long before there was an OSI Model or a major industry focus on interoperability, there were proprietary network architectures. These architectures deviate from the OSI Model to a greater or lesser extent in the functional layering they employ.

In the final analysis, there is not much argument with the overall set of functions that must be performed to achieve interoperability. Most proprietary networks deviate from the Model only in the assignment of functions to layers. For example, there is no particular harm in performing encryption at the Transport layer. It works perfectly well; it just doesn't happen to agree with the OSI Model's assignment of the task.

Systems Network Architecture

For more than three decades, SNA has been the designated communications approach from IBM. SNA was introduced as the communications architecture for the System 370 family of computing products. It provides a master plan for the interconnection of remote users and distributed computing systems in a private network environment.

SNA defines the logical organization of the network. IBM's System 370 hardware and software products are designed in accordance with that logical structure to interoperate within an SNA environment. In certain applications, such as Online Transaction Processing (OLTP) for medium to large business customers, SNA has become the most widely used data network architecture in history. In selected large markets, IBM controls up to a 90-percent share of the total market for computing and data communications equipment.

The earliest SNA networks were single-host environments. Applications ran on System 370 host computers. An SNA program product, Virtual Telecommunications Access

Method (VTAM), ran on System 370 machines and mediated the transfer of data between applications and a high-speed parallel communications bus. The Network Control Program that ran on IBM communications controllers, or front-end processors, took data from VTAM and transferred it to devices over low-speed communications links using the Synchronous Data Link Control (SDLC) protocol. The beta test product environment for SNA was the 3600 Finance Communication System. In this system, cluster controllers connected automatic teller machines to host application programs using SNA software and protocols.

The mainframe-centered, hierarchical nature of classical SNA is not suited to many modern computing environments. In such environments, distributed intelligence takes the form of minicomputers and PCs. Furthermore, in these environments there is a requirement for users to interact with each other, not just with a host application program (such as e-mail). The transmission characteristics of these distributed environments (file transfer) are not well suited to the hierarchical terminal-to-host SNA architecture. Finally, hierarchical SNA would require that a mainframe computer be involved in PC-to-PC transfers, a cumbersome solution at best.

The diagram in Figure 2-4 compares the OSI Model with the layered protocol stack implemented by SNA. Working from bottom to top in the figure, in the lowest two layers (Physical and Data Link), SNA corresponds to the OSI Model faithfully. Beyond that, the

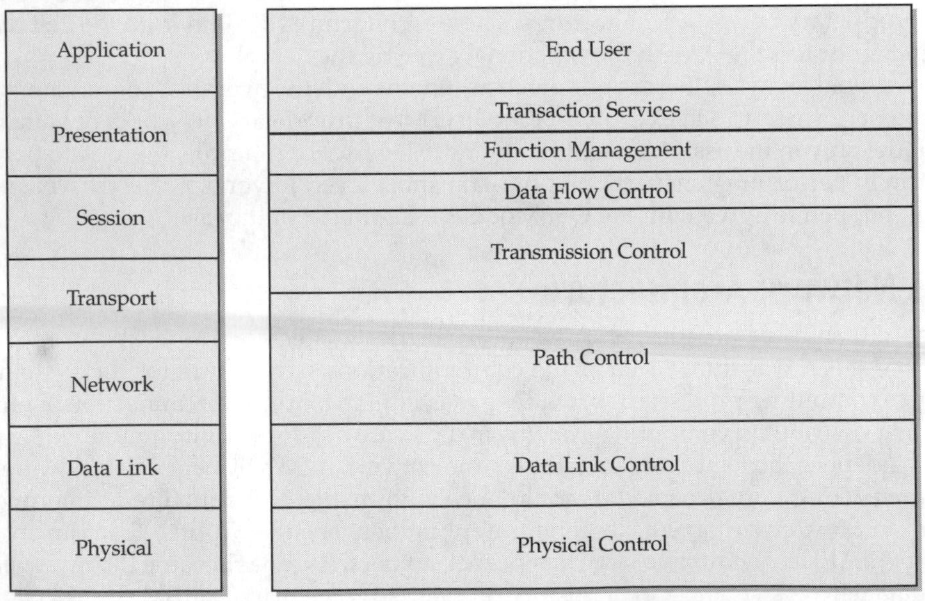

OSI Model	SNA
Application	End User
Presentation	Transaction Services
	Function Management
Session	Data Flow Control
	Transmission Control
Transport	
Network	Path Control
Data Link	Data Link Control
Physical	Physical Control

Figure 2-4.   OSI and IBM SNA model comparison

relationship is fuzzy, at best. While it is not of any practical significance to compare the two systems (since they are very different), it is illustrative as a learning tool to make some broad associations.

The Path Control layer of SNA implements functions common to the Network and Transport layers of OSI. In addition to routing and congestion control (OSI Network layer tasks), Path Control is responsible for establishment of end-to-end (i.e., network attachment unit to network attachment unit (NAU-to-NAU)) network paths. Such end-to-end path establishment is thought of as a Transport layer function in OSI terms.

The OSI Session layer is roughly divided into two layers in the SNA architecture. Transmission Control is responsible for the establishment, management, and termination of NAU-to-NAU sessions. Data Flow Control is responsible for the control of dialog between two NAUs during the session.

The OSI Presentation layer corresponds roughly to the Function Management layer in SNA. Not unexpectedly, Function Management performs various data manipulations for SNA end users. The exact nature of those manipulations will depend on the specific end user in question.

When devices (keyboard/displays or keyboard/printers) are used to access host applications, SNA is a six-layer architecture; end users interface directly to the Function Management layer. When advanced program-to-program communication (APPC) is involved, the SNA architecture implements a set of Transaction Services that correspond to the OSI Application layer. These Transaction Services include program-to-program services such as e-mail and remote file and database access and manipulation.

SNA was developed in the time of proprietary networking, so it was optimized for communication between IBM mainframes. Today the world is very different, and so is IBM's approach to networking. While SNA may still be used by the application, in all likelihood communication between computers is likely to use TCP/IP. Many applications we accessed via the Internet are SNA applications sitting on IBM mainframes. All vestiges of SNA are hidden from the user, who sees only a standard Internet interface.

Internet Protocol Suite

In the late 1980s, many large corporations were thrust into a multivendor computing and communications environment. Such an environment was created by workgroups unilaterally purchasing LANs and associated applications, and by mergers and acquisitions. As users began to need access to the resources of other workgroups, the requirement to interconnect LANs on a company-wide basis emerged.

None of the major computer vendors (IBM and DEC) offered a model for communications that supported connectivity and interoperability among many different vendors' equipment. Corporate IS managers began to search for a model for multivendor interconnection and soon discovered the Internet model and its foundation—the TCP/IP suite. They began building structures based on the Internet model using the protocols developed within that environment. The structures currently emerging in corporations across the United States have been termed *enterprise* networks.

TCP/IP enjoys a significant share of the internetworking market. TCP/IP is a mature technology, having been conceived in the early 1970s. With the standardization by the

Defense Advanced Research Projects Agency (DARPA) and inclusion in the 4.2 BSD UNIX operating system, it has become the de facto standard for most heterogeneous network applications.

Many vendors offer TCP/IP as an optional addition to their own native protocol stack. For example, Digital's proprietary network architecture is the Digital Network Architecture (DNA), but it also supports TCP/IP and OSI protocols. Similarly, Novell supports TCP/IP in addition to its native NetWare protocols.

If the Internet is the network that links us all via TCP/IP, what should we call those *private* networks interconnected by TCP/IP? Experts use the term *intranet* to refer to those portions of corporate and private networks that run TCP/IP. Intranets are created because many companies are concerned about security issues in Internet connectivity. An intranet is a "private internet," using TCP/IP for communications among many different types of systems for access to information servers, but separate and distinct from the Internet. Similarly, an *extranet* is a virtual, extended internetwork of TCP/IP-based networks (usually involving the Internet to some degree), which connects a company with its partners, associates, and most importantly, customers.

On the list of WAN/MAN technologies used in constructing corporate intranets, we could include the Internet itself. Notwithstanding the foregoing observations about security, a number of large ISPs are offering virtual private network (VPN) support on the Internet. These offerings provide secure connectivity among corporate sites and might also feature service-level agreements guaranteeing throughput and packet loss levels. VPNs might one day provide the bulk of corporate communications.

The Internet protocol suite is composed of four layers. These correspond approximately to the OSI Model, as depicted in Figure 2-5, although there is not an exact match between OSI and TCP/IP.

The bottommost layer of the Internet protocol suite (TCP/IP) is the Network Interface layer. It defines the software and hardware used to carry TCP/IP traffic across a LAN or WAN. Most, if not all, network interfaces are supported, including ATM, Ethernet, Frame Relay, Point-to-Point Protocol (PPP), Token Ring, and X.25.

The Internet layer, which employs the protocol IP, provides a connectionless OSI Network layer service. As such, IP does not sequence or acknowledge packets, nor does it provide bit-error detection or correction for the datagram. IP routes are based on the network portion of a 32-bit host address. An IP packet can be up to 64KB long. IP supports a packet fragmentation and reassembly capability, since the size of an actual IP transmission is limited by the underlying physical networks to which the host systems are attached and any intervening subnetworks.

The Host-to-host layer in TCP/IP specifies two protocols: TCP, which defines a connection-oriented service, and the User Datagram Protocol (UDP), which defines a connectionless service. TCP is more commonly used when a reliable connection is required (such as file transfer or terminal services), whereas UDP is most often used in transaction-oriented applications (such as network management or routing information exchanges).

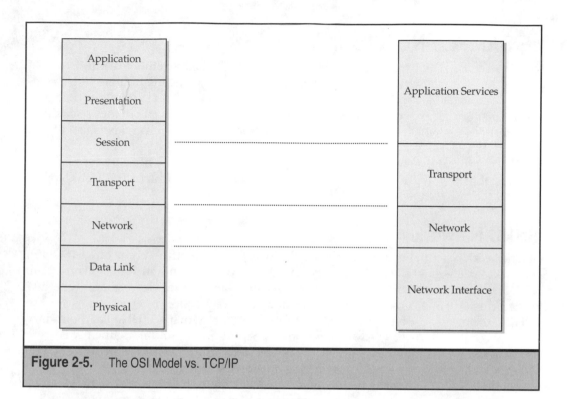

Figure 2-5.   The OSI Model vs. TCP/IP

The uppermost layer of the TCP/IP protocol suite is the Application Services layer. Here we find the definition of some of the basic network services, tools, and utilities that are common to many user applications. In fact, it is these applications that users see; few network users get down to the "bare metal" of the TCP, UDP, and Internet Protocols (IPs).

Important protocols associated with TCP/IP will be examined in future chapters. They are listed here:

▼   **Domain Name System (DNS)**   A distributed database that allows users to refer to programs, devices, and services by easy-to-remember names (for example www.hill.com) rather than numeric addresses.

■   **Hypertext Transfer Protocol (HTTP)**   A protocol that allows for the exchange of World Wide Web (WWW) text, graphical, audio, and video documents over the Internet.

■   **Simple Mail Transfer Protocol (SMTP)**   An electronic messaging service.

▲   **Post Office Protocol (POP)**   A protocol that allows electronic messaging clients to retrieve e-mail stored on an electronic messaging server.

SOME SAMPLE NETWORKS

Networks come in many different forms—from two PCs linked together to a multinational corporation's global network composed of hundreds of servers, mainframes, intercontinental backbone links, and hundreds of thousands of host PCs. It is also likely as time progresses that our concept of the network will shift. Already, there is talk of the *personal area network* (PAN), in which an individual's network-ready devices, such as a personal digital assistant (PDA) and cell phone, interact with each other and the environment.

This section of the chapter offers some perspective on common network types. In the chapters that follow, these examples will allow the reader to place the technologies detailed into a concrete framework.

The SOHO Network

One network type that is growing in popularity is the small office/home office (SOHO) network, which generally includes less than 10 PCs and may not include servers at all. Network resources such as DNS server resolution and e-mail servers are generally located offsite, either hosted by an ISP or at a corporate office. Internet access for the SOHO network is usually provided by cable, DSL, or perhaps ISDN. The boundary between the LAN and the WAN connections is an inexpensive router, frequently costing less than $100. This router may also serve double duty as a firewall to shield the SOHO network from malicious activity originating outside the network.

On the LAN side of the network, either a workgroup hub or low-end switch may be used to provide interconnections between client PCs and the router, and many routers include an integral hub or switch. Due to its simplicity, Ethernet is generally the LAN standard used to wire the SOHO network. Wireless standards such as 802.11b are starting to appear for the SOHO market, eliminating the need for adding LAN wiring in the home. The IP suite of protocols is used for communications on the Internet. In addition to the IP, a requirement to support other protocols used in the corporate network, or to provide local communication in the SOHO network, may exist. In later chapters, we will discuss how this complex environment might be supported.

In some instances, when a small office needs to connect to a corporate environment in a secure manner, some sort of VPN device is either built into the router itself or on the LAN. Figure 2-6 shows an illustration of a SOHO network.

The Small Company Network

Most of the world's corporate networks are considered small to medium-sized companies that generally have some sort of main office branch with one or more remote offices. Employees typically work from the office; however, as we have entered the age of telecommuting, support for SOHO connections may be required.

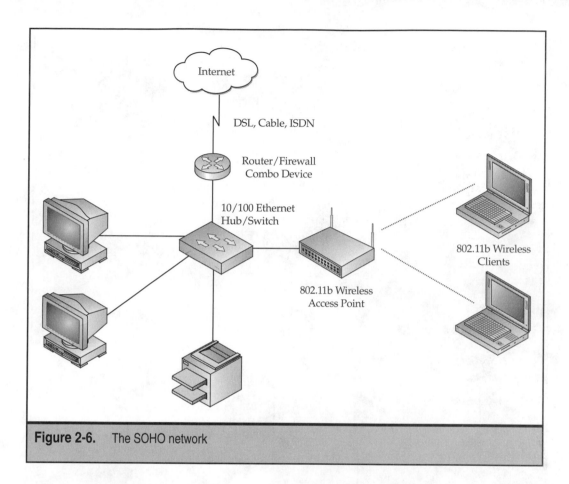

Figure 2-6.   The SOHO network

Figure 2-7 shows the business network connected to the Internet. Increasingly, connection to the Internet is an essential part of doing business. In Chapters 13 through 17 we shall examine the options, including ATM, DSL, frame relay, ISDN, and T-1.

While there are exceptions, the LAN will usually be an Ethernet connection. Its ease of use, low cost, good performance, and ability to support multiple Network layer protocols has made Ethernet the LAN of choice for small and medium-sized companies.

These businesses may support a variety of networking protocols, particularly if in-house networks have existed for a number of years. Many businesses may have one or more of the following networking systems: AppleTalk, Windows NT, and NetWare. Larger companies may have some form of UNIX. In most instances, TCP/IP will be included in the networking approach, if not used as the sole approach.

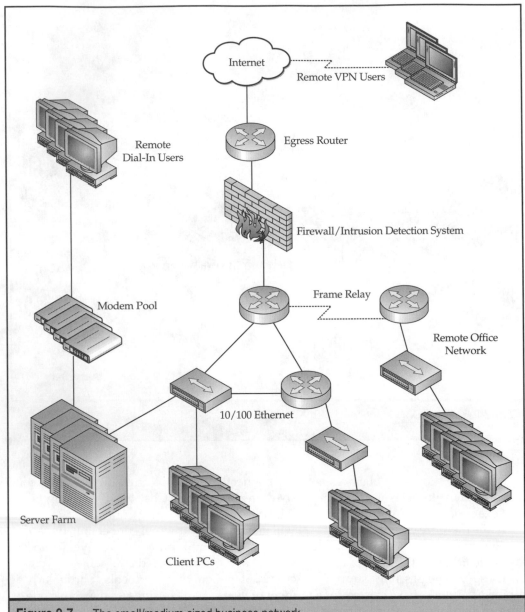

Figure 2-7. The small/medium-sized business network

The resources provided by the company are also much greater than those of the SOHO network, including multiple services that are available to the outside world, such as a WWW server that hosts the company Web site. File, print, database, directory, and other server types may be available to both internal and external users of the company network. To ease the administration of these powerful computers, they may be physically grouped into one or more server *farms*. Access to these servers will be provided by highest bandwidth links and high-capacity switches.

In any network, security is of primary importance. All access to and from the Internet should be secured through the use of firewalls. The most secure installations will employ multiple layers of protection employing firewalls, routers, and intrusion detection systems to ferret out inappropriate behavior. VPN devices and encrypted tunnels may be in use between the main office and remote offices/users, or even on the LAN itself.

Finally, as networks grow in complexity, the need for support grows. Unlike the SOHO system, companies generally use an IT professional to maintain operation of the network. Depending on the importance of the network to the revenue of the company, the network responsibilities may be divided between multiple individuals, with one person responsible for the routing, switching, and hardware infrastructure; another responsible for the maintenance and configuration of the servers and desktop computers; and finally others responsible for database development or Web site design. When advanced or rare skills are needed, outside network/software consultants may be contracted on a case-by-case basis.

The Enterprise Network

The largest and most complex of network types is the enterprise network. These networks are found around the world in the offices of multinational corporations. While a company may have a main corporate headquarters, the network itself may have more than one data center, acting as a regional hub. The data centers would be connected to one another using some form of high-speed WAN; in addition, numerous lower-speed spoke networks radiate from each hub, connecting branch offices, SOHO telecommuters, and traveling employees.

The reliance on computer networks creates some serious challenges for today's corporations. Network reliability and security are essential, particularly when connected to the Internet. Companies must be willing to make significant investments in hardware, software, and people to achieve these goals. Not doing so could be fatal.

As with the medium-sized company, large company networks use a variety of LAN technologies. The most common technology is Ethernet, but other technologies may be found, including Token Ring and Fiber Distributed Data Interface (FDDI). Unlike smaller companies, the large corporate network most likely evolved through the years as technology matured, and as mergers, acquisitions, and new branch offices added new network segments. As such, the enterprise network could best be conceived as many different LAN technologies connected by WAN links.

Many different networking protocols are likely in the corporate network, particularly in older more established companies. They will be supporting many legacy applications and protocols alongside the IP suite. In short, the network is a microcosm of the Internet as a whole, except under the administrative control of one or more IT professionals.

The enterprise network topology is complex. Typically, the WAN links between the hubs of the network will be engineered to operate as a high-speed and reliable backbone connection. Each part of the hub network operates as a transit network for the backbone as well. This means that data from one remote office to another remote office will be routed through one or more hubs. This backbone network may be so large and so well engineered that the hubs will also serve as transit networks for information from other corporations.

Since the enterprise network is composed of many hubs, branch offices, and SOHOs, the internal LAN topology will resemble that of the branch office closely. Information from the backbone will be distributed to the edges of the network and from there will access the LANs in a hierarchical fashion. One remote office sending traffic to another remote office must do so through the backbone because the offices do not share a direct connection (see Figure 2-8).

Because of the complexity, size, and importance of the information on the network to the financial health of the company, staff will be devoted solely to network security on the enterprise network. Users will be strictly policed through the use of passwords, internal firewalls, and proxy servers. Network usage such as e-mail and Web access will be monitored, and well-defined and strict network security polices will be in place and enforced on a regular basis. While branch offices may have a person responsible for the security of that network under guidelines from the main office, some sort of network operations center will monitor the health and security of the network full time from a central location.

Firewalls, proxy servers, and intrusion detection hardware and software will also be in use throughout the network to help provide network security. To protect communications

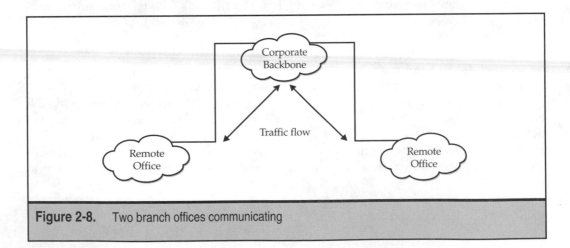

Figure 2-8.   Two branch offices communicating

between hubs and between the remote branch or SOHO user, VPN devices will also be employed. Physically, the network will be secured as well, and access to servers and workstations will be controlled by locks and identity checks whenever possible. Figure 2-9 shows the enterprise network with hubs and firewalls in place.

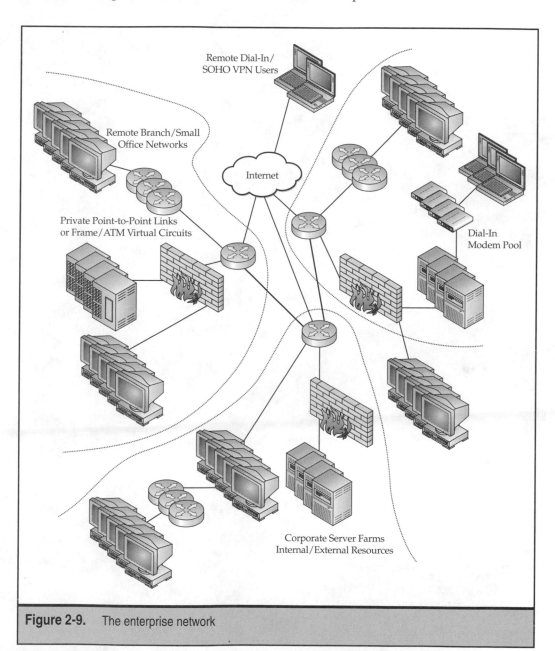

Figure 2-9.   The enterprise network

CHAPTER 3

An Introduction
to the Internet

This chapter provides a basic introduction to the Internet, including a historical overview, an overview of other related networks and organizations, and the structure of the Internet itself. The chapter begins with a history of the original ARPANet and its progeny and discusses the evolution from academic to commercial network, followed by a brief overview of Internet oversight provided by various organizations. It then explores the components and architecture of the Internet. Finally, the chapter examines Internet service providers (ISPs) and Internet access options.

WHAT IS THE INTERNET?

For a large number of people, the Internet is pure magic. Where did it come from? Who owns it? Who pays for all of this? The Internet is a collection of computers (called *hosts*) and networks based on use of the TCP/IP suite—often referred to as the Internet Protocol (IP) suite. Furthermore, the Internet is not actually a network in the usual sense—that is, it is not merely a networked collection of hosts. Instead, the Internet is a number of backbone networks that connect a variety of subnetworks, some of which are publicly funded while others are private or commercially funded.

As a collection of millions of subnetworks, the "owner" of the Internet is difficult to identify. In fact, no single owner exists; although, of course, one can identify the owners of the component subnets. In some ways, the Internet is owned by everyone who uses it, because all members of the Internet community contribute to the network. Similarly, no single entity operates the Internet; various organizations operate their own little pieces, although, again, some central authorities have been defined where necessary for convenience. In addition, individual users and/or organizations may offer information to the community in a number of ways, such as by contributing to central information servers or maintaining the information servers themselves.

A Historical View of the Internet

Where did this "network of networks" come from? In 1957, the USSR launched Sputnik, the first artificial satellite, causing a great deal of anxiety in the Western bloc of allied countries. The fear generated by Sputnik's launch was the catalyst for the United States to put greater emphasis on math and science in schools, and for the U.S. Department of Defense (DOD) to create the Advanced Research Projects Agency (ARPA). ARPA, as the name might suggest, was responsible for advanced research and development projects for the armed services, with control over such areas as U.S. space and missile programs.

The Internet began as an ARPA-funded experiment in the use of packet-switching technology. In 1968, a contract to design and deploy a packet-switching network was awarded to Bolt, Beranek, and Newman. In 1969, less than a year after the award of the contract, the network, called ARPANet, became operational with four nodes: UCLA, Stanford Research Institute, the University of California at Santa Barbara, and the University of Utah. At the heart of the network were the precursors to modern routers, specialized minicomputers called *interface message processors*, that handled the data traffic between sites. The ARPANet spanned the continental U.S. by 1971 and had connections to Europe by 1973.

The initial host-to-host communications protocol introduced in the ARPANet was called the Network Control Protocol. Over time, however, this protocol proved incapable of keeping up with the growing network traffic load. In 1974 a new, more robust suite of protocols was proposed and implemented throughout the ARPANet, based on the Transmission Control Protocol (TCP) and the Internet Protocol (IP). Although the protocol suite is usually referred to simply as TCP/IP, the suite actually comprises a large collection of protocols and applications, of which TCP and IP are only two (albeit perhaps the most important). In 1979, the Internet Control and Configuration Board (ICCB) was chartered to provide an oversight function for the design and deployment of protocols within the connected Internet.

Although TCP was initially defined in 1974, the most common versions of TCP and IP were developed in 1981. In 1982, the DOD declared that TCP/IP would become the new standard ARPANet protocol suite and the official switch occurred on January 1, 1983. An even bigger boost for TCP/IP came from the University of California at Berkeley in 1983, when TCP/IP came bundled with Berkeley's version of UNIX, called 4.2BSD.

In 1983, the ICCB was renamed the Internet Activities Board (IAB) and has since been renamed the Internet Architecture Board; it became a de facto standards organization for the Internet. The IAB was reorganized in 1986 to oversee two major activities: the Internet Research Task Force (IRTF) coordinated research activities related to the TCP/IP protocol suite and the architecture of the Internet; and the Internet Engineering Task Force (IETF) concentrated on short- to medium-term engineering issues related to the Internet.

The development of the Domain Name System (DNS) also occurred in 1983. The DNS was an important development for continued network growth because prior to the development of the centralized name service, all ARPANet hosts had to know the name and address of all other hosts. Just as it's easier to remember 1-800-CALLNOW instead of 1-800-225-5669, customers remember the names of corporate Web sites, not IP addresses. So the DNS, a distributed mechanism for translating mnemonic names that people use into the IP addresses that computers use to deliver data across the network, is absolutely critical for businesses to maintain a presence on the Internet.

During the next year, the ARPANet was split into two components. One component, still called ARPANet, was used to interconnect research/development and academic sites. The other network, called MILNET, was used to carry military traffic and became part of the Defense Data Network (DDN).

In 1988, the Internet *worm*, a self-replicating program that hopped through the network, crashed most hosts in its path. Launched accidentally by Cornell graduate student Robert T. Morris, the worm exploited a known hole in the UNIX *sendmail* program and, because of a programming error, produced disastrous results. While this specific incident didn't make headlines at the time, it probably sounds familiar to today's readers. Morris's program demonstrated the Internet's vulnerabilities years before business became dependent on the network and applications such as e-mail and the World Wide Web (WWW). Indeed, e-mail attacks have become common, as evidenced by the success of the

Bubbleboy, ILOVEYOU, NIMDA, and other well-known exploits—all intellectual offspring of Morris's worm.

Many new tools were introduced to the Internet for the organization of information and automated searching of file archives. The World Wide Web Project was conceived as a way to interconnect these disparate sites in 1992. Mosaic, the first widely used WWW browser, became available from the National Center for Supercomputer Applications (NCSA) in 1993. In 1993, the Internet Network Information Center (InterNIC) project was funded by the National Science Foundation (NSF) as a five-year project to provide information to the Internet community. A directory and database service was maintained by AT&T, and a registration service was controlled by Network Solutions—General Atomics also originally collaborated on the project but eventually dropped out. Also in 1993, the executive branch joined the online revolution when the White House established its presence on the Internet with a domain name of whitehouse.gov.

By July 1994, the Internet had more than 3 million host computers comprising almost 30,000 networks in more than 65 countries, testimony to TCP/IP's good health. Even the U.S. government has taken a fresh look at TCP/IP; an early 1994 report from the National Institute for Standards and Technology suggested that the Government OSI profile should incorporate TCP/IP and drop the "OSI only" requirement. This effectively became the death knell for OSI as a worldwide standard.

Late in 1994, Mosaic Communications was formed. Renamed Netscape Communications, the company went public in 1995 to enthusiastic public response—their selling price at the end of the first day (about $78) was more than two-and-a-half times the asking price at the initial public offering ($28). Certainly the glee for Netscape's stock shown by the market was a precursor of the dot-com boom and bust seen by the turn of the new millennium. By the end of 1995, it was estimated that Netscape supplied 60 to 70 percent of the browsers, with that figure peaking at better than a 95-percent share. Netscape had the mind share in the browser market, which the company was able to leverage for many years to sell other products. However, branding is only half the battle. Microsoft's dominance on the desktop has propelled its Internet Explorer product into dominance in the browser world, which does have an impact on how Web sites are designed and even on the server software they use.

In the early 1990s, the IETF determined that the functionality of IP was limited due to a number of factors, including a decreasing address space and lack of security. The IP Next Generation (IPng) working group was formed, and the next version of IP, version 6 (IPv6), was released in December 1995. Before the protocol was really viable, other simple mechanisms such as Dynamic Host Configuration Protocol (DHCP) and Network Address Translation became common ways to alleviate some of the pressures on the existing address space, so for years there was some reluctance to deploy the new version. Only recently have vendors begun to support IPv6 and address blocks became available, so production networks are starting to use it.

By early 1996, the Internet had nearly 9.5 million hosts, 76,000 of which were named *www.*; in fact, WWW traffic accounted for nearly 40 percent of all traffic on the global Internet. Over the years, many sites have actually dropped *www* from their hostnames, as the Web has become almost synonymous with the Internet as most users' primary application. Whatever the name, Web servers are legion, numbering in the tens of millions, and the native Web protocol, HTTP, accounts for a vast majority of today's traffic.

As the global Internet phenomenon grew, censorship became a major issue; one international Internet provider eliminated a discussion list on its network because one country found the topic offensive. In the U.S., the Communications Decency Act (CDA) ostensibly outlawed pornography on the Internet, but it also outlawed many legitimate discussions, such as those on breast cancer. In protest, many Web sites set their pages to a black background on February 8, 1996, the day the bill was signed into law. By 1997, the U.S. Supreme Court found certain provisions of the CDA unconstitutional on the grounds that it violated the First Amendment right of free speech.

Just over a decade after Morris's worm, it appeared that the Internet still hadn't shored up all its defenses. In 1999, one of the first viruses to be noticed by the public at large was released into the wild. Melissa, a macro virus, spread quickly through corporate e-mail systems, causing administrators in many cases to disable outgoing e-mail to prevent its further spread. Globally, the infection cost companies countless hours and dollars in lost productivity. The ease of creating and distributing viruses, particularly via e-mail, has led to their becoming one of the greatest security threats today.

1999 was certainly a watershed year in many respects. Napster, the controversial electronic music exchange service, hit the scene and made the recording industry aware of the potential threat and opportunity the Internet offers. Consumer acceptance of e-commerce also increased dramatically. One particularly interesting phenomenon was the growth in consumer-to-consumer marketplaces, such as online auctioneer eBay. As more and more people sold goods online, new payment mechanisms were required, giving rise to companies like PayPal. Now the average Internet user can use an easy electronic payment method that is faster than money orders and personal checks, allowing for quicker fulfillment.

With e-commerce becoming more accepted in the mainstream as the twentieth century ended, it was no surprise that e-commerce companies, particularly the "pure-play" (i.e., only online) variety, would become inviting targets. The biggest news in 2000 was arguably not the Y2K bug, but the spate of Distributed Denial of Service (DDoS) attacks that brought Internet giants such as Amazon.com, Yahoo!, and E*TRADE to their knees. The most nefarious aspect of these breaches is that relatively little skill is required to launch them, as most of the code is freely available on the Internet for "script kiddies" to download and run with the push of a button. These incursions continued to plague the Internet. According to some estimates today, thousands of DoS attacks are launched per week.

The Internet still faces issues that are related to its continued growth. Governments around the world have noticed the network and are wrestling with taxation, regulation, and other legal concerns. As businesses have become more dependent on the Internet to generate revenue, they also have become more exposed to the inherent security risks. Home users also rely on this global network to find information, buy products, and communicate, often without being aware of the security and legal ramifications of their activities. However, there is increased awareness of the attendant issues, which bodes well for the Internet's continued vitality.

The Internet and Other Networking Projects

The history of the Internet is tightly intertwined with other computing and networking developments. During the past few decades, many types of networks have rivaled, paralleled, and been connected to the original ARPANet or the modern Internet. Some are still in use today, while some have come and gone. All have contributed to the understanding of how to connect computers and networks in a global community, so it is worth a quick survey of some of these other facilities and projects.

The 1980s was a time of tremendous data networking growth. The Computer Science Network (CSNET), for example, was an NSF-funded network linking computer science departments at major colleges and universities in North America that were not connected via the ARPANet. Chartered in 1981, CSNET formed a gateway with the ARPANet in 1983 and merged with BITNET in 1987, only to be shut down in 1991 in favor of more general Internet access.

BITNET (Because It's Time Network), formed in 1981 at the City University of New York with funding from IBM, was based on IBM's Network Job Entry (NJE) protocols and intended for e-mail and information exchange. The original BITNET (sometimes referred to as BITNET-NJE) thrived until it became apparent that it was growing obsolete in light of the Internet's continued maturation. The Corporation for Research and Educational Networking (CREN), BITNET's managing organization, announced IRA (Internet Resource Access) in 1994. This next iteration of BITNET, called BITNET-II, encapsulated the BITNET protocol inside the IP to facilitate access from BITNET members to the Internet itself. In 1996, CREN recommended terminating use of the network.

Europe also saw its share of networking experiments. For example, the European UNIX Network (EUnet), was created in 1982 to provide e-mail and USENET services throughout western Europe and is now the largest ISP in Europe. The European Academic Research Network (EARN), chartered in 1983, was similar to BITNET and is similarly defunct.

The Japan University Network (JUNET) was formed in 1984 for a purpose akin to EUnet's. Reflecting the growing interest in networking networks, JUNET was connected to CSNET in 1987, bringing Japan into the international networking community. With changes in the Ministry of Posts and Telecommunications' regulations regarding the Internet in Japan, JUNET was shut down in 1994. The success of JUNET spurred the 1986 creation of the Widely Integrated Distributed Environment (WIDE) project, which is still in operation today as a test bed for Internet-related research in Japan.

In 1986, the NSF built a backbone network to interconnect four NSF-funded, regional supercomputer networks and the National Center for Atmospheric Research (NCAR). This network, dubbed NSFNET, was originally intended as a backbone for other networks. The NSFNET continued to grow and provide connectivity between both NSF and non-NSF regional networks, eventually becoming the backbone that we know today as the Internet. The NSFNET, originally composed of 56Kbps links, was completely upgraded to T-1 (1.544Mbps) facilities in 1989. Migration to a "professionally managed" network was supervised by a consortium made up of Merit (a regional network headquartered at the University of Michigan), IBM, and MCI. Advanced Network & Services,

Inc. (ANS), a nonprofit company formed by IBM and MCI, was responsible for managing the NSFNET.

A new NSF backbone network structure, the very high-speed Backbone Network Service (vBNS) went online in 1994. MCI also took over operation of the NSF T-3 network backbone that had been operated by ANS. Later that year, America Online (AOL) bought ANS's commercial T-3 backbone. On April 30, 1995, the old NSFNET was shut down.

1994, the twenty-fifth anniversary of the first man on the moon (July), Woodstock (August), and the Amazing Mets (October), also brought the silver anniversary of the ARPANet (September). And what of the original ARPANet? It had grown smaller and smaller during the late 1980s as sites and traffic moved to the Internet, finally being decommissioned in July 1990.

Following the development of IPv6 in 1995, several entities were created to study the protocol's capabilities and impact and to facilitate its eventual deployment. In 1996, a collaborative test bed network called the 6bone began life as a virtual network, tunneling IPv6 packets over the existing IPv4 Internet. Today hundreds of sites in dozens of countries participate in testing implementations and transition strategies. Another initiative, the 6REN (Research & Education Network), was started in 1998 by the Energy Sciences Network (ESnet) to promote IPv6. Other related organizations include the IPv6 Forum, a consortium of vendors and research networks, and the IETF's IPng Transition (ngtrans) working group.

Today's Internet has been called the "Information Superhighway" by some, while others scoff, preferring the term "Information Super Onramp" to illustrate how the network has yet to live up to its true potential. To foster development of the next, better iteration of the Internet, the Internet2 project was formed in 1996 by 34 research universities, with the first meeting of its members occurring the following year. The consortium currently has more than 180 member institutions working with industry and government on developing new network capabilities and applications, and introducing them to the Internet at large.

The Commercial Internet

Until 1988, use of the NSFNET was limited to organizations that were funded by the NSF. As the benefits and advantages of internetworking became known during the 1980s, several companies started to offer commercial e-mail services. In 1988, MCI and the Corporation for National Research Initiatives (CNRI) formed a gateway between the MCImail service and the Internet, and CompuServe and Ohio State University formed another gateway for CompuServe's commercial e-mail users. With these gateways, any user of MCImail, CompuServe, or the Internet could exchange e-mail. While the e-mail content on the NSFNET was still limited to noncommercial, NSF-related topics, the experiment was a rousing success—many view it as the first crack in the dam holding back commercial use of the Internet. After access to e-mail, the commercial users also wanted file transfer and remote login capabilities.

While historically the Internet was not used for overt commercial purposes, this began to truly change in the 1990s. For example, in 1991 the Commercial Internet Exchange

(CIX) was formed by General Atomics, Performance Systems International (PSI), and UUNET Technologies to promote and provide a commercial Internet backbone service. Today the Internet is an important medium for commerce by a wide range of businesses.

1990 found the Internet at a crossroads that directly impacted most users, particularly those in the commercial arena. The ARPANet and the Internet in the United States had been based on some sort of government funding, particularly from the DOD and NSF. Things dramatically changed in 1994 when the NSF made it known that it never intended to be in the networking business; after eight years, the Internet regional networks were to become self-sufficient and the role of the NSF was to be directed elsewhere. Significantly, in 1994 a process of increased privatization and commercialization of the Internet began.

Privatization of the Internet brought into question what the price structures would look like and how that would affect use of the network. Almost all of today's network and host access to the Internet is billed on a flat-rate basis, rather than on a usage-sensitive basis. Although users might actually save money with usage-based billing, most users opt for the flat-rate option. The "all you can eat" plan still flourishes, with some modification, as Internet access has become a commodity.

Commercialization of the network also resulted in an increase of purely commercial traffic, particularly junk e-mail commonly referred to as *spam*. Corporate and consumer Internet users are bombarded daily with e-mail advertisements for everything from pornography, to mortgages, to the next "sure thing" stock tip. Attempting to stem the tide of unwanted e-mail solicitations, most ISPs publish appropriate use policies that ban such practices, with limited success. Many e-mail applications support server- or client-based message filtering, somewhat reducing the clutter users must sift through, though this does not reduce the amount of network traffic. Unfortunately, law, policy, technology, and age-old "netiquette" appear to be unable to prevent innovative spammers from clogging our e-mail.

Today, one of the major issues facing the Internet is its continued growth. The ARPANet, which began operation with four nodes in 1969 had expanded to a mere 213 hosts by August 1981—hardly a tremendous rate. However, the Internet doubled in size in 1986, quintupled in size in 1987, and nearly doubled in size again in 1995. By the end of the 1990s, a new network was being added to the Internet every 30 minutes! A saturation point has apparently been reached today so growth has slowed somewhat, though it is still vigorous.

Most of the growth in the U.S. Internet is in the commercial arena. But the fastest rate of growth of Internet users is outside the United States—in fact, by 1999, half the online population was outside the U.S., representing almost every country on the globe. The network is becoming less techie-, DOD-, and U.S.-centric, thus drastically changing the profile of the network.

The Internet's greater diversity means greater commercial opportunity, with the network allowing businesses to reach a worldwide audience. There is a limit to the Internet's reach, however, even today. The "irrational exuberance" of the 1990s gave way to the "irrational pessimism" of the new millennium as New Economy companies discovered that Old Economy rules, such as the need for profit, still applied. Regardless, the Internet continues to thrive and e-commerce companies still are viable, albeit chastened. More people and businesses continue to get online, and that is driving the Internet's growing impact on the global economy and how we live at a fundamental level.

There are many studies and surveys about the Internet's growth, changing demographics, and other trends conducted each year. The list here contains several online resources for current information.

▼ **Internet Indicators (http://www.internetindicators.com)**   Semi-annual Internet economy indicators

■ **Netcraft (http://www.netcraft.com)**   Monthly Web server survey

■ **Internet Domain Survey (http://www.isc.org/ds/)**   The Internet Software Consortium's semi-annual DNS survey

■ **Nielsen//Netratings (http://www.nielsen-netratings.com)**   Global count of Internet users, top properties, and other measurements

▲ **CommerceNet (http://www.commerce.net/research/stats/)**   Historical and projected demographic information

Internet Development, Administration, and Oversight

Who runs this huge, ungainly, multiple headed monster called the Internet? The earliest days of the Internet (ARPANet) saw ad hoc administrative centers at the ARPA, the DOD, several major universities, and several other points. One would think that the intervening years would have seen a consolidation of administrative effort, but in fact, the responsibility remains distributed, albeit in a slightly more organized way.

The next list itemizes several of the more prominent Internet oversight organizations. For current information on their various projects, visit the Web sites provided:

▼ Internet Society: http://**www.isoc.org**

■ Internet Engineering Task Force: http://**www.ietf.org**

■ Internet Assigned Numbers Authority: **http://www.iana.org**

▲ Internet Corporation for Assigned Names and Numbers: **http://www.icann.org**

As the U.S. government backed out of funding the Internet, the introduction of commercial ISPs further signaled the loss of a central administration, which, in turn, threatened the process for making Internet standards. In January 1992, the Internet Society (ISOC) was formed with a charter to provide an institutional home for the IETF and the Internet standards process, as well as administrative support for the IAB, IETF, IRTF, and the Internet Assigned Numbers Authority (IANA).

The ISOC controls cooperation and coordination for the Internet, advising and overseeing the activities of many of the other boards and taskforces that hold sway over Internet development, research, and organization. Interested members of the Internet community can join the ISOC to help determine the Internet's future.

The primary body responsible for maintaining control of the Internet's technical side is the IETF. While having existed informally for quite a while, it was defined and chartered in 1986. Its charter gives it control over engineering and development of the protocols that run over the Internet infrastructure. The Internet Architecture Board (IAB) oversees the editorial management and publication of the Internet RFCs and standards, oversees the activities of the IANA, and provides general technical oversight for the IETF. The activities of both the IETF and IAB are coordinated by the ISOC.

A more recent entry into the fray is the Internet Corporation for Assigned Names and Numbers (ICANN). Formed in 1998, ICANN is responsible for coordinating the assignment of Internet domain names and IP addresses. Three organizations aid ICANN in the review and development of policies: the Address Support Organization (ASO), Domain Name Supporting Organizations (DNSO), and Protocol Supporting Organization (PSO). The group has had many luminaries on its board of directors, including Esther Dyson and the so-called "father of the Internet," Vint Cerf. The board also consists of several "at-large" members elected by Internet users. Despite the lofty makeup of ICANN, it has often been (perhaps unfairly) criticized for its lack of mandate.

Commerce is not about to sit around and allow procedures and protocols to be dictated to them. As such, commercial Internet users become members, whenever possible, of as many of the Internet councils and forums as possible. They, along with the research divisions of higher education, drive most of the research that goes into the development of Internet protocols. ISPs, as well, have a vested interest in the development of standards, so they are also present in most working groups.

THE STRUCTURE OF THE INTERNET

We usually think of the Internet in terms of a web, but you can also think of it as a kind of railway system. In much of the world, the gauge of railroad tracks differs from region to region, as well as from country to country. This has led to substantial inefficiencies in both commercial and personal travel/shipping. It would clearly be far better if trains could carry goods and passengers smoothly from source to destination.

When the United States and Canada were developing as nations, a decision was made to use a common track gauge for all service providers. That decision was a monumental one, because it assured that everyone in North America could play on a level field in the railroad business. A small local carrier such as Vermont Railway could provide local service, while at the same time obtaining access to regional carriers via switching points. These carriers could, in turn, get access to national carriers at designated access points. Thus, the small carrier can offer nationwide connectivity, all as a consequence of the common choice of gauge. The gauge of the railroad tracks is a *protocol* (an agreement that facilitates connectivity). The gauge of the tracks is then an *internetworking* protocol, similar in function (if not in form) to the IP.

The Internet itself is structured in exactly the same fashion as is the railroad "internet." A common protocol (in this case IP) allows for the interconnection of local, regional, national, and even global service providers. The equivalent of the railroad switch is the router, where packets are directed from one provider's network to another's, until final delivery to the destination processor.

There are, however, some differences between the Internet and the railroad network, specifically with regard to pricing models. Large ISPs generally agree to carry one another's traffic for free, in an arrangement known as *peering*. Peering works if flows are roughly symmetric, and it is an easier arrangement than a traffic-sensitive approach. Larger ISPs may charge smaller ISPs for connection—typically on a flat-rate basis. In the railroad system, cars are counted, and railroads reimburse one another on a traffic-sensitive basis for providing carriage. There are some who believe that this type of billing algorithm will soon appear in the Internet as well, particularly as asymmetric flows such as those associated with Web-based video applications become more common.

Internet Components

In April 1995, the NSF-funded backbone network was turned off to commercial traffic, causing their regional networks to become self-supporting. Therefore, we have seen increased competition, price wars, expanding territories, and acquisitions and mergers.

A brief description of the major components of the Internet backbone is listed here.

▼ Network access points (NAPs) in four cities: Chicago; San Francisco; Washington, D.C.; and Pennsauken, New Jersey. The NAPs act as interconnection points between the ISPs and NSF-funded networks, providing the physical and logical means for networks to interconnect.

■ Backbones interconnect the NAPs. While traditional network diagrams show the NSF-sponsored vBNS as the backbone interconnecting the NAPs, a number of large telecommunications providers actually connect them.

▲ A routing authority (RA) acts as a routing arbiter to oversee certain aspects of the network traffic, such as ensuring that traffic is fairly and uniformly distributed across the various carriers.

While the NAPs originally received partial funding from the NSF, they were and still are commercial service offerings, which means that competition exists in the NAP market. NSF was particularly aggressive in this arena. Today, WorldCom operates network interconnection points in Silicon Valley; the Washington, D.C., area; and Dallas.

Internet Hierarchy

The components of the Internet fit into a loose hierarchy. While general definitions of the various components are helpful, providers' networks and interconnection profiles vary, and those differences are exploited by providers' marketing and sales offices in an attempt to gain a competitive advantage, resulting in a rather confusing picture of what constitutes typical Internet connectivity. Nevertheless, some similarities remain. At the top of the Internet connection picture are the NAPs. While the NSF originally funded four NAPs, the explosion in the growth of Internet connectivity has resulted in the establishment of providers' own NAPs, which are of greater or lesser significance depending on what technologies are implemented and the number of providers at the NAP. The original NAPs still rank at the top of the list when it comes to size, speed, number of interconnected networks, and prominence in the public mind, but some of the larger providers' connection points are getting close.

One step below the NAPs—and the primary reason for their existence—are the largest of the ISPs, called variously Tier 1 providers, national ISPs, or carrier providers. Originally, a provider was considered Tier 1 if it connected to at least three of the four NSF NAPs. This definition is inadequate today. A Tier 1 provider also typically has private peering arrangements with other networks of similar size—this peering is outside the constraints of the NAP structure for greater efficiency. While there are no hard and fast criteria for belonging to this group, these providers' networks are national (or even global) in reach and have many smaller downstream ISPs and large corporations as customers.

Tier 2 providers, or regional service providers, are smaller versions of the Tier 1 networks. They sometimes have a presence at the NAPs to interconnect with other providers, yet they lack the geographic reach and customer base of the bigger networks. Tier 2 providers often pay one or more of the Tier 1 providers for access to the Internet at the speeds that they (and their customers) require. Tier 2s can leverage their regional focus by offering better access to regional information while still maintaining a relatively large network footprint.

Tier 3, or local, ISPs depend on other ISPs for their access to the backbone of the Internet. While these providers have traditionally been the corporate and consumer market entry points to the Internet, many are merging or being acquired by larger providers who value them for their local access facilities and customer bases. Given that corporate and consumer customers can obtain service from any size of provider, Tier 3s might seem to be at a disadvantage today. However, these providers can usually make up for their lack of network reach with a local presence greater than that of the larger players.

NOTE: Many definitions of the three tiers exist, and the terminology has become quite diluted by providers attempting to differentiate themselves. Perhaps it would be better to rely on a larger set of distinctions that includes network size, but also entails latency, service-level agreements, and other factors. All aspects of a network are prone to marketing manipulation—when more information is considered, a more effective distinction can be made.

The NAPs

As stated earlier, the Chicago, San Francisco, Washington, and New Jersey NAPs established as part of the NSF network reorganization remain the important points of network interconnection. In recent years, though, providers have pushed ahead with the creation of their own NAPs. All that is really necessary to establish a NAP, after all, is the willingness of a number of access providers to interconnect their networks at some physical location.

So what distinguishes a NAP from just any connection between providers? Little, if anything. But typically a NAP is established when at least some of the following conditions are met:

▼ A large number of providers connect to each other at the same location.

■ The connection is made through a high-speed switching system. Often this system is as simple as a number of high-speed routers connected via an Asynchronous Transfer Mode (ATM) or high-speed LAN backbone.

■ Corporate and other large customers of the ISPs connect to them via this site.

▲ Perhaps most important, the ISPs announce the creation of the NAP with the intention of promoting it as an interconnection point in the future.

Today, dozens of locations meet at least some of these criteria, and therefore these sites call themselves NAPs. However, the actual equipment present at a NAP can vary widely, depending on the philosophy of the network engineers who design it, the availability and cost of the equipment, and the speed and number of access links required. However, they usually share the same basic design and include the components listed here:

▼ A backbone network that interconnects the access nodes, typically composed of high-speed ATMs or high-speed Ethernet-based technologies. FDDI implementations still exist as well, though generally are not available to new customers.

■ A number of fast routers that connect via the backbone, each of which aggregates customers' network traffic onto it. Typically these are high-end routers.

■ A route server, which holds a copy of the Internet routing table, makes routing information available to the backbone routers as they need it.

■ Associated equipment, such as data service units (DSUs), wiring hubs, power supplies, and so on.

▲ Peering agreements between providers. Merely having a physical connection at the NAP is only one piece of the puzzle—other networks still need to agree to carry traffic for other networks.

Figure 3-1 shows the configuration of a typical NAP.

Note that the listed components don't include anything about location. The NAPs themselves aren't always elegant. Although the facilities have moved in recent years, the original MAE-East, for example, was located in a very small area in the corner of a parking garage! Today most NAPs have moved to larger and more secure accommodations.

One of the biggest issues with having a large concentration of providers in one location is congestion, both physical and in terms of packet traffic. The NAPs will not be going away any time soon, but many ISPs have looked to alternatives to such access points, often maintaining private connection points with other providers as well.

National (Tier 1) ISPs

Originally only a handful, today many ISPs are recognized as Tier 1 providers, including AT&T, Cable and Wireless, Genuity, Level 3, Sprint, and UUNET (WorldCom). These

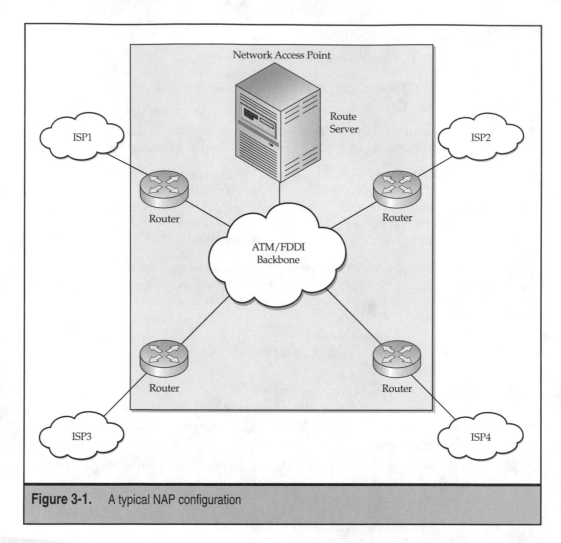

Figure 3-1.   A typical NAP configuration

companies have a number of common attributes that distinguish them from the rest of the crowd:

▼ Their enormous geographic scope means that any one of them can reach nearly any location with their own facilities. All are represented throughout the continental United States and each has overseas connections as well.

■ Their high-speed, redundant backbones mean that large customers with mission-critical data feel relatively safe about using these networks for transit. The large customers typically account for a high percentage of each network's revenue.

■ They provide access to the Internet for a large number of smaller ISPs.

▲ They have significant experience in long-distance data networking—although that experience may or may not have begun with the IP—and therefore have well staffed network operation centers (NOCs).

Regional (Tier 2) ISPs

Regional, or Tier 2, ISPs come from an interesting mix of backgrounds. Some are formed from parts of former NSF-funded networks that have since been privatized, some are run by Regional Bell Operating Companies (RBOCs) as part of their regional data networks, and some are relatively new networks that have been started by entrepreneurs as part of the Internet boom. Whatever their origins, most regional ISPs extend coverage over an area of one to a few states.

These networks typically connect to only one NSF NAP; or, they may access an NAP via one or more Tier 1 providers. Therefore, while their backbone networks are important to their customers, in as much as they will be used for regional access and for access to the Internet, it is often the dial-up and direct access connection points of these providers that are their most attractive features. Many regional ISPs in fact have more dial-up capacity than do the national providers.

Usually these providers are more diverse and less redundant than the national providers. Assuming equal pricing, when a customer chooses to buy access from a regional ISP instead of a national one, he or she is placing a premium on regional connectivity and ease of access, as opposed to national or global reach.

Local ISPs are like regional ISPs, except they usually serve a significantly smaller area, such as a small state or a large metropolitan area. Local ISPs typically connect to a regional or national ISP for worldwide Internet connectivity.

Interconnection

Given the mishmash of NAPs, ISPs of different sizes, and networks of various descriptions, how does an e-mail message get from the computer of a user somewhere in North Dakota to the computers of a giant corporation in Atlanta? And how does a request to see a Web page make it from the desk of an executive to the Web servers of the Yahoo! or AltaVista search engines? In either case, the traffic flow describes a similar pattern. We will describe a possible—and even probable—traffic flow, but one that is by no means definitive for our endpoints.

Figure 3-2 shows a stylized map with several faux ISP networks to illustrate how providers are interconnected.

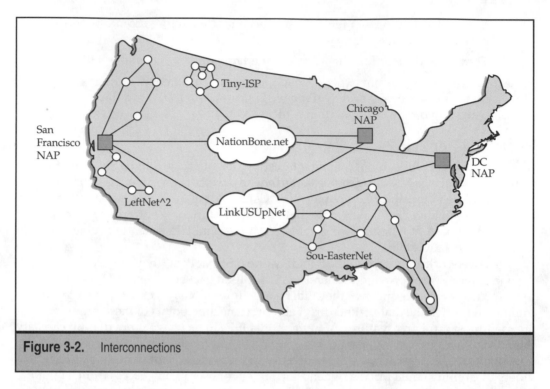

Figure 3-2.   Interconnections

The following steps describe how a user on Tiny-ISP in North Dakota might reach HAI Enterprises's corporate server in Atlanta.

1. A packet leaves the computer of a private user in North Dakota, who obtains access to the Internet from Tiny-ISP. The packet exits the user's modem and travels through the local phone system to one of the modems of Tiny-ISP in Fargo.

2. Tiny-ISP's router examines the packet and determines that the destination network is not within its assigned range of IP addresses; therefore it can be delivered only by forwarding the packet to Tiny-ISP's Tier 1 provider, NationBone.net. Such a conclusion is actually how most of the upstream (i.e., toward the NAP) routing decisions on the Internet are made.

3. The packet travels across Tiny-ISP's leased line to NationBone's point of presence (POP) in Cheyenne, Wyoming. Upon examining the packet, NationBone's routers determine that the destination network is advertised as belonging to LinkUSUp, another Tier 1 provider. This does not mean that the destination network is directly connected to LinkUSUp, just that NationBone hears about it from LinkUSUp. The destination network is Sou-EasterNet.

4. NationBone forwards the packet over its own backbone until it can transfer it to the LinkUSUp network. The nearest exchange point to Cheyenne is the Chicago

NAP, so NationBone sends the packet there. It is in NationBone's interest to get rid of the packet as soon as possible to save capacity on its own network.

5. At the NAP, the high-speed routing system transfers the packet from NationBone's backbone to LinkUSUp's backbone.

6. The first LinkUSUp router to see the packet determines where to send it in the LinkUSUp network. In this instance, the destination network is advertised by a LinkUSUp router that connects Sou-EasterNet to the LinkUSUp backbone. The packet is sent over the LinkUSUp backbone to that router.

7. The router that forms the boundary between LinkUSUp and Sou-EasterNet forwards the packet through the Sou-EasterNet backbone to the router that connects its customer HAI Enterprises.

8. The packet is then sent across the HAI Enterprises network until it arrives at its destination computer.

Note that while we have mentioned the major jumps from network to network in this example, for simplicity we have not documented every instance of a packet arriving at a router. In fact, such a trip could probably not be made in as few router hops as there are steps. Each backbone mentioned needs to route the packet internally to arrive at the exit point that connects to the next network.

Now that the basic concept of interconnections has been introduced, let's look at a real trace of a path through the Internet. A program called *traceroute* is available for every computer platform, from Macintosh to Sun workstation. In its most basic form, the program is simple, command-line software that shows the route that packets take from one computer to another. Figure 3-3 shows the version built into the DOS command prompt that comes with Microsoft Windows-based PCs.

At the C:\ prompt, a user at Hill Associates invokes the command **tracert www.osborne.com**. The software generates traffic using the Internet Control Message Protocol (ICMP), a diagnostics protocol that's part of the IP suite. The results show that packets from Hill Associates to Osborne/McGraw-Hill follow this path from the user PC:

1. To the Hill Associates router, golem.hill.com, which is in Colchester, Vermont.

2. To the border router of Hill Associate's ISP, Cable and Wireless (CW), in Boston. This happens to be via a dedicated Internet connection, as opposed to dial-up.

3. Through three more CW routers, including another one in Boston and two in New York City.

4. To an AT&T router in New York. This router represents a peering arrangement between CW and AT&T, which could be located at a NAP, or more likely at a private connection point.

5. Through AT&T's router network from New York to Chicago.

6. To the Osborne Web site.

```
C:\>tracert www.osborne.com

Tracing route to www.osborne.com [198.45.24.130]
over a maximum of 30 hops:

 1   170 ms   190 ms   190 ms  golem.hill.com [208.162.106.1]
 2   211 ms   200 ms   200 ms  bordercore2-serial6-0-77-77.Boston.cw.net
        [166.48.67.53]
 3   201 ms   200 ms   200 ms  acr1-loopback.Bostonbol.cw.net [208.172.50.61]
 4   210 ms   200 ms   200 ms  acr1-loopback.NewYorknyr.cw.net [206.24.194.61]
 5   190 ms   210 ms   200 ms  cable-and-wireless-peering.NewYorknyr.cw.net
        [206.24.195.230]
 6   190 ms   200 ms   200 ms  gbr3-p50.n54ny.ip.att.net [12.123.1.122]
 7   310 ms   300 ms   261 ms  gbr3-p30.cgcil.ip.att.net [12.122.2.173]
 8   250 ms   261 ms   280 ms  gbr1-p100.cgcil.ip.att.net [12.122.1.154]
 9   250 ms   291 ms   260 ms  ar8-a300s1.cgcil.ip.att.net [12.123.4.169]
10   241 ms   290 ms   361 ms  12.125.173.66
11   310 ms   271 ms   270 ms  198.45.24.244
12   230 ms   250 ms   261 ms  198.45.24.130
```

Figure 3-3.   An example traceroute

Many network administrators name their computers and routers so it is possible to know the geographic, as well as logical, path through the Internet—in many cases, the underlying networking technology (such as ATM) is also indicated. Some versions of traceroute are graphical in nature and even have capabilities to look up information in databases to determine precise locations of nodes in the network.

Traceroute is a useful program for network administrators to diagnose performance and other problems. Some people even use the information to differentiate between service providers—presumably, a network with fewer hops will offer better performance than one with more nodes to traverse (though that isn't necessarily always the case).

CONNECTING TO THE INTERNET

Today a myriad of options for connecting to the Internet are available, including traditional dial-up via analog modem as well as high-speed technologies for the corporation and even at home. In this section, we will take a brief look at the predominant access methods for residential and corporate users. For our purposes, *residential* entails consumer and SOHO customers. A summary of basic connection types is shown in Figure 3-4.

Exactly how a user obtains a connection will depend on the user's intended application, which drives the level of services that are demanded from the ISP. For example, does a customer just want access to e-mail (these folks still exist!), does she want to telework or watch streaming media, or does she want to set up a Web site? Although the options for connecting to the Internet are legion, they can be organized into a relatively simple classification that defines three basic types of access: dial-up, broadband, and dedicated access.

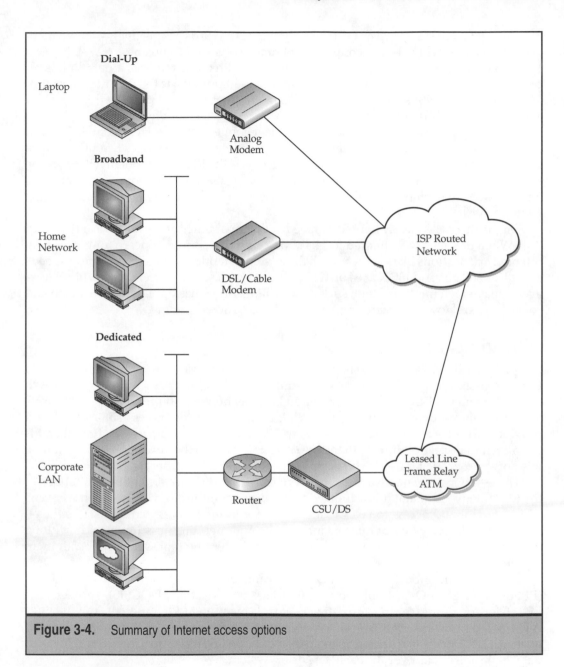

Figure 3-4.   Summary of Internet access options

Dial-Up Access

Basic dial-up access is shown in Figure 3-4, in which the modem is a separate functional device (and not to scale). Most computers, whether of the desktop or laptop variety, have built-in v.90 modems as PCI or PC (formerly PCMCIA) cards, respectively—not to mention Ethernet cards. Dial-up is common for home use as well as business travel.

A dial-up connection is essentially another data link and requires a data-link layer protocol. For Internet access, the most commonly used protocol is the Point-to-Point Protocol (PPP). Many ISPs still support Serial Line Internet Protocol (SLIP), which is older and less capable, but SLIP is fundamentally a legacy protocol today.

Dial-up access is most often supported via a conventional modem connection, but other options are becoming common. These include Integrated Services Digital Network (ISDN) and wireless. An example of wireless access is the Ricochet modem network. The service supports speeds up to 128Kbps in Atlanta, Dallas, Minneapolis, San Francisco, and several other cities, as well as many airport and university campuses across the United States.

Dial-up access is clearly limited in speed to the capabilities of the modem and dial-up link. When modems were limited to 14.4Kbps, this was a serious concern for graphically oriented applications like the Web. With the advent of modems that can operate at speeds as high as 56Kbps and ISDN at 64 or 128Kbps, this was less of a concern for a while. However, today's rich media content, such as streaming video, has upped the ante again and consumers are now looking for high-speed access options at home.

Broadband Access

A typical broadband scenario is shown in Figure 3-4 as a simple home network, most likely Ethernet-based, and a couple of desktop computers. These machines can be connected to each other and the DSL/cable modem via an Ethernet hub—the hub, in fact, could be integrated into the modem or a small router. While called "always on" access, some broadband providers will disconnect idle users, usually to free up scarce IP addresses. Broadband is often asymmetrical in nature, with the greatest bandwidth in the downstream direction, so it is not appropriate for users who want to maintain a local Web server.

These services are not available to all homes, given distance and other limitations of the technology, although it is estimated that around one-third of U.S. homes have some sort of broadband access today. Broadband is not limited to residential users either—with the explosion of corporate reliance on the Internet, business travelers have long tied-up hotel phone lines, so many chains are implementing high-speed connectivity in their rooms.

The dominant protocols used in DSL and cable deployments are PPP over Ethernet (PPPoE) and Data Over Cable Service Interface Specification (DOCSIS), respectively. PPPoE leverages the features of existing protocols including PPP, Ethernet, and the ISP's ATM-based networks. DOCSIS (also known as the Certified Cable Modem Project) is an open set of communications and management specifications to provide high-speed datacommunications over the cable television system.

In addition to the well-known wired services, several wireless broadband options such as satellite and Multichannel Multipoint Distribution Service (MMDS) are also available. All have their advantages and disadvantages and are thus finding specific niches depending on application, geography, demographics, and other factors. Regardless of what method is used, the proliferation of broadband services allows for a wide range of new high-bandwidth applications that are only now beginning to mature.

Dedicated Connections

A generic dedicated connection is also shown is Figure 3-4. The premises equipment includes the computers and other equipment that make up the LAN, as well as the hardware necessary to connect the LAN to the ISP, such as a router and channel service/data service unit (CSU/DSU). We have made a distinction here between traditional dedicated access and today's newer broadband access because services such as frame relay are typically not available in the residential market.

Connection options include leased lines (at speeds ranging from 2400bps to 1.544Mbps), and packet-based data services (such as frame relay or ATM, at speeds ranging from 56Kbps to 45Mbps). Leased lines and frame-relay permanent virtual circuits (PVCs) are the more common methods for dedicated access. Over leased line facilities, the PPP is by far the most common Data Link layer protocol.

Dedicated access provides a way to connect a host computer or network to the Internet 24 hours a day, seven days a week. This type of connection typically employs a permanent link from the user's premises to the ISP's nearest switch; this also offers the user(s) full-time access to the network. Dedicated connections provide the most robust and flexible type of Internet service and tend to be particularly cost effective for large organizations. A dedicated connection is necessary if the user wants to offer any form of online information service, such as FTP or WWW servers.

In contrast to residential users, corporate networks have a few additional tasks that generally have to be done. For example, if a company will be running a Web site to reach customers, a domain name must be obtained. IP addresses also must be assigned to hosts on the network so end users can access the Internet. Some individual must be responsible for administrative issues and security, which is a major concern. In addition, connecting a network to the Internet over a dedicated line could require a local mail relay host and a DNS server.

Internet Services

Tens of millions of Internet users rely on traditional online service providers for Internet access. These online services are usually accessible via dial-up lines using client software supplied by the service provider. AOL and other ISPs give users a simple interface and access to databases of information for such things as sports, entertainment, news clippings, travel, and other topics.

Online services provide full Internet access, albeit mediated. All the online services provide e-mail and Web browsing, usually partnering with either Microsoft (Internet Explorer) or Netscape (Navigator), who provide the browser. Most offer some form of file transfer capability, and fewer offer remote login capabilities.

While the online service providers boast the largest customer base, many Internet users use "pure" Internet service providers (ISPs). AOL and its ilk have in large part countered this migration by offering services and pricing similar to that of the ISPs, while re-emphasizing the content that has been their mainstay for so long. While much content is available for free on the Web, AOL has demonstrated that owning content is still a differentiator, as evidenced by the company's merger with Time Warner in 2001. Although not offering content *per se*, pure ISPs do usually provide some sort of Web portal for quick access to an Internet search engine, and local/regional news, weather, and other information.

The main function of ISPs is to provide connectivity. However, connectivity has become a commodity, so providers are trying to move up the value chain. They will usually provide a wide range of services, including these:

▼ **Web hosting**   Relevant to both corporate and residential users. Many companies lack the core competency to maintain their own Web sites and will outsource this to service providers. For consumers who often use dial-up or asymmetric access, the only way to have a Web site is to use the ISP's facilities. Sometimes hosting is bundled with Internet access. Many hosting companies specialize in e-commerce and might offer better deals than ISPs in terms of cost, storage space, and application support.

■ **E-mail**   Still arguably the most widely used application today. Mail hosting is important for corporate customers, though not all ISPs offer it because the user population can be extremely volatile and therefore difficult to manage. All consumer services offer e-mail accounts, often with multiple mailboxes for a family.

■ **File transfer**   Not the sexiest Internet application by any means. File Transfer Protocol (FTP) is still incredibly valuable as a distribution mechanism, as well as the most common method for posting HTML pages to Web servers. Most providers will make FTP access to Web directories available so customers can easily maintain their Web content.

■ **Domain name service**   Necessary to have a branded corporate, or even personal, presence online. Many ISPs will register a domain name on behalf of the customer as a value-added service. Virtually all offer DNS hosting, both in a primary and secondary (backup) capacity.

■ **IP addressing**   Required to be connected to the Internet. Corporate networks with dedicated access will generally be assigned an appropriately sized pool of addresses for internal servers and clients. Consumers using dial-up or broadband typically receive a single dynamic address via DHCP or other similar mechanisms.

▲ **Security**   Has quickly become one of the most important differentiators in Internet service, not just for corporations but for consumers as well. The recent spate of DoS attacks on premier e-commerce companies has made security issues much more visible. Managed firewall and related security services are popular for corporate customers. ISPs are also at least trying to educate their residential customers about Internet risks and solutions such as personal firewalls—because of liability issues, some providers are bundling security products with their services.

SUMMARY

From its modest beginnings in 1969, the Internet has grown to become a global phenomenon, widely accepted not only by academics, but by corporations and consumers as well. It is not an overstatement to say that the Internet has changed the way we live and do business. We now take the Internet for granted—indeed, a large portion of the world population has come to depend on it for information and commerce. There are many ways to access the Internet, and consumers and businesses have varying needs, but the goal is always the same: to connect to an international community.

PART II

Network Foundations

CHAPTER 4

The Physical Layer

In Chapter 2, we discussed the seven protocol layers of the Open Systems Interconnection (OSI) Reference Model (the Model). Although each layer of the Model is indispensable in performing its particular function for the good of the whole, the Physical layer (at the bottom of the hierarchy) is the only layer that is both necessary and sufficient for intercommunication to occur between two points.

Every system that communicates across a network must first communicate from point A to point B. The role of the Physical layer is to provide a channel by which bits can be transferred either between two points (in the case of *point-to-point* communication) or among multiple points (in the case of *point-to-multipoint* communication). As such, the Physical layer plays a critical role in standardizing such simple-but-critical issues as the use of a particular physical medium, the representation of bits on that medium, the connector standard for that particular interface, and the specific operations by which bits are transferred across the interface.

In this chapter, we will explore various aspects of the Physical layer and its protocols, with a focus on the principles that are common to many Physical layer standards. Thus, we will look at the Physical layer from the following points-of-view:

- ▼ Physical layer architecture
- ■ Communications media
- ■ Signaling and information content
- ■ Timing and synchronization
- ▲ Multiplexing at the Physical layer

Following these discussions, a brief overview of popular Physical layer protocols will be presented as a summary of the chapter.

PHYSICAL LAYER ARCHITECTURES

Relationships at the Physical layer can be described by the "send-receive" characteristics of the communicating parties as well as by their "endpoint status" characteristics. In addition, Physical layer operations can be distinguished as either *serial* or *parallel* in their configurations. We shall discuss each in turn.

Simplex, Half-Duplex, Duplex—What's the Difference?

Figure 4-1 illustrates three possible send-receive configurations in common use at the Physical layer: a *simplex* send-receive relationship, a *half-duplex* send-receive relationship and a *duplex* (sometimes called *full-duplex*) send-receive relationship.

In a simplex send-receive relationship, one partner can transmit and one partner can receive, period. This relationship can never change in the course of communications. Broadcast television is a good example of a simplex Physical layer relationship.

In a half-duplex send-receive relationship, either partner can be the transmitter and either party can be the receiver, but never simultaneously. Thus, the communications facility

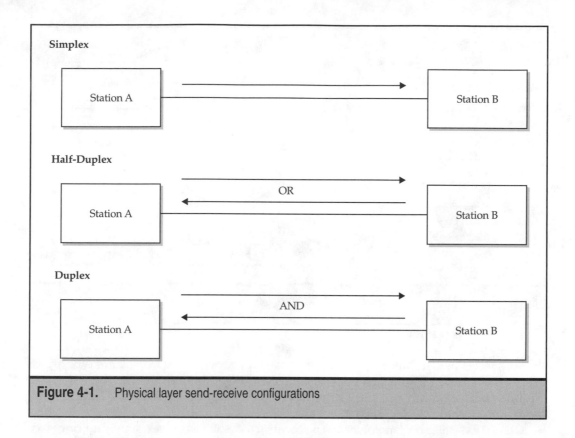

Figure 4-1.   Physical layer send-receive configurations

must be "turned around" for two-way communication to be effected. The way we typically converse on the telephone is a good example of half-duplex communications. Although sometimes we try to talk and listen at the same time, our conversations proceed much more smoothly if we take turns in the speaking process. In human conversation, the protocol for deciding who speaks when is full of nuance (perhaps a brief pause in conversation or a subtle change of intonation). In half-duplex arrangements at the Physical layer, some explicit set of rules must be established to permit communicating parties to determine when it is time to transmit or to receive.

In a duplex send-receive relationship, both sides may transmit and receive at any time (even simultaneously). This relationship predominates in computer-to-computer communication at the Physical layer. If one has access to two communications channels (one for each direction of transmission), duplex communications between machines considerably increases the efficiency of the system.

Point-to-Point vs. Point-to-Multipoint

Figure 4-2 illustrates the basic difference between a point-to-point physical architecture and a point-to-multipoint physical architecture. In the former, two communicating parties

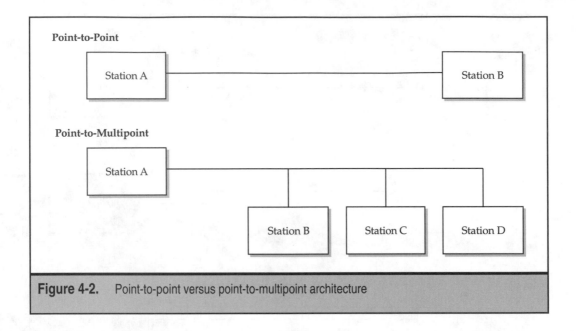

Figure 4-2. Point-to-point versus point-to-multipoint architecture

share a single link. The link may be simplex, half-duplex, or duplex. In the latter type, transmissions from one Physical layer device are received by multiple devices. These links are typically simplex (such as cable TV) or half-duplex (such as 10/100BaseT Ethernet), although they can be duplex (such as SONET).

Other Physical layer topologies can be constructed, such as busses, stars, and rings (see Figure 4-3). These, however, are variations on the point-to-point or point-to-multipoint architectures. For example, the *ring* topology is a circularly, closed-loop collection of point-to-point links. The *star* topology is a set of point-to-point links that radiate from a central hub. The *bus* topology is a true point-to-multipoint topology wherein the signals transmitted by one communicating party are received by all others attached to the physical medium.

Parallel vs. Serial Physical Communication

In a parallel Physical layer arrangement, multiple communications paths lie between communicating parties (see the top half of Figure 4-4). Typically, all but one of the paths is used for data transfer; a single path is reserved for the transfer of timing information between the sender and the receiver. As such, multiple data units (typically bits) can be transmitted simultaneously over the interface. As a result of this characteristic, parallel interfaces typically exhibit high transmission rates. Unfortunately, this advantage is somewhat offset by a phenomenon called *skew*, in which circuits do not respond evenly to a propagated signal. Because each of the physical paths that comprise the parallel channel (typically copper wires) exhibit different electrical characteristics (such as resistance), the bits on each path may behave differently. For example, electrical waves may travel at different relative

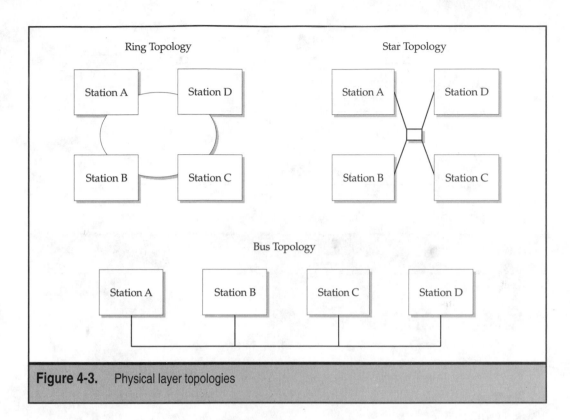

Figure 4-3.   Physical layer topologies

speeds over each of the paths. When this happens, a set of bits placed on the channel at a single point in time at the transmitter may arrive at the receiver at different times. This effect is exacerbated with distance, such that, after traveling a certain distance from the transmitter, the bits may become unrecoverable at the receiver. Due to this skew phenomenon, parallel interfaces are typically distance-limited. For example, IBM uses a parallel Physical layer interface to interconnect mainframe computers and their peripherals. This so-called *bus-and-tag* interface runs at 36Mbps but is limited to a mere 400 feet.

In a serial Physical layer arrangement, a single communications path exists between communicating parties (see the bottom of Figure 4-4). Thus, bits must follow one another down the channel between the transmitter and the receiver. The inability to send data simultaneously down multiple paths tends to limit data rates in serial environments. This is not to say that serial interfaces are inherently slow, but rather to point out that, all other things being equal, parallel interfaces display a speed advantage over serial ones. Of course, the problem of skew disappears in serial environments, making them more suited to long-distance communications. An issue that emerges in serial environments, however, is that of timing and synchronization between the transmitter and the receiver. With only one functional path available, serial approaches must make use of techniques by which timing information can be "extracted" from the data themselves. We will discuss this issue at length later in this chapter in the section "Timing and Synchronization."

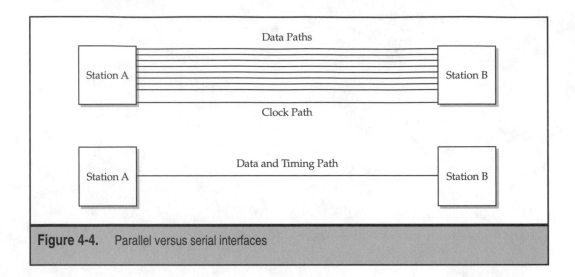

Figure 4-4.   Parallel versus serial interfaces

COMMUNICATIONS MEDIA

Generally, one of four types of media is employed at the Physical layer:

▼   Twisted-pair media (such as telephone cable and networking cable)

■   Coaxial cable (such as cable used in community-antenna TV networks)

■   Optical fiber

▲   Free space (such as that used in wireless communications)

Each of these media differs in terms of installed base, information carrying capacity, noise immunity, ownership of infrastructure, ease of installation, and a myriad of other considerations. For our purposes, Physical layer media can be characterized in the following ways:

▼   By information carrying capacity

■   By noise immunity

■   By ease of installation

▲   By popularity among users

Table 4-1 shows a matrix that compares various physical media along the lines just established.

Twisted-pair cable is manufactured by intertwining two insulated conductors in much the same shape as a DNA molecule (see Figure 4-5). The particular rate of twist and the type of insulation used, among other things, give rise to a number of categories of twisted-pair cable. Generally, the higher the category number, the higher the information carrying capacity of the cable. Because one conductor is now partially surrounded by the

Medium	Capacity	Noise Immunity	Ease of Installation	Popularity
Twisted Pair	Low	Poor	Easy	Very Popular
Coaxial	High	High	Difficult	Unpopular
Fiber	Unlimited	Excellent	Easy	Very Popular
Free Space	Growing	Poor	N/A	Growing

Table 4-1.   Comparison of Physical Layer Media

other, the noise immunity of the pair is greater than if the two conductors were not twisted. Compared with other physical media, twisted pairs offer the least information carrying capacity and relatively poor noise immunity. They are easy to install and are by far the most popular medium for the Physical layer. Twisted-pair wiring for low-speed LAN applications (called Category 3 twisted pair) is suitable for data rates up to 10Mbps over 100 meters. In contrast, high-speed LAN wiring (Category 5 twisted pair) will support data rates of 100Mbps over the same 100-meter span. Most of the twisted-pair wiring in place in LAN environments today is either Category 3 or 5.

In comparison to twisted-pair media, coaxial cable (COAX) offers significantly more information carrying capacity. Figure 4-5 shows that coaxial cable is structured such that one conductor is completely surrounded by the other. For this reason, COAX is particularly immune to electromagnetic noise. Unfortunately, as the information carrying capacity of coaxial cable increases, so does its size and weight. COAX is notoriously difficult to install and maintain. For this reason, it has lost popularity in real-world applications over the past several decades. Rarely seen in LANs anymore, COAX is most prevalent in community-antenna TV networks (CATV) for use in delivering signals across the "last mile" (where the size and weight of the cable can be limited to manageable values).

Fiber-optic media are manufactured essentially from extremely transparent glass (silica), and they work not by conducting waves of electrons like twisted-pair and coaxial media, but by transmitting waves of photons (light). The physical structure of a strand of optical fiber cable is shown in Figure 4-5. The refractive indices of the core and the cladding materials are set in such a way as to promote the phenomenon of "total internal reflection." When one introduces a light wave to the core of the fiber, the principle of total internal reflection ensures that it stays there as it propagates down the fiber. It is sometimes convenient to think of the interface between the core and the cladding as a mirror that "guides" the photon wave down the core of the strand without letting it escape into the cladding. Although some fiber-optic media are limited to relatively low information carrying capacity (such as multimode fibers), the technology has not yet been invented that can fully utilize the information carrying capacity of optical fiber. In modern

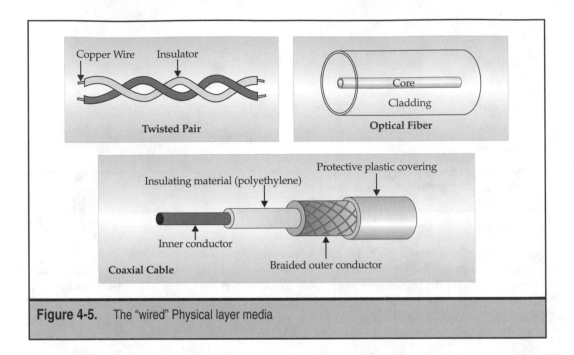

Figure 4-5.   The "wired" Physical layer media

systems, data rates in the thousands of billions of bits per second are commonplace over a single strand of optical fiber. Because light signals are not electrical in nature, fiber-optic communications is particularly immune to electromagnetic noise. Optical fiber is at least as easy to install as twisted pair, which would explain its enormous popularity first in long-haul networks (1980s), metro networks (1990s), and perhaps even residential and desktop networks (2000s).

Free space media represent a special case, in that, unlike twisted pair, COAX, and fiber, this medium is typically owned by the public in the public domain. As such, its use is carefully regulated within a particular geography. In the United States, the Federal Communications Commission (FCC) determines how the electromagnetic spectrum is divided up and assigned to those who wish to use it for communications purposes. For example, the frequencies between 88MHz and 1GHz are reserved for FM radio and broadcast television. Higher frequencies (2GHz–50GHz) are assigned to various terrestrial microwave communication systems. Unlike other media whose installed quantity can simply be increased to meet increases in demand, free space media are limited in fact. The information carrying capacity of a portion of the spectrum increases as the width of the *passband* (the difference between the lowest frequency and the highest frequency in a given spectral band) increases, all other things being equal.

Figure 4-6 shows the electromagnetic spectrum between 10KHz and 1000THz. The figure highlights areas of the spectrum that have been assigned to common communications services (such as radio and television). In addition, the figure illustrates the spectral

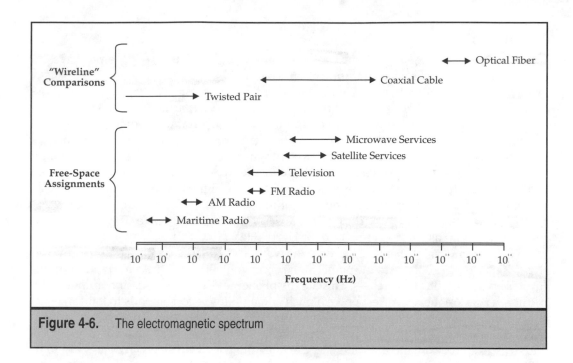

Figure 4-6.   The electromagnetic spectrum

capacity of the other forms of physical media discussed in this section (twisted pair, coaxial cable, and fiber-optic cable) for comparative purposes.

Depending on the particular frequency, free-space media are more or less noise immune. As a general rule, however, free-space media have a noise immunity characteristic closer to twisted pair than to COAX or fiber. At lower frequencies, electrical impairments (such as lightning strikes) are an issue; at higher frequencies, other forms of interference are commonly encountered (such as "rain fade" in microwave systems). Of course, aside from having to obtain a license to operate in an area of assigned spectrum, there are no installation issues that impact free-space media. The popularity of the use of free-space media for communications is evidenced by the explosive growth of wireless technologies (cellular telephones, PCS systems, wireless LANs) during the past several decades. Moreover, technological advances aimed at exploiting more of the available spectrum are occurring at a rapid pace.

SIGNALING AND INFORMATION CONTENT

Signaling refers to the precise method by which information is conveyed across the physical medium. Signaling techniques can be grouped into two general categories: *broadband* and *baseband*. Broadband signaling corresponds roughly to what is commonly thought of as analog in nature, whereas baseband corresponds roughly to digital signaling.

Broadband Signaling

Broadband, or analog, signaling is based on the use of continuously varying waves to convey information across a communications channel. Such continuously varying waves are usually represented by the sine function, and are called *sine waves*. Figure 4-7 illustrates a sine wave and shows two of the three ways in which it can be characterized. As shown in the figure, a sine wave can be thought of as beginning at some midpoint between two extremes (such as voltages). As time progresses, the sine wave approaches and becomes equal to the higher of the two extremes. The wave then decreases through the midpoint once again on its way to the lower of the two extremes before returning once again to the midpoint. This set of excursions (midpoint>highest extreme>midpoint>lowest extreme>midpoint) composes one *cycle* of the sine wave. The number of such cycles that occur over 1 second is said to be the *frequency* of the sine wave. Frequency is expressed therefore as cycles-per-second, or Hertz (Hz).

A second important characteristic of a sine wave is its *amplitude*, which refers to the relative distance between the extreme points in the excursion of the wave. As the points become further apart, amplitude increases. To state the phenomena of frequency and amplitude in musical terms, frequency corresponds to pitch and amplitude corresponds to loudness.

A third characteristic of sine waves is more difficult to describe. That characteristic is *phase*. The phase of a given sine wave is measured relative to another sine wave (a reference) and is expressed as an angular difference between the two. Two sine waves are said to be "180 degrees apart" when, at the same instant that one of them has reached its positive-most point, the other has reached its negative-most point. Figure 4-8 illustrates relationships between pairs of sine waves along the amplitude, frequency, and phase dimensions. The first sine waves in the figure occur at different frequencies. Those in the middle of the figure occur at different amplitudes. The sine waves at the right side of the figure occur at different phases; in fact, they are 180 degrees apart. Notice that for the waves that display a phase difference, the other two characteristics (frequency and amplitude) are equal.

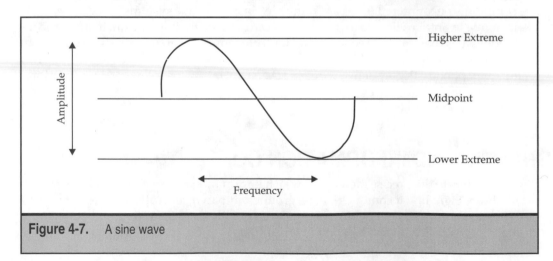

Figure 4-7.   A sine wave

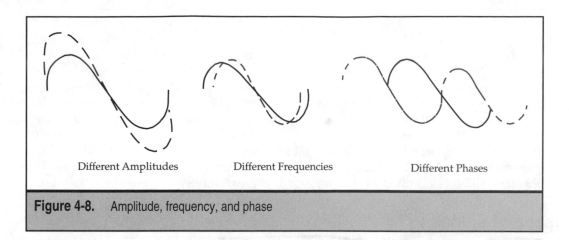

Figure 4-8.   Amplitude, frequency, and phase

The point of this discussion is to emphasize that only three characteristics of sine waves are available to be manipulated for the purposes of conveying information over a communications channel. Thus, if a communications channel requires the use of broadband signaling, and the information to be conveyed is digital in nature (0s and 1s), some mechanism for "impressing" the digital information onto a sine wave must be designed. A device that is designed to convey digital information over a broadband communications channel is a *modem*. Modems manipulate one (or a combination) of the three characteristics of a sine wave to signal the presence of either a 0 or a 1 on the channel. Historically, modems have manipulated amplitude (amplitude shift keying), frequency (frequency shift keying) or phase (phase shift keying). Currently, most modems manipulate both amplitude and phase by a technique called *quadrature amplitude modulation* (QAM).

Figure 4-9 illustrates the operation of a frequency shift keyed modem. The broadband communications channel in this example is a telephone line. The line can accept only frequencies that fall roughly between 300 and 3300Hz. To convey digital information across

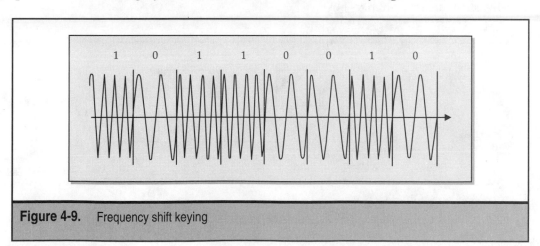

Figure 4-9.   Frequency shift keying

such a channel, we first select two discrete frequencies that fall within the channel (such as 800Hz and 2400Hz). Then we arbitrarily assign one frequency (the lower one in the figure) to represent a digital 0, and the other frequency (the higher one in the figure) to represent a digital 1. By switching between the two frequencies, depending on whether a 0 or a 1 must be conveyed to the receiver, the transmitter *modulates* the incoming bit stream to be compatible with the broadband nature of the channel. The receiver, in a process called *demodulation*, looks at the line, notes which of the two frequencies it is currently receiving, and converts the frequency back to a 0 or a 1, whichever is appropriate. Whether a modem manipulates frequency, amplitude, or phase, the principle of modulation at the transmitter and demodulation at the receiver are the same. The specific technique used in practice will be determined by the bandwidth of the channel and the desired data rate.

Baseband Signaling

Baseband, or digital, signaling is based on the use of discrete states to convey information across a communications channel. These discrete states are typically represented as pulses of some sort (such as voltage) and are thus often called *square* waves. Figure 4-10 shows a square wave by which two discrete states (0 and 1) can be represented each by a different voltage (plus and minus 12 volts in the figure).

Many different baseband signaling schemes have been developed over the years; a few of the most popular are shown in Figure 4-11. At the top of the figure is a *unipolar* signaling scheme, in which a digital 1 is represented by a +5 volt state and a digital 0 is represented by a no voltage (ground) state. This scheme was widely used in early transistor-transistor logic (TTL) circuitry. In the middle of the figure is a bipolar signaling scheme, in which a digital 1 is represented by a –12 volt state and a digital 0 is represented by a +12 volt state. This scheme is widely used today in a Physical layer protocol known as EIA-232-E. At the bottom of the figure is a bipolar return to zero (BPRZ) scheme in which digital 0s are represented by a no voltage (ground) state and digital 1s are represented by alternating 3-volt pulses. This latter scheme, sometimes called alternate mark inversion (AMI), is used on T-1 lines by carriers and customers alike.

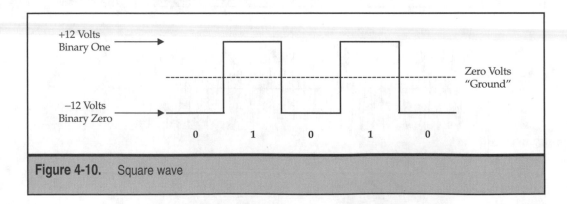

Figure 4-10.   Square wave

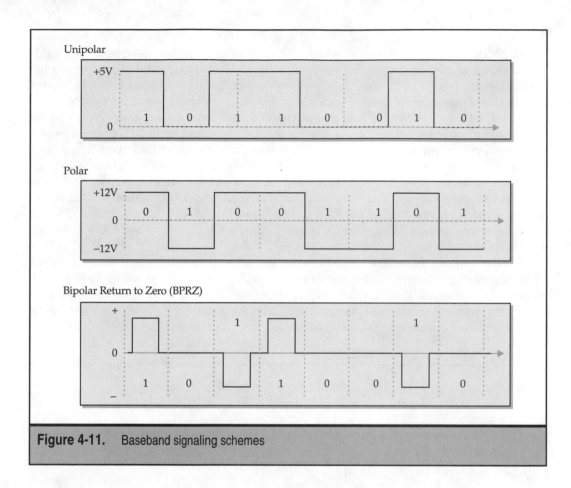

Figure 4-11.   Baseband signaling schemes

As we have seen, each of these baseband signaling schemes has either seen widespread use in the past or is in widespread use today. A major determinant of the choice to use one baseband scheme over another has to do with the ease with which timing and synchronization is supported by the technique. We shall return to this issue later in the chapter on "Timing and Synchronization."

Broadband vs. Baseband and Analog vs. Digital

Thus far, we have used the terms *broadband* and *baseband* to refer to the actual signaling technique used on a communications channel. Thus, a telephone line can be viewed as a broadband channel and a T-1 line can be viewed as a baseband channel. The information content to be conveyed over a given communications channel can be distinguished along similar lines as either analog or digital. These distinctions give rise to an important concept at the Physical layer, namely that of a physical DTE and a physical DCE.

A DTE (data terminal equipment) generates information in the form of data to be conveyed across a communications channel. The information content generated by a DTE can be either analog or digital. The purpose of the DCE (data circuit terminating equipment) is to receive the data from the DTE in whatever format it is generated and to convert that data into a format that is compatible with a given communications channel.

Figure 4-12 is a four-cell matrix that attempts to put these terms into perspective. In the upper-left quadrant, the information is in analog form and must be conveyed over a broadband communications channel. While it is not immediately obvious that any conversion of information takes place in this instance, it is usually the case that some conversion needs to occur. Take, for example, analog information from a human voice that must be conveyed over a telephone line (a broadband channel). The information arrives in the form of sound pressure waves, but the channel requires electrical representation of those waves. A microphone serves as the *transducer* between the analog source and the broadband channel.

In the lower-left quadrant, digital information must be conveyed over a broadband communications channel. This is the classic example of a computer using a phone line for communication. As shown, a *modem* (DCE) is required to effect the requisite conversion.

In the upper-right quadrant, a stream of analog information must be conveyed over a baseband communications channel. A common example of this is the use of a T-1 facility

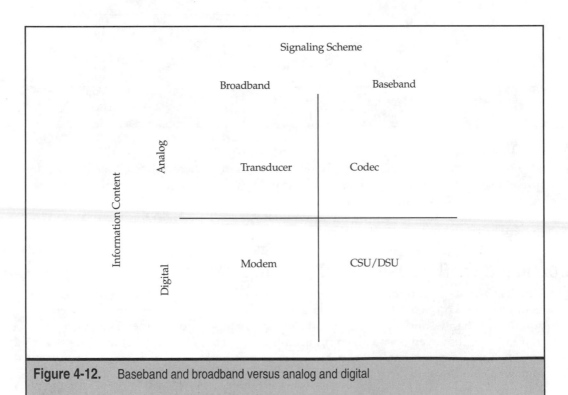

Figure 4-12. Baseband and broadband versus analog and digital

(baseband) to carry analog video information. In this case, a device called a *codec* (from the contraction of the terms "coder" and "decoder") functions as the DCE. The exact method by which a codec affects this conversion is beyond the scope of this chapter, but it will be covered later in Chapter 10.

Finally, in the lower-right quadrant, it is not immediately obvious that a conversion function is necessary. In this case, digital information must be conveyed across a baseband communications channel. Conversion is often required in such cases, because the signaling scheme used by the DTE is not directly compatible with the one required by the channel. For example, if the DTE uses a polar signaling scheme such as EIA-232 and the channel requires some other scheme such as BPRZ in the case of T-1, a conversion is required. The particular DCE that affects this sort of a conversion is called a channel service unit/data service unit, or *CSU/DSU*.

In summary, the DCE plays a vital role in the implementation of the Physical layer. Through the use of different types of DCE functions, analog or digital information can be made compatible with either a broadband or baseband communications channel.

Signaling Constraints and Channel Capacity

A given communications channel has a finite capacity to carry information, even though in some instances (such as a single-mode optical fiber), that capacity is not yet known. The information carrying capacity of a channel is related to its bandwidth and the prevailing signal-to-noise ratio on the channel in a precise way. We shall describe these relationships here.

The first determinant of the amount of information that a given channel can convey is the maximum signaling rate that can be supported by that channel. A signaling event can be thought of as a transition in a signal, whether that signal is broadband or baseband in nature. For broadband systems, the change from one frequency to another or from one amplitude to another both indicate *signaling events*. In baseband systems, the transition from one discrete state to another (for example, from +12 volts to –12 volts) is also a signaling event. In fact, such signaling events are given a special name; they are called *bauds*. Therefore, the maximum signaling rate supported by a given channel is equivalent to the maximum baud rate supported by that channel.

In 1924, working at Bell Telephone Laboratories, Harold Nyquist discovered a fundamental relationship between the bandwidth of a channel and the maximum baud rate that it could support. Nyquist stated that the baud rate cannot exceed twice the bandwidth of the channel. Thus, for a 3000Hz telephone channel, the baud rate cannot be greater than 6000 baud. For an analog television channel (6MHz), the baud rate cannot exceed 12Mbaud.

In many signaling schemes, each signal transition on the channel corresponds to the conveyance of a single bit of information from the transmitter to the receiver. With such 1-bit-per-baud schemes, the bit rate and the baud rate are equivalent. Many signaling schemes, both broadband and baseband, are able to convey more than 1 bit during each signaling event. Such multiple-bit-per-baud schemes therefore are associated with data rates that exceed their baud rates. Signaling schemes such as these are typically classified in terms of how many bits of information are conveyed by each signal transition. Thus, *dibit* (2 bits per baud), *tribit* (3 bits per baud) and *quadbit* (4 bits per baud) are commonly encountered. We shall proceed to examine an example of a common dibit encoding scheme called *differential phase shift keying* (DPSK).

If a demodulator can recognize the difference between only two phases of a sine wave, it can only convey 1 bit of information with each phase transition. To construct such a modem, we can arbitrarily assign the digital value of 0 to one phase (e.g., 0 degrees) and the digital value of 1 to the other (e.g., 180 degrees). Each time the modem shifts between 0 degrees and 180 degrees, a single bit of information is conveyed. Suppose that same demodulator was sensitive enough to distinguish the phases in the middle (i.e., 90 degrees and 270 degrees). In being able to recognize four different phases of a sine wave, we can assign each phase a dibit value. For example, 0 degrees of phase shift can indicate the digital value "00," 90 degrees of phase shift can indicate the digital value "01," 180 degrees of phase shift can indicate the digital value "10," and 270 degrees of phase shift can indicate the digital value "11." Expanding upon this theme further, we can extrapolate that, if a demodulator can recognize eight different phases, we can encode 3 bits of digital information on each phase transition. For a demodulator that can recognize 16 different phases, the encoding of 4 bits of digital information on each phase transition is possible. Note that by increasing the sensitivity of the demodulator, we can increase the data rate without a corresponding increase in the baud rate. Seemingly, this progression could go on forever were if not for one very large limitation, namely noise on the channel.

In 1948, also working at Bell Labs, Claude Shannon, noted as the Father of Information Theory, carried Nyquist's work one step further. Shannon observed that Nyquist's theorem applied only to noiseless channels. Such channels do not exist in the real world. All channels operate in the presence of noise, and that noise sets a limit, not on baud rate *per se*, but rather on the number of bits that can be encoded on a particular signaling event. Shannon's theorem shows that, for a typical telephone line (with a signal-to-noise ratio of about 30 dB), the maximum number of bits that can be encoded in a single signaling event is about 5. Thus, the maximum bit rate achievable under such circumstances is approximately 30Kbps (6000 baud multiplied by 5 bits per baud). Interestingly, modems that are able to operate at bit rates higher than this (such as the 56Kbps V.90 modem) were initially referred to as "anti-Shannon modems." Rest assured, however, that these modems do not invalidate the work of Nyquist and Shannon. Rather, they use techniques of data compression and forward error correction to increase apparent bit rates beyond the limits imposed by the work of the two Bell Labs scientists.

TIMING AND SYNCHRONIZATION

For intelligent communication to take place at the Physical layer, the transmitter and the receiver must be synchronized—that is, the receiver must know when to sample the channel to detect the presence of a valid signaling event. If the receiver samples too soon or too late, erroneous data will result. In such cases, 0s may be falsely interpreted as 1s and vice versa. The following discussion explores timing issues in both parallel and serial Physical layer environments.

Timing in Parallel Physical Interfaces

The issue of synchronization in parallel Physical layer environments is actually quite simple. Referring back to Figure 4-4, we see that, among the multiple communications paths between the transmitter and the receiver, the clock path is used to synchronize the two communicating parties. Using the clock circuit, the transmitter explicitly indicates to the receiver when valid data are present on the data circuits. If data rates and interface distances are held within limits to ensure that skew does not occur, the information supplied by the clock circuit is adequate to avoid erroneous sampling of the data circuits.

Timing in Serial Physical Interfaces

When only one path is available for communication between a transmitter and a receiver, the issue of synchronization becomes more complex. There are basically two ways to synchronize a transmitter and a receiver in a serial environment. One relies on implicit timing between communicating parties and is called *asynchronous* communication, while the other is based on the exchange of explicit timing information between communicating parties and is called *synchronous* communication.

Asynchronous Communication

The use of the term *asynchronous* to describe timing arrangements in which the transmitter and receiver do not exchange explicit timing information is quite unfortunate. To exchange intelligent information, a transmitter-receiver pair must *always* be synchronized with each other. The term *asynchronous* refers to the relative period of time over which a transmitter and receiver must be synchronized, which, in an asynchronous environment is typically very short (usually a single character from a known code set such as ASCII). Thus, the term *character-oriented* better describes asynchronous communications.

Figure 4-13 shows the method of transmission of a single character (an uppercase Z from the ASCII code set) in an asynchronous environment. In such environments, which are characterized by people typing on keyboard devices, the time between the transmission of successive characters (the inter-character interval) is variable, hence the origin of the term *asynchronous*. Once a key has been pressed, however, the data emitted from the keyboard typically consists of a 10-bit entity (a start bit, 7 data bits, a parity bit, and a stop interval). To be received without error, these 10 bits must be sampled at the proper time by the receiver. In other words, on an intra-character basis, the transmitter and the receiver must be synchronized. Ideally, the receiver should sample each bit at precisely the mid-bit interval so that maximum accuracy can be achieved. At a data rate of 1200bps (a common rate for asynchronous transmission), each bit appears on the channel for approximately 830 microseconds (1/1200bps). Thus, to sample each bit precisely, our sampling interval should be about 830 microseconds. The issue then, is in knowing when to begin the sampling process. The *start* bit supplies this information. The beginning of the start bit is signaled by a transition on the line from some positive voltage state to a state of no voltage (i.e., ground). It is perhaps easiest to think of the start bit as a "wake-up call" to the receiver. When the receiver sees the transition that signals the arrival of the start bit, it simply counts to half of the bit interval (415 microseconds in our example) and takes its

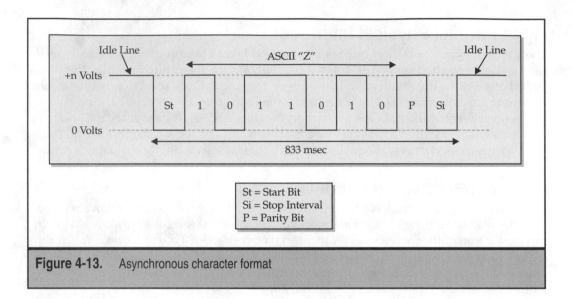

Figure 4-13.   Asynchronous character format

first sample. To read the remaining nine bits in the character, the receiver now simply counts out nine equal periods of 830 microseconds each, sampling the line at the end of each count. In this fashion, the transmitter and the receiver stay synchronized during the transmission of the character without the need to exchange explicit timing information. Of course, the asynchronous method depends on a fair degree of accuracy in the free-running clocks at the transmitter and the receiver. In addition, asynchronous communications works well only at low data rates, where bit intervals are relatively long. At a data rate of 45Mbps, each bit is present on the line for only 0.002 microseconds. Such short-bit intervals render asynchronous techniques, with their implicit timing mechanisms, totally inadequate for accurate Physical layer communication.

Synchronous Communication

When the number of bits to be conveyed between the transmitter and the receiver exceeds some small number (e.g., 10), it becomes a requirement for the communicating parties to exchange explicit timing information for synchronization purposes. On a serial link with only one path for communication, such timing information must be "embedded" in the data themselves. Such communication, wherein a timing relationship must be maintained for long periods of time (i.e., for many bit times) and where the transmitter supplies timing information as a part of the data stream is called "synchronous."

Referring back to Figure 4-11, we can see that baseband signaling schemes can be used to convey not just data, but also timing information between a sender and a receiver. The key to using baseband signals for synchronization purposes is to make sure that frequent transitions (1 to 0 and vice versa) occur in the data stream. In fact, the best synchronization pattern that exists is alternating 1s and 0s. In that case, every bit in the stream provides clocking information to the receiver. The problem in relying on state transitions for the maintenance of synchronization is that, in some cases, long strings of either 0s or 1s must be sent. Looking at both unipolar and bipolar signaling schemes (see Figure 4-11, top and middle), we observe that during a long string of either 0s or 1s, there are no line transitions to synchronize the receiver. The bipolar return to zero scheme (see Figure 4-11, bottom) has an advantage in this regard. In BPRZ, 1s are not represented by a particular line state, but rather by bipolar transitions in the stream, long strings of 1s are acceptable if not preferable. Long strings of 0s still cause problems with BPRZ encoding. They must be avoided when using this signaling scheme.

We mentioned previously that BPRZ encoding is used on T-1 circuits and, in fact, the T-1 standard states explicitly that no more than 15 consecutive 0s may occur across a T-1 span. If that number is exceeded, the line is likely to lose synchronization. Mechanisms have been developed to ensure that strings of more than 15 0s never occur on a T-1 line (although further discussion is beyond the scope of this chapter).

MULTIPLEXING AT THE PHYSICAL LAYER

The concept of multiplexing at the Physical layer is illustrated in Figure 4-14. A multiplexer is typically shown as a forward-pointing triangle. As shown in the figure, a multiplexer combines many narrowband (or low bit rate) channels into a single wideband (or high bit rate) channel. The main advantage of multiplexing is the notion of *pair gain*, wherein one can save on physical facilities and their associated costs. Multiplexing was first introduced in telephone networks to carry large numbers of calls between switching offices using relatively few physical facilities. Multiplexing can be done with broadband signals or with baseband signals. In the former case, the operation is called *frequency division* multiplexing; in the latter case, it is called *time division* multiplexing.

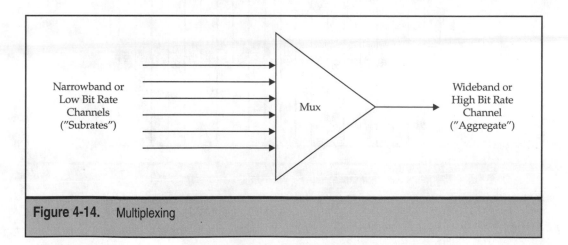

Figure 4-14.   Multiplexing

Frequency Division Multiplexing

Frequency division multiplexing (FDM) is shown at the top of Figure 4-15. In FDM, a wideband channel is divided "horizontally" into a number of independent channels. In the figure, a wideband channel with 40KHz of capacity is divided into 10 subchannels, each with a capacity of 4KHz. Because a single telephone conversation uses approximately 4KHz of bandwidth, such a multiplexer can carry ten conversations over the same physical facility previously used to carry only one. The technique by which FDM is accomplished is called *carrier modulation*. Each subchannel supported by the multiplexer operates a carrier

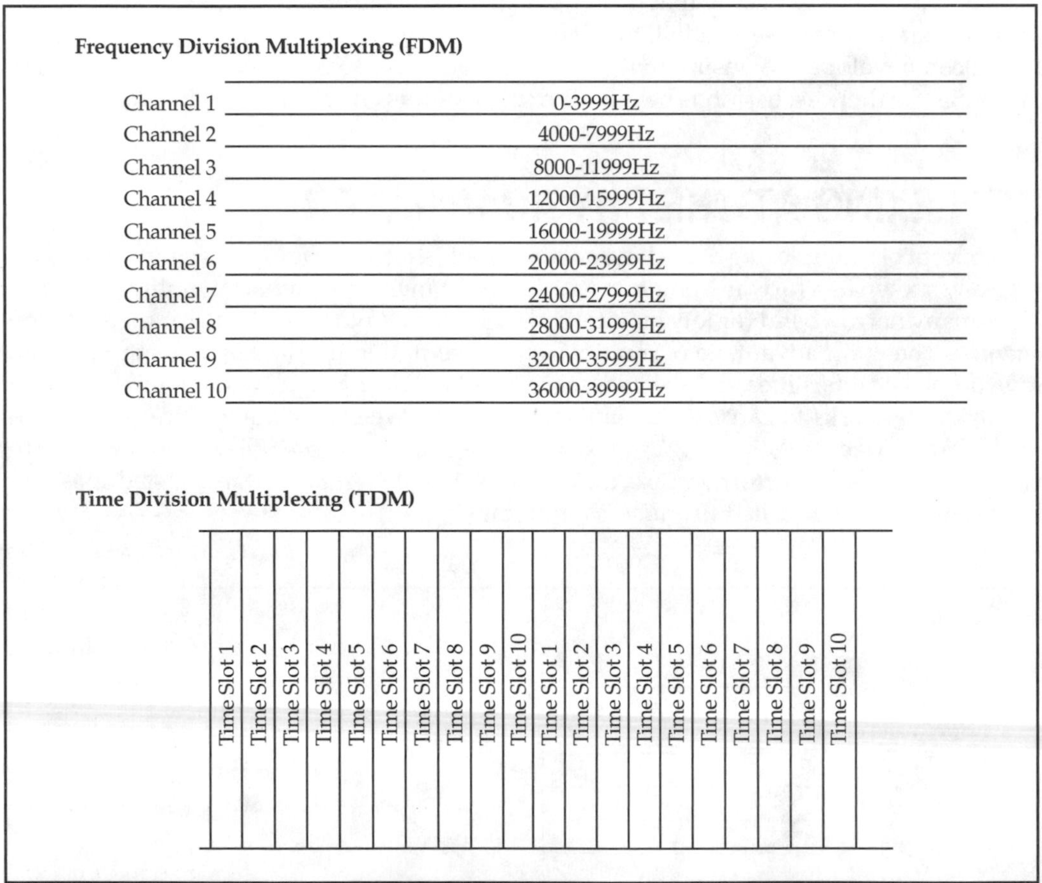

Frequency Division Multiplexing (FDM)

Channel 1	0-3999Hz
Channel 2	4000-7999Hz
Channel 3	8000-11999Hz
Channel 4	12000-15999Hz
Channel 5	16000-19999Hz
Channel 6	20000-23999Hz
Channel 7	24000-27999Hz
Channel 8	28000-31999Hz
Channel 9	32000-35999Hz
Channel 10	36000-39999Hz

Time Division Multiplexing (TDM)

Time Slot 1 | Time Slot 2 | Time Slot 3 | Time Slot 4 | Time Slot 5 | Time Slot 6 | Time Slot 7 | Time Slot 8 | Time Slot 9 | Time Slot 10 | Time Slot 1 | Time Slot 2 | Time Slot 3 | Time Slot 4 | Time Slot 5 | Time Slot 6 | Time Slot 7 | Time Slot 8 | Time Slot 9 | Time Slot 10

Figure 4-15.   Frequency division multiplexing versus time division multiplexing frequency exactly in the center of the subchannel (e.g., 2000Hz, 6000Hz, 10KHz, 14KHz, and so on in our example). At the transmitter, the narrowband signal to be multiplexed into a given subchannel is modulated by a carrier frequency and then slotted into the system in the appropriate place. At the receiver, a subchannel is isolated and then demodulated to remove the carrier frequency. What remains is the original narrowband signal that was modulated at the transmitter.

FDM is the technique by which many radio stations share a single portion of the spectrum. When you tune your radio, you are adjusting the circuitry to isolate and demodulate a particular carrier frequency. The result is a demodulated signal that is interpretable by the human ear. FDM technology is widely used by many communications systems in current use, including cellular telephone systems, satellite broadcast systems, and most CATV systems.

Time Division Multiplexing

The spectral energy of a baseband signal (i.e., a square wave) cannot be contained in the frequency domain. Although a detailed explanation of this fact is beyond the scope of this discussion, suffice it to say that a square wave is composed of a fundamental sine wave component and an infinite number of "harmonics" that bear a mathematical relation to that fundamental. Each harmonic is of a higher frequency and a lower amplitude than the fundamental. As such, square waves contain infinitely high frequency components. When the telephone network began to make use of baseband signaling techniques to improve voice quality, a new method of multiplexing had to be deployed. FDM was simply not possible with baseband signals.

Time division muliplexing (TDM) was developed to permit the combination of multiple low bit rate data channels onto a single high bit rate channel. The concept of TDM is illustrated at the bottom of Figure 4-15. The high bit rate channel is divided "vertically" into a number of independent time slots. Each low bit rate channel is assigned a particular time slot on the high bit rate aggregate. When its particular time slot becomes available, each low bit rate input is permitted to use the entire bandwidth of the aggregate channel for the duration of that time slot. The next slot is used by another low bit rate input, and so on. TDM is used extensively in systems that use baseband signaling. A prime example is in the terrestrial telephone network where TDM can be found both as an access technology (e.g., T-1 access) and as an interoffice trunking technology (e.g., SONET).

FDM and TDM can be combined. That is, a subchannel in an FDM system may be further subdivided into multiple channels using TDM. Digital cellular telephone networks operate in exactly this fashion. To return to our FM radio analogy, when you tune your radio to a certain channel, you will note that the channel is further subdivided in time. For about 3 minutes, you hear one song, and for the next 3 minutes, you hear another song. This is an example of FDM channels being further subdivided by TDM technology.

PHYSICAL LAYER PROTOCOL EXAMPLES

Hundreds of Physical layer standards have been developed over the years. All of them, however, share certain characteristics:

▼ A Physical layer protocol must specify the mechanical relationship between communicating parties (DTE-to-DCE or DCE-to-DTE). The sort of connector(s) to be used and the positions of leads on those connectors are example mechanical considerations.

■ The standards must specify the electrical characteristics of an interface. The voltages (or frequencies, amplitudes, and phases) used to represent 0s and 1s must be agreed upon.

■ The rate(s) of data exchange and the maximum operational distance of the interface are typically specified in some fashion.

■ Functional characteristics of the interface, such as which leads on the connector are responsible for performing the functions carried out over the interface, must be defined.

▲ Procedural characteristics of the interface must be specified. For example, if a Physical layer protocol is designed to support dial-up operation, the sequence of events required to dial a remote telephone number must be specified.

Detailed information on even a few of the most popular Physical layer protocols would fill the pages of an entire book. Thus, the aim of the following discussion of two popular Physical layer standards is intended to present the general characteristics of such protocols.

EIA-232-E

Once known as RS-232-C, the EIA-232-E Physical layer standard has enjoyed popularity as a standard for numerous years. EIA-232-E is found in both synchronous and asynchronous environments in which data rates do not exceed 20Kbps (note that over short distances, EIA-232-E can be driven to data rates of about 56Kbps). A DTE-DCE interface is commonly used between low-speed data terminals (such as PC COM ports) and modems for communication over telephone lines. Although not specified in terms of distance per se, the maximum permissible distance between the DTE and the DCE is about 50 feet using EIA-232-E (the use of "low capacitance" cable can extend this distance substantially). EIA-232-E specifies the use of the DB-25 connector (a 25-pin connector), and can support up to 25 individual leads on the interface. Rarely are all 25 leads used in practical applications, however. In fact, IBM popularized the use of the DB-9 connector (a nine-pin connector) for support of EIA-232-E in its PC products. The most common leads and their functions are shown in Figure 4-16.

The functions of the "transmit" and "receive" leads are self-explanatory. Note that the use of the terms *send* and *receive* are from the perspective of the DTE. The "data terminal ready" and "dataset ready" leads are used to indicate, in both directions, that the

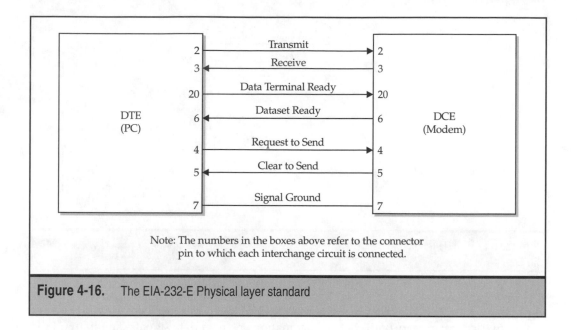

Figure 4-16.   The EIA-232-E Physical layer standard

equipment on either side of the interface is powered on an operational. "Request to send" and "clear to send" are used to effect line turnaround in half-duplex environments. The DTE signals its intention to transmit by a transition on the "request to send" lead. If the physical facility is available for transmission at the time, the DCE indicates this condition by a transition on the "clear to send" lead. Transmission then ensues.

Note that every lead on the EIA-232-E interface is referenced to a common "ground" lead. For this reason, EIA-232-E is called an *unbalanced* interface. Another Physical layer standard, EIA-422-A specifies *balanced* operation in which each lead is associated with its own separate "ground" lead. As a result, EIA-422-A can support higher data rates over longer distances than EIA-232-E. The configuration of EIA-232-E shown in Figure 4-16 is for asynchronous, leased-line communication. When used in synchronous environments, other leads are used to support synchronization between the DTE and the DCE. As a final note, EIA-232-E is said to be an *out-of-band* Physical layer interface because the only signals that flow over the send and receive leads are data signals. All control signals are given their own separate leads; control signals never flow across the send and receive data leads.

ITU Recommendation X.21

The X.21 Physical layer standard is popular outside of North America in what are called *circuit switched data networks* (CSDN). X.21 specifies a synchronous serial interface between a DTE and a DCE that comprises eight leads. The interconnection of the DTE and the DCE across an X.21 interface is shown in Figure 4-17. X.21 is implemented using a

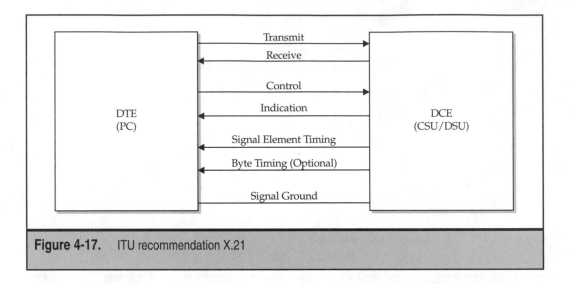

Figure 4-17.   ITU recommendation X.21

15-pin connector. It is considered an *in-band* Physical layer interface because the transmit and receive leads can carry both control information and user data signals. The status of the control and indication leads determine whether a particular signal on the transmit or receive leads should be interpreted as control information or user data. For example, when the control lead is in one state, the information on the transmit lead is dialing information (such as a telephone number). When the control lead is in the opposite state, information on the transmit lead is user data.

SUMMARY

This chapter has explored the fundamental characteristics of the Physical layer of the OSI Reference Model. Of all of the OSI layers, only the Physical layer is both necessary and sufficient to support communications between systems.

Many agreements must be reached at the Physical layer to ensure intelligent communication between (and among) end stations. From the architecture of the interface (point-to-point versus point-to-multipoint, serial versus parallel) to the medium over which communication is to occur (twisted pair versus fiber) to the method for signaling (broadband versus baseband) to the method of multiplexing that is to be used (FDM versus TDM), the Physical layer is rife with complexity. We have attempted here to foster an appreciation of that complexity in the view of the reader. Armed with the information in this chapter, you can now embark on your own study of the myriad of Physical layer standards and documents that pertain to the modern world of telecommunications. The information delivered in this chapter will help you understand concepts that follow in subsequent chapters.

CHAPTER 5

Data Link Layer Protocols

This chapter provides a look at the protocols and services offered by the Data Link layer of the OSI Reference Model. The Data Link layer has evolved the most from its original definitions. Today, just as 30 years ago, the data link protocols provide a set of services that are critical to data communications. These protocols are never flashy and are sometimes totally ignored (catchy phrases such as Packets Over SONET, or Packets Over Glass, neglect to mention that a Data Link layer protocol is necessary and used). In this chapter, we explore the traditional functions and services offered by the Data Link layer, and we visit a few traditional data link protocols that implement these functions. We also review the evolution of data link services and visit a number of modern data link protocols.

DATA LINK LAYER FUNCTION

The principal function of the Data Link layer is to ensure error-free transmission over a physical transmission path. The Data Link layer cannot prevent errors from occurring in the physical path; rather, it recognizes errors have occurred and corrects them. The error messages sent among entities at the Data Link layer are usually called *frames* (see the sidebar "Data Link Protocol Terminology").

Errors can be handled in different ways, depending on the characteristics of the physical path. *Forward error correction* relies on additional information being carried in the frame, so that the receiving station can repair an erroneous frame. This approach is appropriate for transmission paths that are prone to error or that have significant delays—transmissions from space probes are likely candidates for forward error correction schemes.

The most common approach to error correction is for the station that receives an erroneous message to discard the "bad" frame and wait for an error-free copy. Depending on the agreements passed among communicating Data Link layers, frames might also be considered in error if they arrive out of sequence. Some contemporary Data Link layer protocols do not perform error correction; instead, they perform only detection and discard. They assume that higher layer protocols will handle errors, as necessary. Protocols adapting this approach rely on a high-quality transmission path with a low bit error rate, such as optical fiber.

A basic principle of the OSI Model is that a given layer does not "worry" about services provided by lower layers. As such, no layer above the Data Link layer is required to do anything to ensure that the correct bits arrive. The sending and receiving Data Link layer stations use control information to ensure that only "correct" information is passed to higher layers. The sending station adds control information to the beginning and end of the data (most Data Link layer protocols function in this manner), and then sends the frame to the Physical layer. The receiving station uses the control information to ensure that the frame has been received correctly.

Header information typically includes the receiving station's address, a frame sequence number, and an acknowledgment number. An error detecting code is normally included in the Trailer field.

DATA LINK LAYER SERVICES

To deliver error-free data to the higher layers, the Data Link layer must support a set of services that defines how communications takes place, how information is transferred, and how errors are handled.

▼ **Line Access Control**   Line access control determines which station can transmit next. This is easy for two stations on a full-duplex link. When more than two stations are in use on a full-duplex link (such as multipoint) or any number of stations are in use on a half-duplex line, however, transmission needs to be controlled carefully.

■ **Frame Synchronization**   The Data Link layer is responsible for providing synchronization at the frame level—that is, it determines the beginning and end of each frame. The Physical layer is typically responsible for maintaining bit synchronization.

■ **Data Delimiting**   The Data Link layer adds control information to the transmitted frame. The receiving station must be able to determine where the Control field ends and the Data field begins. Control fields contain error detecting/correcting codes, sequence numbers, acknowledgment numbers, addresses, and other information.

■ **Transparency**   Special characters or bit patterns indicate the beginning and end of frames, as well as the beginning and end of fields within frames. If the Data field contains the special character or bit pattern, the receiving station could interpret its meaning incorrectly. This cannot be allowed, so such characters (or bit patterns) must be made "transparent" to the receiving station.

■ **Error Detection and Recovery**   Using a combination of sequence numbers and an error detecting or correcting code, the Data Link layer protocol ensures that frames with errors are recognized and not delivered to higher layers. Recovery is via retransmission for error-detecting codes. Timers are used to ensure that all transmitted frames are received.

■ **Flow Control**   Sometimes the receiving station must be able to cut off the transmitter; the receiver may be too "busy" to accept new frames. At the Data Link layer, flow control enables the receiver to tell the transmitter it is not ready, and to later indicate its willingness to accept more frames.

■ **Addressing**   On two-station connections, addresses can be used to separate commands from responses. Addresses are required on multipoint links with more than two stations to denote the intended receiver and sometimes to identify the sender as well.

▲ **Connection Establishment and Release**   After all physical connections are in place and the equipment is running, the Data Link layer must be activated. Similarly, a mechanism to deactivate a data link connection must exist.

Data Link Protocol Terminology

During the discussion of the functions of a generic data link protocol, we referred to *frames* as the unit of information transfer. You should note that many protocol definitions use different terms for the same aspect of their operation, and other aspects and terminology may also differ.

For example, the Open Systems Interconnect Reference Model refers to the Data Link Layer Protocol Data Unit (DLPDU) as the unit for information transfer. This unit is defined in the ITU-T standards X.200 and X.212. The information field is called the *Service Data Unit*, and the various control fields that manage the link make up the Protocol Control Information.

IBM developed the first bit-oriented data link protocol, called Synchronous Data Link Control (SDLC). It was modified eventually and adopted by the ISO and ITU-T as the High Level Data Link Control (HDLC) protocol. HDLC (or SDLC) has many subsets, including Link Access Procedure-Balanced (LAPB). These protocols all call the DLPDU a *frame*.

However, in IBM's character-oriented Binary Synchronous (BISYNC) protocol, the DLPDUs are called *blocks*. A large text block can also be called a *message*; control information, such as an acknowledgment, is often called a message as well. In DEC's Digital Data Communications Message Protocol (DDCMP), all DLPDUs are called *messages*. Finally, local area networks (LANs), especially those defined by the IEEE, call their Media Access Control (MAC) PDUs *frames*.

Keep in mind that protocols are merely agreements that facilitate communication. We must often adapt our terminology to communicate with others, but the concepts of these terms remain similar.

Line Access Control

Line access control is determined by the type of physical circuit connecting the stations and the number of stations connected. Physical links can be full-duplex (two-way simultaneous) or half-duplex (two-way alternate). Full-duplex links allow two communication stations to transmit at will, whereas on a half-duplex link, the stations must take turns. Some protocols (such as BISYNC) force two-way alternate communication even on a full-duplex link.

Two link configurations are possible: *point-to-point* (only two stations) or *multipoint* (more than two stations). In a multipoint configuration, one of two modes of operation is possible: *hierarchical* and *peer-to-peer*. In the hierarchical mode, one station is designated the *primary* and the others are *secondaries*. In the peer-to-peer mode, a distributed access control protocol is in operation between the stations. When a station wants to "talk," it must obey the access protocol.

The possible data link arrangements are as follows:

▼ **Two-Way Simultaneous Point-to-Point**   If the link is full-duplex and point-to-point, the line access control issue is moot.

■ **Two-Way Simultaneous Multipoint**   Stations on a half-duplex, point-to-point link must agree to and follow a procedure for taking turns.

■ **Two-Way Alternate Point-to-Point**   This procedure is often accomplished by passing a "wand." Only the holder of the wand can transmit. In practice, the wand is either a defined message or a bit, or bits, within a message sent between stations. Alternatively, a station wanting to transmit could monitor the line for data (from the other station). If no data is sensed, the line is assumed to be idle and the station transmits.

▲ **Two-Way Alternate Multipoint**   Multipoint configurations on a half-duplex link are handled in a similar way to those on a full-duplex link. However, the primary station *cannot* transmit after it has polled a secondary. It must wait for the secondary to return the poll. (A *poll* is used to ask the station whether it has anything to send to the primary station.)

Other Line Access Controls

A discussion of line access controls would not be complete without mentioning those methods employed by LAN data links. Token Ring, Fiber Distributed Data Interface (FDDI), and token bus LANs use variations of the distributed polling technique, called *token passing*. These networks are half-duplex multipoint links operating in a peer-to-peer fashion. A station wanting to transmit waits until an electronic "token" is passed to it by one of its peers. After sending its data (or after an established maximum time) the token is passed to another station on the link. This orderly passing of the token allows each station an opportunity to transmit, but no single station dominates the others, as in primary/secondary environments.

An alternative to the distributed polling mechanism is a contention-based system. As the name implies, contention-based methods find stations contending for a common half-duplex multipoint link. When a station needs to transmit, it checks first to see whether the link is already in use by another station by simply "listening" to the link. If the link is in use, the station defers its transmission so as not to cause a "collision" that would corrupt both its own and the other station's data stream. If the link is idle, the station transmits its data.

Frame Synchronization

Frame synchronization defines how a receiver determines the beginning and end of a data link frame. This data link service distinguishes between the data content and idle lines. Two data transmission methods are defined for communications: asynchronous (character oriented) and synchronous (frame oriented).

Asynchronous Frame Synchronization

In the asynchronous protocol, only one character (usually, an 8-bit character) is sent at a time. The character is preceded by a start bit and followed by one or two stop bits. A 1 bit is transmitted to indicate an idle line. The start bit (0 bit) is sent to indicate that a character is being transmitted. The first bit of the character to be sent is usually the least significant bit, with the others transmitted in order of significance. After the character has been sent, the line is returned to the 1 state for 1- or 2-bit intervals. This interval is called the *stop interval* but is often referred to as *stop bit*, since it usually is 1 bit long.

Synchronous Frame Synchronization

Most modern data link protocols use synchronous transmission. A larger block of bits (a frame) is sent as a unit. In this environment, frame synchronization is performed by the designation of a set of bits as the start of a frame and another set as the idle pattern. The frame synchronization is performed when the receiver recognizes a non-idle bit pattern. In two older protocols, IBM's BISYNC and Digital's DDCMP, the transmitter must send at least two SYN characters prior to the start of a frame. The receiver, in turn, must see at least two consecutive SYN characters to recognize that a frame is coming. In most modern protocols, synchronization is accomplished by agreeing to a bit pattern that indicates when a frame is coming. This special flag indicates both the start and end of a frame.

Data Delimiting

The Data Link layer sends the Physical layer the message passed down by higher layers, encapsulated with its own control information. The receiver must be able to strip off the control information, process it, and if no errors are detected, pass the data up to higher layers. Therefore, mechanisms are needed for determining where the control information and the data information start and stop. Several approaches are possible; the most common are *character-oriented*, *byte count*, and *bit-oriented*.

Data Delimiting Methods

In character-oriented protocols (such as IBM's BISYNC), a special character is used to designate the start and end of the data.

In byte count protocols (such as DDCMP), we place the byte count of the data in a fixed-length Control field. The start of data is determined to be a fixed number of bytes from the start of the frame. The position corresponding to the end of data is calculated from the byte count. Frames are called *messages* in DDCMP.

In bit-oriented protocols (such as HDLC), the data is surrounded by fixed-length Control fields. A flag denotes the beginning and end of a frame. If the Control fields are of fixed length, the positions corresponding to the beginning and end of the data can be calculated.

Transparency

Interpretation problems occur if the data happens to contain any of the special characters or flag patterns used to delimit fields. For example, if the flag pattern is found within the data, the Data Link layer has no choice but to interpret it as the end of the frame. This is, of course, an incorrect interpretation.

To alleviate this problem, the transmitted data stream is altered so that the flag pattern can never occur. The alteration has to be reversible so the receiver can distinguish between a flag to be interpreted as a delimiter and one that is simply part of the data stream.

For bit-oriented protocols, the alteration is in the form of *bit stuffing*. That is, if the flag pattern contains six consecutive 1 bits, the data stream must not contain six consecutive 1 bits. Within the data stream, the transmitter stuffs an extra 0 bit after any five consecutive 1 bits. The receiver, upon seeing five consecutive 1 bits, examines the sixth bit. If it is a 0 bit, it is discarded (this is the stuffed bit). If it *is* a 1 bit, a flag pattern has been found indicating the end of the frame.

For character-oriented protocols, the alteration is usually character stuffing or insertion. The idea is the same: if a "special" character appears in the data stream, a specified character is inserted before it.

Error Detection and Recovery

Error detection is one of the Data Link layer's primary reasons for existence. Within the scope of error detection are two general categories: a Data Link layer must be able to detect bit errors within a frame, and it must be able to determine whether all transmitted frames have arrived.

To determine whether all the delivered bits are correct, an error detecting code is transmitted in a Control field. This code (usually a Cyclic Redundancy Check, or CRC) is calculated using the data and header fields at both transmitter and receiver. If the receiver's calculation matches what the transmitter sent, the bits are assumed to be correct.

For connection-oriented data link protocols, sequence numbers are used to ensure that all frames arrived in the correct order. If, for example, frame 6 arrives before frame 5, at least the receiver knows that frame 5 is required. It is up to the individual protocols to specify if out-of-sequence frames are accepted or rejected.

Sequence numbering alone cannot detect the fact that an isolated frame failed to arrive. Thus, where guaranteed delivery is an important issue, transmitter timers are also employed. The transmitter gives the receiver a specified amount of time units to acknowledge receipt. If the receiver fails to respond in the allotted time, the transmitter resends the frame or polls the receiver for an indication of its status.

Cyclic Redundancy Check

CRC, the most common error-detection method, involves appending a code (most commonly 16 bits) to a computed frame. The bits of the frame are treated as coefficients of a polynomial. Thus, the bits 1101 represent $X^3 + X^2 + 1$. The sender and receiver agree on another polynomial used to generate the CRC. Both the high- and low-order bits of the generator must be set. Using polynomial division, Modulo 2, the generator is divided into the frame and a remainder calculated. If the generator is 17 bits, the calculated remainder is 16 bits. The remainder bits, or the CRC, are transmitted after the frame bits.

The receiver treats the frame bits, plus the CRC bits, as a polynomial and performs the same division. If the receiver computes the remainder to be 0, the frame is error-free. If the remainder is not 0, one or more bit errors have occurred.

CRC is typically implemented in hardware and causes little delay. In most settings, CRC achieves better than 99-percent accuracy (more than 99 percent of errored frames are correctly detected as errored).

Forward Error Correction

The most dramatic approach to the problem of error correction is *forward error correction* (FEC). This technique includes additional information with the transmitted frame, enabling the receiver to not only detect the presence of an error but to correct it.

FEC is attractive under certain conditions. For instance, when the channel delay is extremely long, retransmission of flawed information leads to very low throughput. Similarly, if the channel is extremely noisy, the retransmission of erroneous information merely leads to a second erroneous transmission. Finally, in a situation where there is no reverse channel for the transmission of acknowledgments, forward error correction is necessary. Such a situation might arise in deep ocean probes or space vehicles for which transmission equipment might add undue weight or might unnecessarily increase size. Although none of these situations occur frequently over the telephone network, forward error correction is still often used. Many modems use trellis-coding modulation (TCM) to correct errors before the data is transferred to the DTE.

Detect-Retransmit

The alternative to forward error correction is the *detect-retransmit* technique used in synchronous communication. In this scheme, the transmitter supplies only enough redundant information for the receiver to detect an error. When the receiver detects an error, it does not know which bit or bits are in error—it merely discards the synchronous frame and asks for a retransmission. This scheme is much more efficient than forward error correction because it requires less redundant information. For this reason, it is the error control mechanism used in all connection-oriented synchronous data link protocols for normal data communications.

This error control scheme has several requirements. For example, all frames must be sequence numbered so the receiver can identify which frames to retransmit. In addition, the receiver must periodically transmit positive as well as negative frame acknowledgments. This provides for the case in which entire frames are lost, rather than simply arriving with errors. Examples of detect and retransmit error correction schemes are the stop-and-wait and the pipelining schemes, go-back-N and Selective retransmission.

Stop-and-Wait   Figure 5-1 illustrates a stop-and-wait protocol. The transmitter sends a data block and waits for an acknowledgment of that block before the transmission of a second block. IBM's BISYNC protocol is an example of a stop-and-wait protocol.

Stop-and-wait data link protocols are now obsolete in the synchronous environment. Nevertheless, many legacy systems continue to use them.

Pipelining   Pipelining was developed to overcome the shortcomings of stop-and-wait protocols. Pipelined protocols (such as IBM's SDLC) transit frames continuously (up to the agreed "window size" number), without waiting for acknowledgments. Acknowledgments arrive out of synchronization with the transmitted frames. For example, the sender might be transmitting frame 4 when the acknowledgment for frame 0 arrives. This process eliminates the effect of channel delay, but has a built-in Achilles heel: sequence

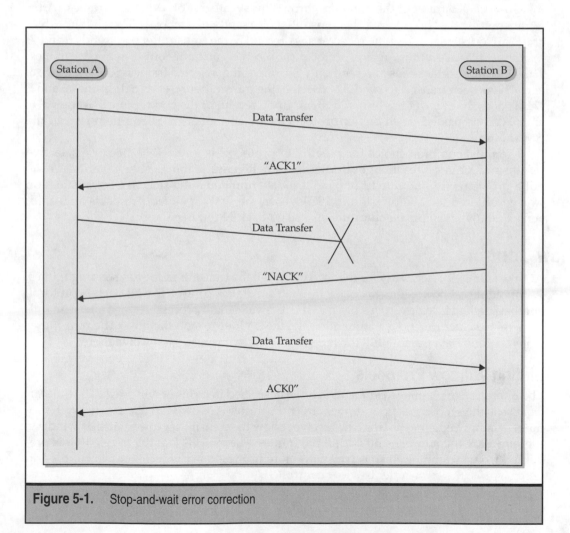

Figure 5-1.   Stop-and-wait error correction

numbers must be used to identify the data frames. Whether the receiver sends a positive acknowledgment that the frame was good or a negative acknowledgment to resend the frame, the sequence numbers are integral to this process. Without them, pipelining would be impossible.

Go-Back-N or Selective Retransmission   In a pipelined protocol, the receiver throws away "bad" frames. What does it do with the "good," but out-of-sequence ones that follow? It can either throw away out-of-sequence frames or keep them. A different protocol scheme deals with each choice. Regardless of the decision, an out-of-sequence frame always informs the receiver of a potential problem: a frame has not been received. At this point, the receiver can choose to let the transmitter know of the problem or simply wait for the transmitter to time out. Either way, the transmitter eventually learns of the problem. In the go-back-N strategy, the receiver throws away all out-of-sequence frames. After retransmitting the "bad" frame, the transmitter also sends all later ones. Thus if frames 5, 6, and 7 are outstanding and the receiver says resend 5, the transmitter also sends 6 and 7.

In the selective retransmission protocol, the receiver can buffer out-of-sequence frames. Thus the transmitter needs to resend only the offending, rejected frames. So if frames 5, 6, and 7 have been transmitted and 5 is rejected, the transmitter resends 5. If frames 6 and 7 were good, the receiver has buffered them. Upon receipt of the good frame 5, frames 5, 6, and 7 can be passed up to higher protocol layers and an acknowledgment sent to the transmitter.

The most common choice for pipelining protocols is go-back-N, because selective retransmission requires more buffering and intelligence at the receiving station. While not a problem with the chip technologies today, implementing selective retransmission was considered too complex in the 1970s when the first pipelined protocols appeared; consequently, most pipelined protocols use go-back-N retransmission.

Flow Control

At times it may be necessary for the receiver to tell the transmitter to stop sending frames. Withholding acknowledgments is one mechanism of flow control. In a go-back-N protocol, another method is to discard frames that the receiver cannot process. A third alternative allows the receiver to send a command to the transmitter to stop the flow. The transmitter stops sending data frames until the receiver indicates it is ready to receive more.

Sliding Window Protocols

The operation of pipelined protocols is often described in terms of windows. The *transmit window* is the set of sequence numbers the transmitter is allowed to use at any given moment. Acknowledgements from the receiver allow the transmitter to advance its window, thereby allowing it to transmit additional frames. The effect is that the transmitter is prevented from swamping the receiver with more frames than the receiver can handle—in other words, the transmitter is "flow controlled."

Connectionless Data Link Protocols

Data link protocols may operate under either a connectionless or connection-oriented model. So far, we have been focused on the connection-oriented version.

Connectionless data link protocols do not establish a logical connection between stations prior to transmitting data as do their connection-oriented counterparts. In an effort to simplify data link operations, connectionless implementations detect transmission errors (using detection codes such as CRCs) and discard errored frames. No attempt is made by the Data Link layer to retransmit errored or discarded frames. Detection of missing data and any retransmission becomes the responsibility of higher layers (such as the Transport layer).

Connectionless data link protocols do not place sequence numbers on transmitted data. Without sequence numbers, sequential delivery of frames cannot be assured. However, if transmission errors are extremely infrequent, users of the connectionless service will rarely experience data loss. In practice, the use of sequence numbers on each frame transmitted adds to the processing burden placed on the data link entities. Removing this burden typically frees stations to process more frames per second, increasing overall throughput.

Environments typically exhibiting low bit error rates tend to utilize connectionless data link protocols in the hope that errors will be infrequent and that the trade-off in error detection is more than compensated for by an increase in throughput. Most LAN data link protocols operate as connectionless data links.

In a go-back-N protocol, the maximum transmitter window size is the Modulo 1. The size of sequence numbers places an upper limit on the transmitter's window size. In selective retransmission systems, the maximum window size is Modulo 2.

NOTE: The *modulo* of the protocol is the range of the sequence numbers. In most data link protocols, a 3-bit or a 7-bit sequence number is used. The length of the sequence number allows counting from 0 to 7 (Modulo 8) and 0 to 127 (Modulo 128), respectively

Addressing

Data Link layer protocols operate over both point-to-point and multipoint physical links. Many data link protocols, such as SDLC, provide for multipoint configurations in which one station, called the *primary*, is in charge of the link. In such a configuration, an address is required to ensure message delivery to the correct *secondary* station. In this system, an address is contained in frames to the primary, so the primary can determine the sender. Data link addresses should not be confused with network addresses; network addressing is a separate issue. Data link addresses are commonly referred to as *hardware* addresses.

In LAN settings, all stations are generally considered peers (that is, there is no primary/secondary distinction). Since any station can transmit to any other, the Address field of a frame contains both destination and source addresses.

Connection Establishment and Release

A connection-oriented Data Link layer protocol must have a defined starting point. This is determined by software rather than hardware.

Stations that sequence data frames to ensure sequential delivery must have a known starting point for these sequence numbers. Stations must also agree on a sequencing modulus. One station initiates the logical connection by sending a defined Start command to the intended recipient. The format for this command and which station(s) may send it are defined by the specific protocol. Typically, the other station returns a connection acknowledgment before the transfer of data begins. The start-up process can also clear error conditions and restart the Data Link layer software. While not often used, Data Link layer protocols also have a defined stop frame. The Stop command causes a software halt rather than a hardware stop.

LEGACY DATA LINK PROTOCOLS

In the early 1980s a number of data link protocols were in use that performed most of the functions defined by the ISO for the Data Link layer. These protocols were included in the major computer architectures in use at the time—IBM's mainframe environment and Digital Equipment Corporation's minicomputing environment. A sampling of these legacy protocols are described in this section and include the HDLC, BISYNC, and DDCMP. These three protocols are also examples of bit-oriented, character-oriented, and byte count protocols, respectively.

HDLC

The HDLC specification describes a family of data link protocols, all bit oriented with a similar frame structure. Actual implementations of a protocol pick and choose options from the full specification, somewhat like ordering from an *a la carte* menu. To facilitate this, the HDLC standard has defined subsets of the procedures to allow for implementations that do not necessarily require all the capabilities of the process. The standard defines three procedure classes that correspond to the three modes of operation: Asynchronous Response Mode, Normal Response Mode, and Asynchronous Balanced Mode. With this method of designating station capabilities, a system implementer can choose stations compatible with one another and with the implemented system. In the following paragraphs, descriptions of similar and derivative protocols (SDLC, LAPB, LAPD) are provided.

Synchronous Data Link Control

SDLC is a data link protocol designed by IBM in 1974 and used in IBM's Systems Network Architecture (SNA) at the Data Link layer. It was the first of the bit-oriented protocols, and was thus a precursor to HDLC. SDLC allows for either two-way alternate (half-duplex) or two-way simultaneous (full-duplex) exchange of data.

Link Access Procedure Balanced

The X.25 Data Link layer protocol in most common use is the LAPB. While not a part of the original 1976 version of X.25, LAPB was added to X.25 in 1980 and enhanced in the 1984 version. It is considered the protocol of choice for X.25.

LAPB is a subset of the ISO HDLC protocol. It utilizes the Asynchronous Balanced mode.

LAP for the D-Channel

To allow users to connect to an Integrated Services Digital Network (ISDN) in a standard fashion, the ITU-T developed an ISDN interface. The second level of this interface, Link Access Procedures for the D-channel (LAPD), deals with how a user's device sends data to an ISDN over one or more connecting data links. It, too, is a subset of HDLC and uses the Asynchronous Balanced Mode of operation.

HDLC Frame Format

This frame has five basic parts:

▼   **Flag pattern**   Used to delimit the data and signify the beginning and end of the frame. A flag is an 8-bit combination: 01111110 (0, six 1s, 0).

■   **Address**   Used to separate commands from responses in both point-to-point and multipoint protocols. Also used to identify the secondary station in multipoint.

■   **Control**   Identifies the type of frame sent and contains some error checking information (sequence numbers).

■   **Information**   Contains upper-layer data. This field may or may not be present, depending on frame type (for example, some acknowledgment frames do not contain user data).

▲   **Frame Check Sequence**   Contains CRC bits for error detection.

Figure 5-2 shows the format of an HDLC frame.

BISYNC

Released by IBM in 1962 for transmission between IBM computers and synchronous terminals, BISYNC (sometimes also abbreviated BSC) became a de facto standard for a two-way alternative character-oriented data link protocol. Devices on a BISYNC link communicate with an application in a host machine. There is no device-to-device communication.

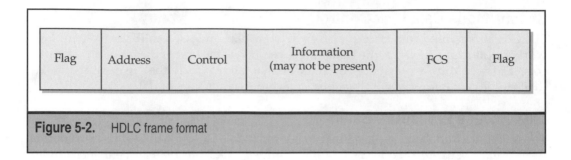

Figure 5-2.   HDLC frame format

BISYNC can operate in point-to-point or multipoint configurations on either full- or half-duplex lines. However, as it was originally designed for a half-duplex transmission facility, communication is always two-way alternate. BISYNC supports three character codes: ASCII, EBCDIC, and 6-bit Transcode. At line initialization time, the character set is chosen. All devices attached to a single line use the same character set. Control characters are used to synchronize devices, delimit frames, and control data interchange.

BISYNC is still a widely used data link protocol in terminal/mainframe and automatic teller machine (ATM) environments, with over 3 million devices in use.

BISYNC is a stop-and-wait Automatic Repeat Request (ARQ) Data Link layer protocol. In BISYNC, frames are called *blocks*, but some can also be called *messages* (for example, there are acknowledgment messages and text blocks). No sequence numbering is used. Instead, two positive acknowledgment messages are used—ACK0 and ACK1—to enable the transmitter and receiver to coordinate their efforts.

In BISYNC, one station is always designated as the control, or *primary* station. On multipoint links, the primary station polls or selects secondary (tributary) stations. A *poll* is used to ask the station whether it has anything to send to the primary station. A *select* is used to inform the secondary station that the primary station has a message for it. In a multipoint environment, up to 32 tributary stations can be attached to a single multipoint line, and up to 32 devices can be attached to each tributary. Associated with each station are two unique addresses: a Station Poll Address (SPA) and a Station Select Address (SSA). These enable the tributary station to differentiate between a poll or select sequence.

DDCMP

The Data Link layer protocol used in Digital Equipment Corporation's DECNet network protocol is DDCMP, an example of a byte count Data Link layer protocol. DDCMP is flexible in that it supports both parallel and serial physical transmission paths. It also supports both synchronous and asynchronous transmission. DDCMP is seldom seen outside DECNet networks, and it is almost never found on commercial equipment other than DEC computers (now owned by Compaq). It can be found supporting specialized equipment's Data Link layer needs. These applications are almost always found in the scientific and engineering areas.

DDCMP supports the following three basic message types:

▼ **Data**   A DDCMP data message contains information from higher layers. DDCMP data messages contain a control part and an information part. Both parts are checked by their own CRC. Data messages are sequenced using Modulo 256 numbers.

■ **Control**   A DDCMP control message contains supervisory information. Control messages contain no data.

▲ **Maintenance**   A DDCMP maintenance message contains data but no sequence numbers. For example, a maintenance message is used to download a secondary station from a primary to a begin operation. No specific use is defined by DDCMP.

Structure of DDCMP Messages

The structure of DDCMP messages is very specific. Data messages contain information of variable length. The length is specified by a value contained in the Control field of all data messages at the beginning of each message type. Every Control field is fully CRC checked, and any message that also contains data has a separate CRC that checks all data. DDCMP does not have a problem with transparency. The Control field is of fixed length and contains a Count field that defines the length of the Data field. DDCMP then reads the next N bytes as data. No attempt is made to look for any special characters, and any character found is treated as data.

EVOLUTION OF SERVICES

If the Physical layer is error-prone, the Data Link layer provides a desirable, if not necessary, set of functions. As the quality of the Physical layer improves, the need for all of the Data Link layer functions lessens. The time required to ensure that "all" messages are received may not be worth the effort if the probability of error on the physical paths is negligible. It may then be reasonable to check for sequencing errors only at the destination node, rather than at the intermediate nodes. The delivered service of the data link has shifted from the delivery of all frames error free to the delivery of bit error free frames. If frames are "lost in the mail," so to speak, a higher layer protocol is responsible for the recovery of the lost information. By the elimination of recovery functions, the Data Link layer becomes faster and simpler.

In the years since the ISO defined the functional breakdown of the OSI Model, networking has evolved, and the functions defined for the Data Link layer have evolved as well. Of the eight functions originally defined, a few have been dismissed and a few others added. Current data link protocols can be separated into three areas: protocols used in point-to-point environments, protocols used in a LAN, and protocols used in (wide area network) WAN service offerings. The data link functions are addressed for each of these areas.

Point-to-Point Protocols

The interconnection of two devices on a physical link for the purpose of sending information is the simplest of arrangements. Therefore, the data link functions required for this scenario can be reduced to the simplest set. The Serial Line Internet Protocol (SLIP), developed for Internet access, is the embodiment of this simplicity. The second-generation protocol for Internet access, Point-to-Point Protocol (PPP) seems overly complex in comparison. The fact is that PPP has become the standard for two communicating devices connected on a serial link. As such, all sorts of bells and whistles have been added to the basic functions of the protocol.

Serial Line Internet Protocol

If ever there was a simple protocol, SLIP is it. Although considered a Data Link layer protocol, the only Data Link layer services it provides are framing, transparency, and data delimiting. Absent are addressing, error detection and correction, and all of the other Data Link layer functions that provide error-free communication.

SLIP is a byte-oriented protocol concerned with moving IP packets across a point-to-point serial link. The SLIP software precedes the Network layer packet with an ESC character (0xDB, not the same as the ASCII ESC character) and follows it with an END character (0xC0)—this is its data delimiting service. The appearance of an ESC or END character in the packet would confuse the receiver, so ESC characters are replaced with the two-octet sequence 0xDB-DD (ESC followed by 0xDD), while END characters are replaced with 0xDB-DC (ESC followed by 0xDC)—this provides for data transparency. The complete specification is in STD 47/RFC 1055.

SLIP's simplicity made it easy to implement, and it was widely used for many years. However, SLIP is a limited protocol; it has no error detection or correction capabilities, and the Internet Protocol (IP) from the TCP/IP protocol suite is the only protocol intended for use with SLIP. To meet the increased demands of multiprotocol networking, the PPP was developed and has essentially replaced SLIP for use over serial lines.

What is PPP?

The PPP is a data link standard from the Internet Engineering Task Force (IETF). The protocol is described in RFC 1661; many extensions are covered in other IETF RFCs (such as RFC 1990 and RFC 2125).

PPP is a peer-to-peer protocol that assumes the use of a point-to-point full-duplex physical facility (switched or dedicated). PPP operates serially over such links in the two-way simultaneous mode. A PPP connection must be established independently by both peers, as options for each direction of transmission can be set separately.

Interestingly, PPP does not perform many of the functions commonly associated with the Data Link layer of the OSI Model. Although PPP performs CRC to detect bit errors, no attempt to correct errors is made. PPP simply discards frames containing bit errors. An upper-layer protocol is responsible for recovery from such frame discard. PPP does not perform flow control operations. Again, PPP simply discards any frames that cause buffer overflow. Finally, PPP does not perform frame sequencing or sequence checking. PPP assumes the use of a Physical layer that performs sequenced delivery of data units. Basic PPP does add a function to the Data Link layer: the ability to identify the transported Network layer protocol. The PPP frame contains a protocol ID field. This function was not necessary in the strict OSI Model, because all layers were independent. As multiprotocol networks evolved, this function was determined to be a necessary component.

PPP Frame Format

The format of PPP frames is shown in Figure 5-3. As can be seen from the general format, PPP is a derivative of the ISO HDLC protocol.

PPP frames are delineated by the occurrence of the standard HDLC flag pattern (01111110). The address field is not used by PPP and is encoded to the "all stations" address (all 1s). Because PPP does not perform sequencing, sequence checking, or acknowledgement, the control field is always encoded 0x03.

NOTE: 0x03 is the control field that defines an unnumbered information (UI) frame, the simplest of information transfer capabilities of a bit-oriented protocol. A UI contains no sequence numbers or acknowledgements.

The 16-bit protocol field indicates the protocol being carried in the information field of the PPP frame. Many protocols, including PPP control protocols, layer 2 protocols, and layer 3 protocols, can be carried by PPP. For a complete list of the protocols that can be carried by PPP, refer to the home page of the Internet Assigned Numbers Authority (IANA) at **http://www.isi.edu/in-notes/iana/assignments/ppp-numbers/**. The protocol field gives PPP the ability to multiplex information from multiple protocols over a single data link.

The information field of the PPP frame can vary in size from 0 octets to some upper limit imposed by the network administrator. The maximum size is negotiated upon initialization of a PPP link; the default value is 1500 octets, which is large enough to take in the maximum sized frame in an Ethernet LAN.

Flag 0x7e	Address 0xff	Control 0x03	Protocol	Information	FCS	Flag 0x7e
8	8	8	16	≥ 0 (Octet-Aligned)	16	8

Figure 5-3.   The PPP frame format

Although PPP incorporates a frame check sequence (FCS) derived from a CRC operation, the FCS does not drive an error-correction process. Rather, frames with bad FCSs are simply discarded by PPP. The default CRC mechanism yields a 16-bit remainder (CRC-16). Other mechanisms (such as CRC-32) may be used in place of the default.

PPP Operations

PPP performs three general types of operations.

▼ **Link control operations** are responsible for the establishment, management, and release of the PPP link. Link control uses the Link Control Protocol (LCP) to perform its tasks.

■ **Authentication operations** are optional and, when used, permit PPP peers to identify themselves to their counterparts at the other end of the link. Two protocols, the password authentication protocol (PAP) and the challenge handshake authentication protocol (CHAP), are available to perform such authentication.

▲ **Network layer control operations** establish and terminate Network layer interactions over the PPP link. For each Network layer protocol supported by PPP, a separate Network Control Protocol (NCP) is defined. Once a Network layer has been established over PPP, data transfer using that Network layer commences.

PPP Session

Figure 5-4 shows the basic steps in establishing and terminating PPP communications sessions.

Although many protocols at many layers are acting at the same time, the operation of PPP is relatively straightforward:

1. The physical connection must be in place before any PPP link operations. This physical connection can be either a dedicated or dial-up connection.

2. The two PPP hosts now exchange PPP LCP messages to establish the PPP connection, which involves configuring various link options. In some cases, authentication will also be provided. Authentication differs from identification in that it provides an independent method of verifying that the user has bona fide access to this system.

3. The hosts exchange PPP NCP messages for each protocol that they want to use over this connection. NCP messages are exchanged to establish a logical connection for a given Network layer protocol. For IP networks, the NCP would be the IP Control Protocol (IPCP). One of the functions of IPCP is to assign the IP addresses to be used over the connection. IPCP provides a mechanism for dynamically assigning IP addresses. In this way, an ISP could manage a pool of IP addresses for its dial-up customers.

4. After the NCP messages are exchanged, the two hosts can exchange data. Additional NCP messages are exchanged to terminate the logical Network layer connection. Multiple Network layer connections can be open at the same time, which might commonly occur in today's multiprotocol environment. NCPs are currently defined for AppleTalk, DECNet, IP, NetWare, VINES, and bridging.

5. When all Network layer data exchanges are completed, additional LCP messages are sent to terminate the PPP logical connection.

6. If it is a dial-up connection, the Physical link connection will be terminated.

Another variant of PPP is called the Multilink Point-to-Point Protocol (MLPPP). This specification defines an inverse multiplexing scheme, allowing a high bandwidth application to utilize multiple low-speed, point-to-point links between a pair of hosts. MLPPP is defined in RFC 1717.

The Bandwidth Allocation Protocol (BAP) is an extension to MLPPP. BAP is described in IETF RFC 2125. BAP supports dynamic allocation of individual PPP links within an MLPPP group. Allocation and deallocation of individual links within an MLPPP group is controlled by the Bandwidth Allocation Control Protocol (BACP).

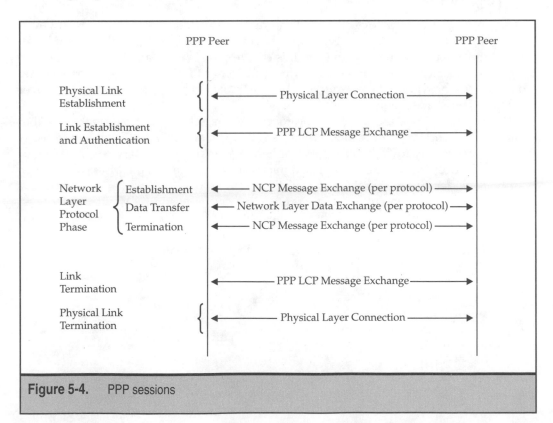

Figure 5-4.   PPP sessions

As you can see, PPP is quite a complex protocol. In fact, you may be wondering why we have spent so much time discussing this protocol. The reason is simple. PPP is one of the most widely used Data Link layer protocols in dial-up and leased-line connections in corporate networks and the Internet. In the case of a corporate network, the multiprotocol capability of PPP may be very useful. Frequently corporations use more than one set of networking protocols—for example Hill Associates used to support both NetWare and TCP/IP protocols, and PPP allowed both traffic types to be carried over the same physical link. The device at the far end of the link knew how to handle the information as the protocol type was identified by the PPP session, making PPP an important protocol for remote access to a corporate network and for WAN links between sites.

LAN Protocols

LAN protocols transport a large portion of the corporate data traffic. A number of LAN protocols have existed over the years. Most LANs today are of one of two implementations, Ethernet or Token Ring. These LAN protocols are simply Data Link layer protocols, and we shall examine them in detail in Chapter 15.

WAN Protocols

WAN protocols act as the "over-the-road trucker" for the data communications environment. They provide *any-to-any* connectivity over very large geographic areas. Unlike most of the other protocols examined in this chapter, the WAN data link protocols have been sold as carrier-based services, allowing a customer to purchase WAN connectivity from a service provider. These services allow the transport of any type of Network layer traffic between corporate locations.

The original public service WAN protocol, X.25, is actually a three-layer stack that combines Network layer functions as well as Data Link and Physical layer functions, as described in Chapter 16. As service provider networks matured, the service guarantees of the X.25 protocol stack were viewed as unnecessary overhead as the physical network became considerably more reliable; migration from analog to digital was the primary reason for this improvement. Two new protocols, *frame relay* and *ATM*, were developed based on the assumption that the Physical layer was essentially error-free. They significantly reduce the processing required by the network and are often referred to as *fast packet* protocols. These protocols are covered in detail in Chapters 17 and 18.

CHAPTER 6

Switching

This chapter explains the role of switching in communications. *Switching* permits the exchange of information between two stations without requiring a dedicated, permanent connection between the stations. Instead, each station is linked to a switched network that conveys the information for the station.

This chapter provides a definition of switching basics and the two major types of switching: circuit switching and packet switching. A brief overview of various packet network examples is explored. Finally, the chapter discusses where circuit and packet switches are found today, along with a discussion of their applications.

SWITCHING BASICS

Switching uses temporary connections to establish a link or to route information between two stations. Two kinds of switches exist in today's networks: circuit switches and packet switches. Circuit switches provide end-to-end connection between two stations. The most recognizable circuit-switched network is a telephone network. Packet-switched networks, a form of "store-and-forward" switching, involve delays as each packet is handled by each switch. An example of a packet-switched network is the Internet.

Multiplexing is a factor in both circuit and packet switching communication networks. Multiplexing entails combining multiple streams of information, or *channels*, onto a common transmission medium—in other words, it allows a number of separate signals to be transmitted simultaneously. The signals are combined in a multiplexer device.

To maintain the integrity of each signal on the channel, signals can be separated via time division multiplexing (TDM), frequency division multiplexing (FDM), or space division multiplexing (SDM). SDM, the least familiar technique of the three, provides multiple fixed-bandwidth channels that travel via multiple physical paths (pairs of wires). A good example of SDM is the use of a 25-pair cable to carry the conversations of 25 individual users from a customer's premises to the local telephone company's central office. The other types of multiplexing will be discussed in the next section.

Circuit Switching

Circuit switching was developed primarily for voice telephony (telephones). Fixed-bandwidth connections are assigned to the communicating parties by a call establishment process, and they remain assigned to those parties for the duration of a call until a termination process releases the facilities used for the call. Thus, these circuits behave as if they were end-to-end dedicated connections for the duration of the call. Once established, circuit switching networks involve no delay in transmitting the information, except for propagation delay on the facility.

When all available connection paths in a circuit-switched network have been assigned to users (when network congestion exists), the network responds by blocking new users until paths are made available to serve them, as indicated by a *fast busy* signal in the telephone network. The dedicated connections provided by circuit switching are useful in many communications applications, but they are required by delay-sensitive applications, such as voice communications.

For example, Figure 6-1 shows two circuit switches, each with two telephones connected via a local loop. Each local connection is dedicated to a single user. When a user makes a call from telephone A to a user at telephone C, the trunk between switches is seized for the connection and is thus dedicated to A and C until the two users are finished talking. Even though telephones B and D have their own dedicated local facilities, neither can make a call for the duration of A and C's conversation in this scenario, because the trunk is being used. After A and C complete their call, the trunk will become available again.

Of course, most phone networks don't work this way—they have multiple paths between switches. Rather than running many pairs of copper, each with a single conversation, phone companies have deployed connection facilities capable of carrying several phone calls at the same time.

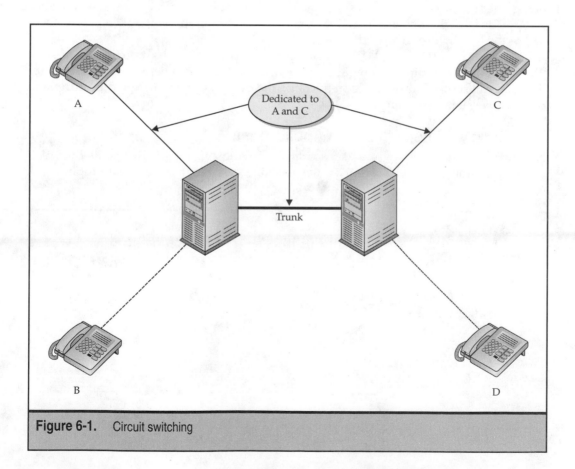

Figure 6-1.   Circuit switching

Two multiplexing approaches, as mentioned previously, are used in circuit switching. FDM is useful for transporting a number of narrowband signals over a single wideband facility using carrier modulation techniques. Because the frequency spectrum of a broadband signal can be confined to a given passband by filtering techniques, FDM is commonly employed to carry such signals. Figure 6-2 shows a facility with a 24KHz bandwidth being divided into six, 4KHz passbands. Each passband is capable of carrying a single voice conversation whose bandwidth requirement does not exceed 4KHz. One way to express the FDM relationship is that each pair of users gets "some of the bandwidth all of the time."

Today, most phone company facilities are digital, and TDM is used to transport multiple digital signals over a single wideband facility. Because the frequency spectrum of a digital signal cannot be confined, such signals are amenable to multiplexing only in the time domain. In TDM, a single wideband facility is split into many time slots. Each pair of users is assigned a time slot at call establishment time; that time slot remains dedicated to them for the duration of the call. Figure 6-3 shows a facility with 250KHz bandwidth being divided into six, 8-bit time slots. Each time slot is capable of carrying a digital voice sample for a single conversation. TDM technology can be said to give each pair of users "all of the bandwidth some of the time."

Note that in FDM and TDM, dedicated paths are provided to users. This is consistent with the aim of circuit switching. In addition, the relationship between users is established at call setup time (that is, a frequency band or a time slot is assigned). Addressing during the call is thus implicit.

Real-Time vs. Interactive Data Applications

No matter the multiplexing method employed, circuit switching is well suited for real-time applications such as voice, but it is not generally suited for *interactive* data applications

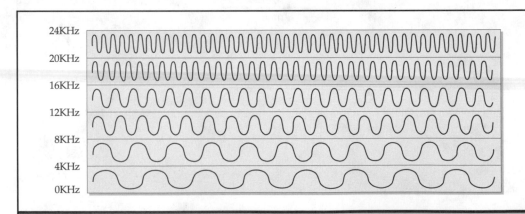

Figure 6-2.   Frequency division multiplexing

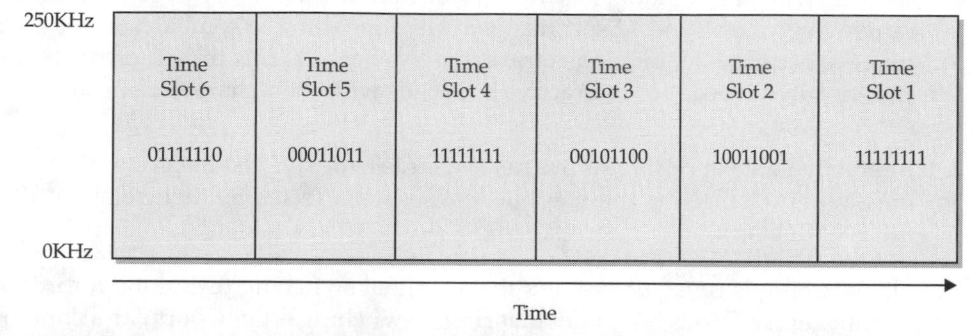

Time Slot 6	Time Slot 5	Time Slot 4	Time Slot 3	Time Slot 2	Time Slot 1
01111110	00011011	11111111	00101100	10011001	11111111

250KHz

0KHz

Time

Figure 6-3. Time division multiplexing

(as opposed to packet switching, discussed next). Such applications are sometimes referred to as "bursty" data applications—the actual data transmissions are much shorter in duration than the idle time between those transmissions. Said another way, if the "think time" in a data application is longer than the transmission time, we are dealing with an interactive data application.

Interactive data applications are typically characterized by human involvement of some sort (such as a human at a cash register interacting with a credit verification program in a computer). As data applications, they usually involve digital signaling procedures (such as EIA-232 interfaces). The holding times for interactive data applications are usually short (30 seconds to 2 minutes), especially when compared to voice application holding times (such as 5 to10 minutes). Finally, interactive data applications are generally delay insensitive. The meaning of a credit verification message does not change regardless of when it arrives.

Common examples of interactive data applications include electronic funds transfers such as automatic teller machine transactions and point-of-sale credit verifications. Inquiry/response applications may qualify as interactive if they meet the criteria. In fact, even file transfers can sometimes be considered interactive, providing they are short in duration.

The use of circuit switching in interactive data applications may be considered inappropriate for a number of reasons:

▼ The low line utilization (transmissions shorter than gaps between transmissions) associated with interactive data are inconsistent with the provision of dedicated paths. As circuit-switched connections are typically charged on a time-sensitive basis, this inefficient use of those connections is both wasteful and costly.

- ■ Circuit switching technologies are typically associated with call setup times ranging between 5 and 12 seconds per call (depending to some extent on the number of switching nodes traversed by the call). This time represents significant overhead for interactive data calls with their characteristically short holding times.

- ▲ Interactive data applications are not sensitive to delay. The meaning of messages exchanged in these applications is not affected by their relative times of delivery.

Circuit switching is considered a less-than-optimal switching technique for interactive data applications. That is not to say that circuit switching is not a popular technology for transport of interactive data, or that noninteractive data applications ("bulky" data applications) are not well served by circuit switched technology.

Packet Switching

Packet switching is relatively new as compared to circuit switching. In 1962, Paul Baran described packet switching in a report titled "On Distributed Communications Networks." In contrast to the circuit switching approach, which entails a dedicated, real-time channel between parties, packet switching is a real-time, shared technology.

Packet switching segments long messages into short entities called *packets* with an upper limit on the packet length. This *packetizing* process increases the addressing overhead associated with a message in the network, because every packet must carry a copy of the destination address. In packet switching, communicating parties share facilities to get better utilization at a lower cost. This sharing requires that data objects be addressed and that switches perform routing on each packet.

Packet switching is a form of *store-and-forward* technology that involves delay, because each packet must be handled by each switch. Messages are relayed from a source to a destination via shared network nodes that store messages until a path is available to the next node. Inside a store-and-forward switch, each link has an associated buffer pool. Messages are stored in the inbound buffers while the routing algorithm chooses the appropriate output link. To do this, the routing algorithm examines the addressing information associated with each buffered inbound message and then selects, on the basis of a table lookup, the correct outbound buffer in which to place that message for transmission. After the appropriate link is identified, the messages may be stored in an outbound buffer until time becomes available on the appropriate outbound line.

In effect, the switching function of a store-and-forward switch is a simple buffer swap based on the message's addressing information and on information contained in the routing tables. The processing time associated with address interrogation and table lookup induces delay in the path between input and output, in addition to propagation delay.

The two major varieties of store-and-forward switching are *message switching* and *packet switching*, which differ substantially in their internal workings and applicability to different environments. In message switching, messages are sent in their entirety from beginning to end. Because message length is unpredictable, external storage devices (such as disk drives) may be used to store messages awaiting output to the next node in the network. Message switching is highly efficient for the noninteractive transport of long messages, because addressing overhead is insignificant compared to message length. Short messages tend to suffer in a message switching network, however, because they must compete for resources with long messages. The quintessential example of a message switching network is the telegraph network (such as Western Union Telegraph).

Package switching transmission is much more efficient than that of message switching. Error checking occurs for each packet at each node in the network prior to forwarding. If an error occurs in packet number 100, for example, only that packet is retransmitted and not the entire message. This reduces the time lost to retransmissions for long messages. In addition, by breaking messages into discrete packets, a given packet can be forwarded while a node is receiving the next packet. This "overlapping" of the store-and-forward process on a packet basis reduces the overall delay incurred on a message basis.

Packet switching nodes use random access memory (RAM) rather than disk space for storing packets. This is possible because the upper limit on packet length (usually 4KB) makes buffer management a predictable task. The use of main memory buffers for packet storage reduces switch delay in these networks. A packet's upper limit size, coupled with the use of RAM for buffering, makes packet switching a real-time interactive technology.

In packet switching, data is transmitted from source to destination over shared physical facilities using statistical multiplexing techniques. Figure 6-4 shows a simple packet-switched network employing statistical time-division multiplexing (STDM). Unlike circuit switching, the packet-switched network can be shared by multiple users for multiple conversations.

In TDM, digital signals from a number of users are transported over a single wideband facility by time slicing the available bandwidth of that facility. In STDM, however, time slots are assigned to users on a first come, first served basis. User messages are queued in a pool of buffers that empty into time slots as slots become available on the shared medium. Thus, any user can be assigned to any time slot at any time, and time slots may vary in length. Because time slots are not dedicated to pairs of users at call establishment time (implicit addressing), each message must carry its own explicit addressing overhead to identify it at the receiver.

The yield of STDM technology is efficient use of bandwidth on the shared facility. If enough traffic is being generated by users, no time slots should be "empty" on the medium, as exists in TDM with its dedicated time-slot arrangement. STDM networks are designed to support peak traffic loads, so at times, some links are idle. In addition, STDM tends to provide bandwidth to those users who need it most (users with a lot of traffic to send). STDM technology can, however, introduce delay for users, as their messages may have to wait in queue for a time slot to become available. For this reason, STDM is typically inappropriate in delay-sensitive applications, such as voice applications.

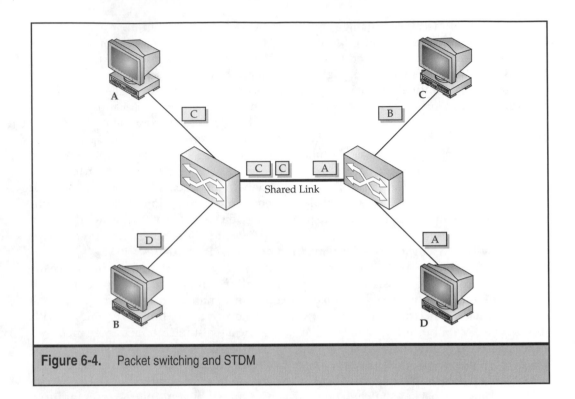

Figure 6-4.   Packet switching and STDM

Types of Packet Switching Networks

Packet switching networks come in a couple of configurations. The oldest technology is termed *datagram* technology. In a datagram network, no call establishment procedure is required; an end user merely begins submitting packets to the network at any time. Packet routing occurs on a packet-by-packet basis in these networks and, as a result, multiple packets belonging to the same message may take different routes through the network. Packets may thus arrive at their destination out of sequence. To route each packet independently, each datagram must contain a full destination address in its header information.

There is no internal sequencing or acknowledgment of packets in a datagram network. Packets may get "lost." In fact, many datagram networks use "packet discard" as their congestion control technique! All responsibility for message integrity is placed with higher layers of protocol in the end user machine (usually the Transport layer). Datagram networks are sometimes referred to as "unreliable" or "connectionless" networks because of their less-than-optimum approach to congestion control.

A good analogy for a datagram network is the postal service. To send a letter to an end user, no setup is required. We simply drop the letter, with a full destination address attached, into the postal network. Regardless of the fact that multiple letters may be sent to the same destination over a period of time, there is no way to lessen the addressing overhead. Each envelope must bear a full address. The post office exerts its best effort to deliver each letter in a timely fashion. However, because there is no guarantee of sequentiality, letters can certainly be delivered out of order. Indeed, the post office offers no guarantee of delivery at all. If the user wants guaranteed delivery, he/she must purchase a higher priced service.

The only way to ensure correctness in the postal scenario is to establish a procedure for end users to acknowledge receipt of letters from the other end. That is, correctness is an end user (upper layer) responsibility. If a recipient must devote considerable processing time to each arriving letter, it is conceivable that the recipient can be overrun by information from the sender. It is the recipient's problem to slow the flow of traffic; flow control is not possible in the network itself. One stops communication in the postal situation by not sending any more letters. Since there is no connection in the first place, there is no reason to tear the connection down.

In contrast to the datagram method of packet switching, many packet networks today provide *virtual circuit service*, a notable exception being the Internet. In such a network, a call-establishment procedure is required before the user can begin submitting packets. A special group of *call setup* packets are defined for the purpose of circuit switching. After a network path (or virtual circuit) is established, subsequent packets follow one another along the established route until such time as the circuit is released. Thus, packets arrive at their destination in sequence.

The full destination address is required for only the call setup packets in a virtual circuit network. Subsequent packets use a short identifier (called a *logical channel number* or *virtual circuit number*) to refer to the pre-established path. This technique results in a savings of overhead in virtual circuit networks.

Internal acknowledgment of packets is a characteristic of virtual circuit networks; this permits the provider of virtual circuit service to guarantee the delivery of each packet. If, for some reason, a loss of data occurs in a virtual circuit network, the network signals such failures to the user by one or more special packet types. Thus, virtual circuit networks are often referred to as *reliable* or *connection-oriented* networks.

The voice telephone network is a connection-oriented structure. The analogy should not be carried too far, however, because the voice network is a circuit network, not a store-and-forward structure. Nevertheless, there are important similarities between the voice network and connection-oriented store-and-forward nets.

In a voice network, prior to transmitting a voice to another user on the network, a path to that user must be established. Once the path is established, signals are delivered in sequence, and no further addressing is required. All the overhead associated with addressing and sequencing is "buried" in the setup phase. To terminate the connection, an explicit action is required; we cannot simply stop talking. Furthermore, we are well advised to take this specific stop action, since there is a "connect-time" charge made as long as the connection is in place.

Figure 6-5 shows a call request packet for establishment of a virtual circuit. The call request packet serves to correlate the full destination address with the short identifier (logical channel number) that is then used by subsequent packets over that virtual circuit. In this fashion, subsequent packets need never again make reference to the full address, thus saving overhead in the packet header.

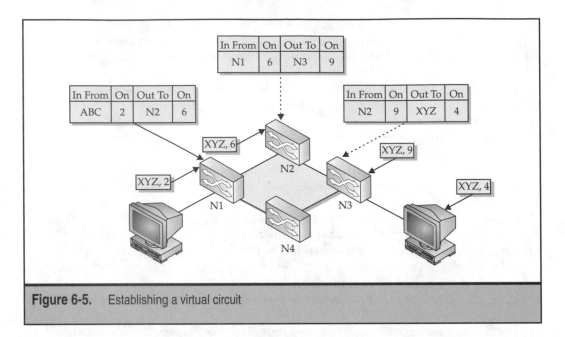

Figure 6-5. Establishing a virtual circuit

The figure shows the virtual circuit establishment process as a call request packet makes its way between two computers (addresses "ABC" and "XYZ") attached to the network. In this case, we trace the process for a call originating at "ABC" and terminating at "XYZ."

"ABC" uses its connection to Node 1 (N1) to submit a call request packet with the destination address "XYZ" and the short identifier "2." Node 1 performs its routing function, which states that the best way to "XYZ" is through Node 2 (N2). Node 1 now uses its connection to Node 2 to forward the call request packet with a new short identifier, "6." Node 2 repeats the routing process, using its connection to Node 3 to forward the call request packet with yet a new short identifier, "9." Finally, as Node 3 has a direct connection to "XYZ," it uses that connection to forward the call request packet with the short identifier "4." Note the independence of short identifiers on each link along the path of the virtual call.

The call request packet, on its hop-by-hop trip through the nodes, causes each node to make an entry in a virtual circuit table. Table entries for Nodes 1, 2, and 3 are shown in the figure. These table entries constitute the virtual circuit, in that subsequent packets for this call simply follow the established chain of short identifier designations through the network (2-6-9-4). Packets flowing in the other direction from "XYZ" to "ABC" simply follow the tables in the reverse order (4-9-6-2). Thus, the virtual circuit is full duplex.

When the users decide the virtual circuit is no longer needed (when one of them decides to hang up), a special *clear request* packet is issued. This clear request causes the nodes to erase their virtual circuit table entries for that call and the call ceases to exist.

While we have discussed setting up and tearing down virtual circuits, they can be (and often are) permanently left connected to prevent delay when sending data. These are known as *permanent virtual circuits* (PVCs); it is interesting to note that most frame relay networks offer only PVC service.

No one packet switching technology is "best" under all circumstances. Clearly, both types of network exist and have their uses. The underlying network technology is chosen based on a mix of application requirements, availability, cost, security, ease of use, and other factors. For example, frame relay and ATM are the mainstays of the connection-oriented world today and are often used for high-speed wide area network (WAN) connectivity. The Internet Protocol (IP), which is connectionless, is the primary Network layer protocol in use and is necessary for Internet access. IP is increasingly a major part of the corporate backbone as well. Interestingly, the underlying access to the Internet service (or other IP networks) often is frame relay. Thus, connectionless and connection-oriented protocols are not mutually exclusive—rather, they can complement each other.

The moral of the story is: There are applications for which each type of service is preferred. The choice within a given network should be dictated by consideration of the type of traffic that is most prevalent in the given setting.

PACKET NETWORKS

Packet networks are used for a variety of applications—typically data-oriented applications. Packet-switched networks provide services corresponding to the lower three layers of the OSI Model (Physical, Data Link, and Network—see Chapter 2). Thus, packet switching networks provide connectivity services. While it is true that packet switching can provide connectivity among devices of different manufacturers, interoperability among such pieces of equipment is not guaranteed. The issue of interoperability (i.e., upper-layer compatibility) remains the responsibility of the end user.

Most packet networks today are based on open standards, but some proprietary approaches are still in operation. In the following sections, we explore some examples of connectionless and connection-oriented packet implementations.

Connectionless Packet Networks

One of the earliest packet networks, the original ARPANet, employed a hybrid of datagram and virtual circuit technology. Internally, the ARPANet was a datagram network. At the destination network node (called an *Interface Message Processor*, or IMP), packets were acknowledged and resequenced on behalf of the user. Users of ARPANet were thus provided with connection-oriented service.

Today's Internet is also a hybrid of sorts. At the Network layer, the IP is in fact a datagram implementation providing connectionless service. However, many Internet service provider (ISP) backbones use Asynchronous Transfer Mode (ATM) technology at the Network Interface layer (combining layer 1 and 2 capabilities), which is based on virtual circuits.

The Internet is being used increasingly for such applications as voice and video, which traditionally use circuit-switched facilities. Using the Internet as a WAN with virtual private network (VPN) methods is also extremely popular today. A great deal of work is focused on providing Quality of Service (QoS) via the Internet, which often entails making the network act more connection-oriented. For example, many providers use *Multiprotocol Label Switching*, a method of combining layer 2 and layer 3 capabilities so that predefined paths through a routed network are created to perform similarly to virtual circuits.

One reason the Internet is attractive as a convergence platform, bringing voice and data together on a single packet network, is that it actually does provide applications in addition to connectivity. The IP is really part of an entire suite that can run on top of any network interface, connectionless or connection-oriented, and defines upper layers for interoperability and user services such as file transfer and e-mail.

One final connectionless network example is Digital's DECNet. There is still some interest in the DECNet architecture even today, particularly for integrating newer Linux boxes into existing DECNet or Compaq's OpenVMS operating system environments. DECNet provides pure datagram service at its Network layer. DECNet's Transport layer, however, provides *logical link* (connection-oriented) service to the layers above.

Connection-Oriented Packet Networks

An immense installed base of IBM's Systems Network Architecture (SNA) is used by businesses today. In its original incarnation, the SNA was based on virtual circuits to provide a reliable data networking environment. Interestingly, the Path Control (Network) layer of SNA uses full origin and destination addresses in each packet, not unlike a datagram network. With the IP suite being the de facto standard for internetworking today, IBM has evolved with the times and provided solutions for integrating SNA with IP. In fact, the vast majority of SNA sites use TCP/IP in some form today.

ITU-T Recommendation X.25 specifies an interface over which connection-oriented service is provided. X.25 can interface either to datagram or virtual circuit packet networks. In its heyday, X.25 was *the* approach to building public packet-switched networks. Today, its importance has been diminished with the rise of frame relay, ATM, and IP networks. Chapter 16 provides a a discussion of X.25.

One popular fast packet technology is *frame relay*, discussed fully in Chapter 17. Originally designed to replace more costly leased lines, frame relay is a virtual-circuit approach to packet switching. The service is generally implemented using PVCs—switched virtual circuits (SVCs), while uncommon, are possible. In addition to its use in the WAN, frame relay is often used as an Internet access method.

A second fast-packet approach is generically known as *cell relay*. ATM, covered in Chapter 18, is similar to frame relay but was designed not only to carry data but also voice and video. ATM has been envisioned since its definition as a convergence platform, bringing together traditional packet and circuit applications in a virtual circuit environment. Corporate network services using ATM are becoming common, though ATM has formed the basis for most ISP Internet backbones for years.

SWITCHES

To employ any form of switching, you, of course, need a *switch*. These devices are found in many environments and can vary widely in configuration and capability. While not meant to be an exhaustive treatment of the subject, this section presents some examples of where and when you might employ switches.

Circuit Switches

The trend for most communications appears to have all applications ultimately collapsing onto a single packet-switched network, such as the Internet and/or ATM. However, as previously discussed, circuit switches are well suited for certain applications and are still extremely common. The primary place to find circuit switches today is in a telephone network.

Many people think of voice communications as being the domain of the phone companies. While these traditional providers of voice services are certainly the most visible users, circuit switches can also be found in the telephone network enterprise—in the form of a private branch exchange (PBX).

PBXs are often used by medium-sized to large businesses as part of a private telephony network. Rather than connecting each individual phone to the phone company's network, some number of outside lines is aggregated at the PBX to be shared by the corporate users. The PBX can provide many features, including voice mail, integrated voice response (IVR), and simplified dialing (such as four-digit extensions). PBXs can support analog and digital phones.

A myriad of PBX vendors exists in the marketplace, including Avaya (Lucent), Mitel, and Nortel. Ironically, with the great interest in IP telephony, many of the traditional leaders in the realm of circuit switching now have products that "IP-enable" PBXs, or even replace traditional PBXs with completely IP-based products. That being said, few customers are throwing away their investment in circuit switches just yet.

Perhaps the most obvious place to find circuit switches is in the Public Switched Telephony Network (PSTN). Here we find a switching network, or fabric, that connects lines and trunks together. The voice switch is typically found in a building called a *wire center*, or *central office*. This facility houses not only the switch, but power and battery equipment, carrier equipment, frame terminations for private lines, and test boards as well. Phone company switches are substantially larger and more powerful, and they support more features than the average corporate PBX but operate on the same basic circuit switching principles.

The players in this environment are Lucent, Nortel, and Siemens. Their circuit switches are found in many carrier networks around the world. Not only do the phone companies operate such switches, but sometimes large private corporations will maintain their own switches, perhaps in a large campus environment. For example, the Boeing corporation has its own network of Lucent 5ESS switches, which are highly reliable, carrier-class products.

Although less commonly used today, the Circuit Switched Public Data Networks (CSPDN) is another place where you could find circuit switches. Sometimes known as "Switched 56" in the United States, this digital switched service operates at 56Kbps and was originally used for leased line backup in a dial-on-demand routing (DDR) environment. Integrated Services Digital Network (ISDN) is also a digital circuit-switched service that provides two TDM channels for voice or data transmissions and a third smaller channel for signaling. It is interesting to note that it took the explosion of Internet use (packet-based) in the 1990s to allow ISDN (circuit-based) to flourish. ISDN service can be offered by the same central office switches used for traditional telephony, loaded with ISDN software.

Packet Switches

LAN switching, sometimes called layer 2 switching, is a hot topic in today's data networking marketplace. LAN switches operate in "cut-through" mode. Cut-through defines a technique in which header information is processed and sent along the appropriate path while the data and trailer information are still being received. Contrast this to store-and-forward processing, which involves buffering a complete transmission before any processing or forwarding actions are taken. The primary benefit of LAN switches is that they greatly reduce delay, thus improving performance of the LAN. Vendors of LAN switches include 3Com, Cisco, Enterasys (Cabletron), Extreme Networks, and Nortel. LANs are covered in Chapter 14.

ATM switches often are the basis for provider backbones, and sometimes for interconnecting LANs and Internet access. ATM integrates voice and data onto one common network, so the switches have a complex role. They need to deal effectively with QoS issues associated with supporting multiple applications on a single facility. In the network, the typical backbone switch has numerous ports for premises access. One or more trunk ports are used to access other ATM switches in the network as well. And non-ATM interfaces provide internetworking with other public network services like frame relay. ATM switch vendors include Alcatel, Fujitsu, Lucent, Marconi, and Seabridge (Siemens).

IP routers are a form of connectionless packet switch, which you will find in LANs and WANs. They store packets, examine addressing information, make routing decisions, and forward the packets to the next router. The era of gigabit (and terabit) routers has arrived. Vendors of such high-speed routers include Avici, Cisco, Lucent, Juniper, and Foundry Networks.

Interestingly enough, a number of layer 4 switches are now starting to appear on the market. These devices peer into the Transport layer to determine the transport protocol and user application (such as File Transfer Protocol [FTP] versus Hypertext Transfer Protocol [HTTP]). The applications, as well as the target network, can now influence the forwarding decision, which can help a company set up and comply with service level agreements (SLAs) based on application type. In fact, some vendors are moving all the way up the stack to layer 7 for load-balancing Web servers and to provide better QoS for IP-based voice applications. It's not enough for a packet switch simply to forward packets in this era of converged networks.

SUMMARY

Switching, whether circuit or packet based, is the foundation of most communications today. Regardless of which approach is used, the intent is to more efficiently use network resources. Applications have traditionally dictated the form of switching implemented, although today we see a trend toward a single, packet-switched network carrying voice and data. At the heart of these networks are the switches themselves, which can be found in enterprise and carrier networks—really everywhere voice and data are being transmitted.

CHAPTER 7

The Network Layer

A number of widely deployed Network layer protocols are on the market, each with its own characteristics, strengths, and weaknesses. Most networks today are connectionless and use the Internet Protocol (IP); however, it has not always been so. Many networking protocols have come and gone, both connection-oriented and connectionless. In this chapter, we will briefly examine Path Control, a connection-oriented protocol in IBM's Systems Network Architecture (SNA). We will also examine IP in detail. The chapter concludes with brief overviews of several common layer 3 protocols.

ROLE OF THE NETWORK LAYER

The Network layer routes packets or messages through an interconnected network. To provide this service, the Network layer protocol must specify the structure of the Network layer address, the format of a data unit (such as a packet), the types of service provided to the upper layers (such as connectionless, connection-oriented), mechanisms for requesting and obtaining certain Quality of Service (QoS) characteristics, and mechanisms for making routing decisions; it must also control congestion.

The complexity of the Network layer directly impacts the complexity of the next layer, the Transport layer. If the layer 3 protocol can guarantee delivery and sequence, there is little need to build capabilities into layer 4 to handle errors and sequence. However, if the layer 3 protocol is *best effort*, offering no guarantees, the layer 4 protocol is probably complex, to ensure that correct information is delivered to the application.

CONNECTION-ORIENTED NETWORKS

Connection-oriented network protocols are frequently far more complex than those found in connectionless networks, because more protocols are required to set up, maintain, and tear down the connections. In addition, the connection orientation allows us greater control over the traffic flow, providing the opportunity to grant strict QoS and security capabilities in ways the connectionless world can only envy. Even so, it is the connectionless world of IP that dominates the networking landscape today. Interestingly, the majority of the work being done to extend the capabilities of IP, such as Multiprotocol Label Switching (MPLS; see Chapter 23), make IP networks behave in a connection-oriented fashion, which proves that there is always room for reinventing the networking wheel.

Public data networks, such as X.25 and frame relay, are often considered to be layer 3 networks. In the context of the customer network, these protocols operate at layer 2, not layer 3. While they may have many characteristics of a layer 3 protocol, such as addressing and packet transport, they simply offer a richer appearance of connectivity than the underlying physical network. The corporate network will employ a layer 3 protocol *on top of* the connection-oriented wide area network (WAN) service.

PATH CONTROL

SNA is a particularly complex network architecture, as is SNA's Network layer structure. The SNA Path Control layer provides both Network and Transport layer functionality, but we will concentrate here on the Network layer aspects of the Path Control layer. Due to the complexity of the protocol, combined with the loss of popularity of SNA, our discussion is limited to a high-level treatise on Path Control. For anyone desiring to delve into more details, several good SNA texts are available.

Path Control is often described as a *statically defined* connection-oriented network. While the tables in an SNA network *are* static, as you shall see, the route selection between two end points is anything *but* static. Many people pooh-pooh SNA because it is complex and appears cumbersome—some may say clumsy. But SNA is a networking scheme that was developed to support very large networks with high availability, at least 5 nines (99.999% uptime), and tightly controlled performance for all users. There has never been another networking architecture capable of supporting the number of users with the degree of control and quality of service provided by SNA. In fact, IP folks are working hard to reengineer and extend IP to add some of the capabilities found in SNA (not that they would view it this way).

In Path Control, the connection is referred to as a *virtual route*, a logical association between two end nodes. Multiple virtual routes may extend between any two end nodes. There are several classes of virtual routes; the class to which a user is attached depends on the class of service requirements that are determined by the user and application profiles. A virtual route corresponds to the concept of a Network Service Access Point. Higher layers obtain Path Control services by attaching to a virtual route.

A number of explicit routes may be associated with any given virtual route; the explicit route for any given session is selected by the explicit route control function in the source node. This selection is based on the virtual route in question (on the class of service required) and on current load conditions. The explicit route manager performs some degree of load balancing by selecting different explicit routes for different sessions.

After a packet is launched, the explicit route manager attaches an explicit route number to the setup packet and sends it to the next node on the explicit route. The explicit route manager in this node uses a static table to determine which node is next in line and which transmission group to use to connect to that node. Since all packets for the session follow the identical explicit route, SNA is correctly identified as a *virtual circuit network*.

A transmission group between two nodes can contain multiple links, although all the links in a transmission group are of similar type—for example, all satellite links or all terrestrial T1 links, and so on. Since different packets on the same session can use different links within a transmission group, packets can get out of order during transmission. It's up to the receiving node to put the packets in order after they arrive.

Here's an example scenario. Suppose that Packet 1 goes over Link 1, and Packet 2 goes over Link 2. Packet 1 suffers a hit on the link and must be retransmitted, while Packet 2 gets through unscathed. Packet 2 is now "ahead" of Packet 1. Resequencing occurs at every node to ensure that packets are always delivered in sequence.

This is the high-level view; you can imagine what the details look like. Path Control provides a great degree of flexibility in using the underlying network resources. While the details are beyond the scope of this book, we hope you have gained at least a sense of the power provided by connection-oriented networks.

CONNECTIONLESS NETWORKS

A variety of connectionless Network layer protocols are available; we have elected to limit our discussion to one, IP. If you understand IP, it is easy to transfer your knowledge to other protocols you may encounter. While the details may be different, the underlying functionality will be similar.

In connectionless networks, each packet contains enough information to identify its ultimate destination. Each node makes a forwarding decision based on the final destination. This is in contrast to connection-oriented networks, in which the nodes are forwarded based on a virtual circuit number, and outside the setup message there is no information explicitly identifying the final destination. In connectionless networks, the act of forwarding packets is frequently referred to as *routing*. Routing has two distinct parts: the building of the route (forwarding) tables, and the act of forwarding (routing) the individual packets.

Routers and Routing

Routing is the process of getting a single IP packet from source to destination in a connectionless network. The device that performs the packet transfer is known as the *router* (although many other terms have been used over the years, *router* has gained universal acceptance).

While a router has many functions related to the operation of a network, its primary function, not surprisingly, is routing information from point A to point B. Consider the simple network diagram shown in Figure 7-1. Without delving into the format of network addresses, we see that each router, numbered 1 through 4, is connected to four Ethernet networks, labeled Networks A, B, C, and D.

If a computer (a *host* in Internet parlance) located on Network A wanted to send a packet to a host located on Network C, the packet would first be sent to Router 1. Router 1 would look up the destination of the packet and determine where to forward the packet. The routing table contains entries for all networks known to the router. Assuming that Router 1 has an entry for Network C, Router 1 simply consults its table and forwards the packet to the appropriate interface.

While some variation exists among routers of various manufacturers, every routing table contains key information that is used to make these forwarding decisions. In the

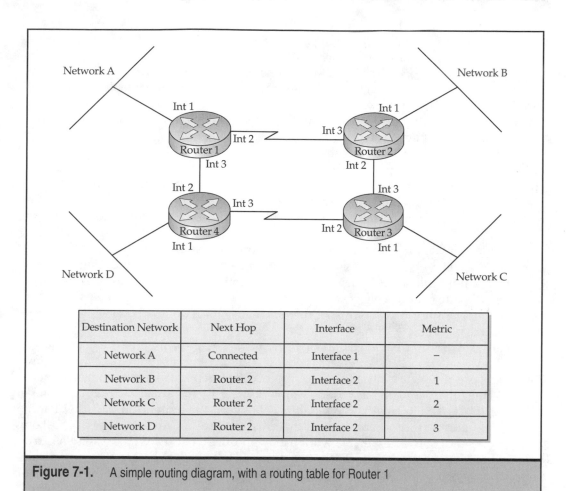

Figure 7-1.   A simple routing diagram, with a routing table for Router 1

example routing table shown in Figure 7-1, you can see a stylized version of these fields: Destination Network, the Next Hop (the next router), the Interface used to reach that next hop, and a Metric value.

The destination network is straightforward; it represents a network the router *knows* about. Routers communicate with one another using a routing protocol. Routers exchange information about their network view so that each router learns what other networks are out there. The routing table in Figure 7-1 shows an entry for each of the networks in the diagram that Router 1 might want to reach. In this example, if Router 1 has a packet destined for Network C, the table's Network C entry is the focus. Each entry in the routing table for destination Network C also contains a next hop. This means that while Router 1 is not directly connected to Network C, the network can be reached by

Router 1 forwarding the packet (or hopping) to Router 2. The interface entry, or protocol, specifically tells Router 1 how to reach Router 2.

This sample network is simple. On examination, you can identify two routes from Network A to Network C, yet only one route is indicated in the routing table. The routing protocol determines which of multiple paths is taken. Every routing protocol uses its own set of metrics to identify the *cost* of reaching a particular destination. Metrics can be very simple, such as the number of hops (intermediate routers), or they can be complex, using multiple parameters, such as load, to determine the route.

In our example, if the metric for Router 1 to Network C was based simply on the hop count, or number of routers (two), two routing options could be included in the table. In a given manufacturer's implementation, however, some form of tie breaker must exist so that only one entry makes its way to the routing table. In this case, the Router 2 route is selected.

As links between networks, routers are important management points in the network. They are excellent locations to use for collecting network information, such as traffic patterns, link utilization, transmission and reception errors, and host and network statistics. The information collected at the router can be sent to a management station for compilation. This information can assist network managers in troubleshooting the network and resource planning.

Finally, routers may be called upon to *fragment* packets—that is, they break up a large packet into smaller packets. Fragmentation is a necessity because routers can use many different types of network interfaces that support different packet lengths. In the example network, the Ethernet segments hosting Networks A, B, C, and D can each support packets up to 1500 octets in length. The serial links between the routers, on the other hand, may be configured to accept only packets that are a maximum size of 576 octets. A host on Network A sending to a host on Network B may produce a packet of 1500 octets. When a router receives that packet and realizes that the serial link it has available for sending will support only 576 octets, it must fragment the original information into three smaller packets. The three smaller packets will then be passed through the other routers as independent packets and reassembled by the destination host on Network B. The intermediate routers do not reassemble the packets because they may need to be fragmented somewhere further along in the network, and constant reassembly and fragmentation would be a waste of resources.

Direct and Indirect Routing

We normally associate routers with the devices that actually perform the routing; however, any IP-enabled device is capable of performing routing. At the least, these devices are capable of routing to hosts attached to their network. In Figure 7-2, Host 10 is directly attached to Network A, and it is able to route packets to any other host on Network A. This is known as *direct* routing.

If Host 10 were to try to send something to a host on Network B, it would need the assistance of a router. In this case, it would forward the packet to Router 1 for delivery to the destination Network B. The process of passing control of delivery to another device is known as *indirect* routing. Let's explore how this process works.

IP Address Structure

Although Chapter 19 offers an in-depth examination of IP addressing, it may be helpful to introduce you to this topic now to help you understand how routers make forwarding decisions.

IP addresses are strings of 32 bits represented in dotted-decimal notation for ease of readability. An example of an IP address would be *192.168.1.10*. This address is divided into two portions: the network portion and the host portion. To determine where in this string of 32 bits the network portion falls, a subnet mask is commonly used. A 24 bit subnet mask would imply that the first 24 bits of the address were the network (192.168.1) and the remaining bits represented the host portion (10). This structure is similar to the hierarchical structure that represents telephone numbers, as described in Chapter 1. As with the telephone network, the combination of the network and host address is unique.

The router is interested only in the network portion of the address. The only time the router is concerned with the host address is when the router is directly connected to the network containing the destination host.

When an upper layer in an end station generates a packet to be sent to another host, the Network layer on the sending host needs to answer this question: "Can this packet be directly routed or must it be indirectly routed?" The Network layer examines the destination IP address and compares that address to the network portion of its own IP address.

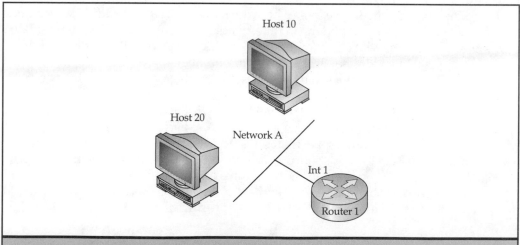

Figure 7-2. Direct routing from Host 10 to Host 20

If the network portion of its address matches the network portion of the destination IP address, the sending host knows that the destination exists on the same network segment, usually a local area network (LAN).

If a match is found, the packet is directly routed to the destination host. The sending host needs to find the data link address of the receiver on the local network segment. IP addresses, being Network layer addresses, are *universal* addresses, which means that hosts from many different Data Link layers can communicate with each other as long as they share the same Network layer IP address. A host with a destination IP address on the local network needs to determine what is the specific Data Link layer address to send the packet to a destination in the same network. So how does it know, or find, the data link address of the destination?

Some Network layer protocols configure themselves so that some relationship automatically exists between the layer 2 addresses and the layer 2 addresses. For example, in Novell's NetWare protocol IPX, the address comprises two parts: the network and host portions. In a LAN environment, the host portion defaults to the data link address, referred to as the Media Access Control (MAC) address, making it easy to find the destination host.

Unfortunately, this is not the case with IP, which needs a process (protocol) to help it identify (resolve) the MAC address of the destination. In IP, the address resolution protocol, or ARP, is used. This protocol takes a known IP address and sends a request to the LAN, asking, "Will the machine configured with this IP address please reply with your layer 2 address?"

Each host is configured with the IP address of the nearest router that is able to reach other networks. This router is also known as the *default gateway*. If a host determines that the packet destination is located on a different network, the packet will be forwarded to the default gateway to be indirectly routed.

To examine this process, let's work through a few simple examples, using a network similar to the sample network we looked at early in this chapter and adding some minor modifications. Now that you understand the basics of routing and how a routing table works, we can dispense with the simplicity of Network A, Network B, Host 10, and so on, and start using real IP addresses and real network numbers.

In Figure 7-3, the links between the routers have been given network IDs. Every port on a router defines a network. If a router has two ports, as these routers do, they will be connected to two different networks. Until this point in our discussion of routing, the networks between the routers have been irrelevant. Not any more! Each port on the router is listed with the portion of the IP address that represents the network. For instance, one of the networks connected to Router 1 is identified as Network 10. This means that of the four octets of the IP address, the first octet is a 10; the other three octets, such as 1.1.1 or 2.2.2, represent hosts on that network. So a complete IP address for a host on Network 10 connected to Router 1 would be the network 10 plus the host address 1.1.1, concatenated to 10.1.1.1.

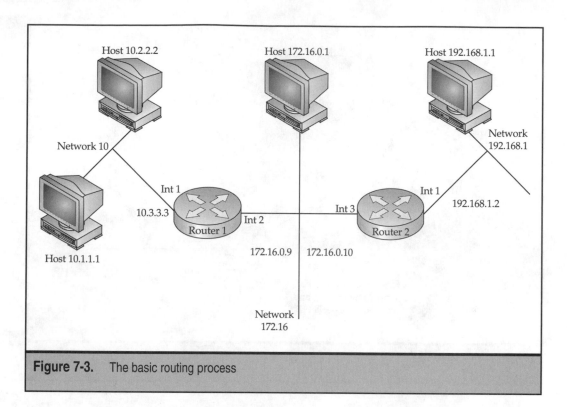

Figure 7-3.   The basic routing process

Notice also that the length of each network section is different. (This is an important element that will be explored in more detail when IP addresses are discussed in Chapter 19.) Consider the network connected to Router 2. The network portion of this address uses three octets—192.168.0. That leaves one octet for the hosts, be it 1 or 11 or 111. A complete IP address for a host on the network connected to Router 2 would be the network number 192.168.0, combined with the host number 1, for an IP address of 192.168.0.1.

With this information, we can now begin to route. For example, in Figure 7-4, the Network layer of the host with IP address 10.1.1.1 has received instructions to send a packet to the host with IP address 10.2.2.2. The first thing that Host 10.1.1.1 does is determine whether the destination IP address is on the same network as itself. To do this, the host compares its network number, 10, with the network number of the destination, which is also 10. Therefore, the sending host concludes that the destination is somewhere on the same network segment as itself.

How is Host 10.1.1.1 so sure that the receiving host is on the same network? Remember that each of the 4.2 billion IP addresses available for use can be assigned only to one

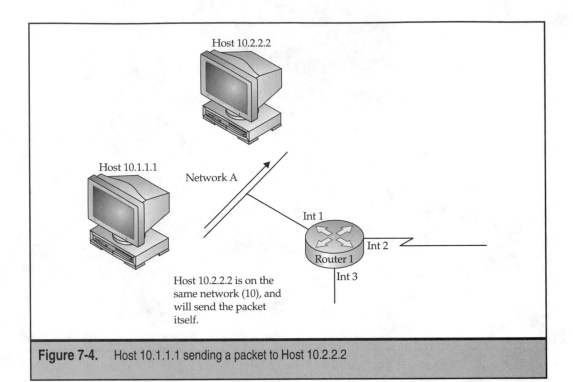

Host 10.2.2.2

Host 10.1.1.1

Network A

Int 1

Int 2

Router 1

Int 3

Host 10.2.2.2 is on the
same network (10), and
will send the packet
itself.

Figure 7-4. Host 10.1.1.1 sending a packet to Host 10.2.2.2

machine. Likewise, each network number can be assigned only to one network. From the point of view of Host 10.1.1.1, the only place it is going to find another machine with the same network number of 10 is on the network it is attached to. So it is with some confidence that Host 10.1.1.1 sends out an ARP that makes the request: "Will the host machine configured with IP address 10.2.2.2 please respond with its MAC address?"

Upon hearing this request, Host 10.2.2.2 dutifully responds with its MAC address, and Host 10.1.1.1 will then have the information that it needs to send the packet to the Data Link layer for insertion onto the local network segment. Figure 7-5 illustrates this packet exchange.

In the next example (Figure 7-6), suppose Host 10.1.1.1 has a packet to send to Host 172.16.0.1. An examination of the network address reveals that the destination is not in the same network as the sending host. Host 10.1.1.1 is configured to send any packets requiring indirect routing to its gateway router, Router 1. To deliver the packet to Router 1, Host 10.1.1.1 needs to use the MAC address of the router whose IP address is 10.3.3.3.

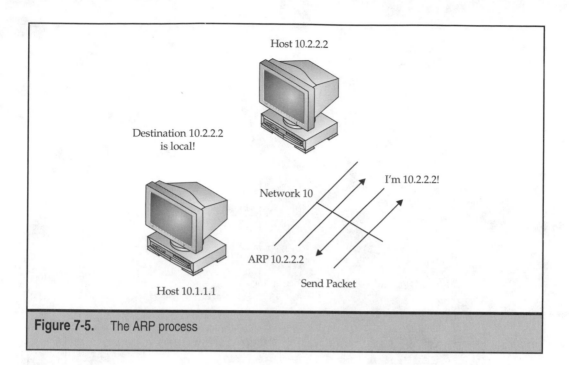

Host 10.2.2.2

Destination 10.2.2.2
is local!

Network 10

I'm 10.2.2.2!

ARP 10.2.2.2

Send Packet

Host 10.1.1.1

Figure 7-5. The ARP process

If the MAC address is unknown to Host 10.1.1.1, it sends an ARP to obtain the MAC address of the router at 10.3.3.3 and the router responds appropriately. The packet destined for 172.16.0.1 is then sent to Router 1 using its MAC address.

Routers route packets according to the information in their IP routing table. To understand what Router 1 will do with the packet destined for 172.16.0.1, let's take a look at its routing table.

Router 1 realizes that it is connected to network 172.16 and therefore assumes that Host 172.16.0.1 must be part of that network. Router 1 will then implement its own direct-routing routine and send an ARP request looking for the destination host configured with the IP address 172.16.0.1.

Now, using the hosts discussed and the routing table for Router 1 as previously illustrated, let's examine what would happen if Host 10.1.1.1 were to send a packet to the destination Host 192.168.1.1.

Much of the process should be familiar to you by now. Host 10.1.1.1 will compare its network number to the destination network number and realize that the destination is on a different network. The default gateway will be ARP'd, if necessary, so that the packet can be forwarded to Router 1.

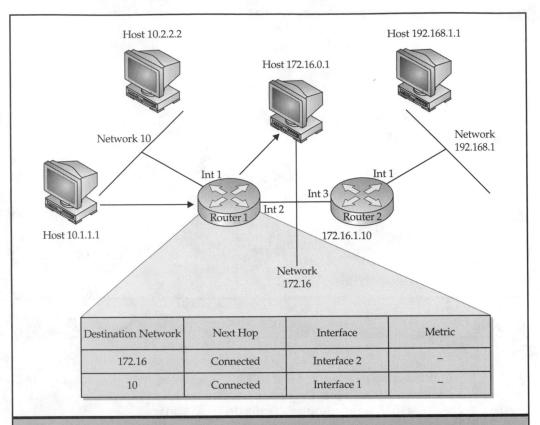

Figure 7-6. Host 10.1.1.1 sending a packet to 172.16.0.1

When Router 1 is in possession of the packet destined for 192.168.1.1, it will check its IP routing table. However, the current table does not include an entry for the 192.168.1 network. Router 1 will be forced to implement one of the most fundamental rules of IP routing: If a packet is received and the routing table contains no information about the destination network, throw the packet away. In this example, Router 1 will discard the packet and notify the sending Host 10.1.1.1 that the network is unknown.

NOTE: Hosts build ARP tables containing recently learned data link addresses to reduce the number of ARPs sent on an individual network. These addresses are purged periodically if no activity is logged with a device represented by an entry. This allows for devices to move around in a network. For more discussion on this topic, see Chapter 19.

Static Routes

As the preceding examples have demonstrated, both hosts and routers use a combination of direct and indirect routing to deliver a packet to its destination. As long as a destination host has an entry in the router's routing table, the router is able to forward the packet appropriately.

But how are routing tables constructed? The simplest way to build routing tables is to configure specific routes manually. Routing from manually configured tables is referred to as *static* routing, because manually configured routes are fixed.

Entire networks with fairly complex topologies could be created using only static routes. The problem with static routes, though, is that *they are static*. Manually entered routes will not change to reflect any changes in the network topology. Consider our original four-router network, shown in Figure 7-7.

In this network, Router 1 can reach Router 2 via two paths. The most direct way would be to send packets via Interface 2; a logical static route would be for Router 1 to send all packets destined for 192.168.1.1 via Interface 2 to Router 2. But if the link between these two routers were to fail, Router 1 would continue to use the static route, and packets would be sent onto a malfunctioning network and never reach their final destination, even though another route exists—via Routers 4, 3, and 2.

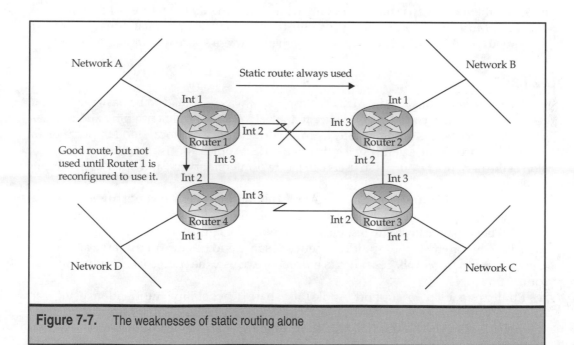

Figure 7-7.   The weaknesses of static routing alone

Despite this disadvantage, static routing is a viable and easy-to-configure solution to many routing problems and is being used in one way or another on most IP networks today. In a network of only a few routers, or networks without many redundant paths, static configuration of routes is straightforward. As the network grows and routers and redundant links are added, though, it becomes difficult for an administrator to keep up with all the possible routes for destination networks using just static routes. In complex networks, dynamic routing protocols are needed to update routers if topology changes in the network. (In Chapter 19, we will explore a number of IP routing protocols.)

EXAMPLES OF PROTOCOLS AND COMMON USAGE

While IP may be the best known of the Network layer protocols, it is by no means the only protocol. Following is a discussion of other Network layer protocols. All the protocols offer the same basic services; the implementations vary according to the specific network architectures.

Most of the following protocols were designed to accommodate communications between hosts in a high-bandwidth LAN environment. Unlike IP, they are proprietary—built by the manufacturers of LAN hardware or software to support only their products. Therefore, use of these protocols is restricted either by licenses from the corporation that invented them or because they work only while using the manufacturer's hardware and software. Most of the protocols also suffer from limitations in scalability due to their inability to combine the network portion of their address space into larger networks.

AppleTalk

Like most applications for the Mac OS, AppleTalk was designed to be easy for the end user. Adding a new computer or device to the network is simple with plug and play, as any new device would automatically learn its address, network locale, and the location of any routers providing a gateway to other networks. AppleTalk also uses a graphical version of the network resources so users can locate printers, servers, and other users simply by clicking on an icon.

To provide this functionality, the AppleTalk addressing scheme is divided into a 16-bit network prefix and an 8-bit host suffix. This allows for more than 65,000 networks with 253 hosts (some hosts are reserved for network functions) on each segment. Because there is no hierarchy to the AppleTalk network space and because of the large amount of network traffic AppleTalk generates to ensure ease of use, the protocol is ideally designed for use on LANs.

Routers enabled to support AppleTalk share reachability information using the Routing Table Maintenance Protocol (RTMP). The RTMP requires that routers share their routing tables every 10 seconds. This short timeframe makes RTMP responsive to network changes for a distance vector routing protocol, but with a heavy portion of LAN bandwidth consumed for router updates.

AppleTalk's use is waning, as most Mac hosts are also generally running IP to communicate with other operating system platforms and the Internet.

DECNet Phase IV

For much of the early days of networking, Digital Equipment Corporation (DEC) was one of the giants of the computer world and the primary competitor of IBM in the early 1980s. When IBM introduced SNA in 1974, DEC soon followed suit with DECNet. By early 1980, DECNet had evolved into the proprietary protocol DECNet Phase IV. In the mid '80s the OSI Model had begun to gather momentum, and DEC, in an effort to comply with this model, introduced DECNet Phase V. Phase V was different enough from Phase IV that many companies never migrated their network to the newer protocol. There are still some DECNet Phase IV networks to be found, despite DEC disappearing when Compaq purchased it.

DECNet networks are divided into three major components. The first is the end node, personal computers or minicomputers running the DEC operating system. These devices do not perform any type of routing. The second, a DECNet node, is a computer capable of functioning as a router; it knows how to reach other routers and end nodes not within its network area. The final component is the logical area, which is a collection of end nodes and nodes that share the same Network layer address.

The DECNet addressing scheme divides each device into a 6-bit area (network) prefix and a 10-bit node suffix. This rather small address space limits DECNet networks to 63 areas and 1024 nodes within each area. Unlike other Network layer protocols that we have discussed, DECNet assigns one address per device. Therefore, a router with six interfaces, for example, will still use one area address and one node address. IP, on the other hand, consumes an address per interface. The same six-port router in IP would use six IP addresses.

DECNet uses its own routing protocol. In Phase V the routing protocol was the basis for Intermediate_System to Intermediate System (ISIS), an ISO standard for routing in connectionless networks. ISIS is frequently found in ISP networks and is similar to Open Shortest Path First (OSPF), described in Chapter 19.

IPv6

Most of the protocols described in this section have had their time and have seen their influence gradually erode in favor of the Internet Network layer: IP. The one exception included in this list is the IP version 6 (IPv6). In the early 1990s, the sudden and rapid growth of the Internet led to concerns that the current addressing structure would not scale to meet projected growth patterns. In 1995, a new version of IP was introduced. IPv6 vastly increased the number of hosts allowable in the addressing space, ensuring that enough IP addresses would be available beyond the foreseeable future. At the same time, the hierarchical nature of the addresses was increased, allowing more efficient routing of the millions of new networks that would be formed.

Early design considerations of IPv6 (the current version of IP) were also reevaluated. When IPv4 was introduced, flow control, quality of service, and security were not major considerations. In today's Internet, however, they are of utmost importance, and IPv6 addresses each of these concerns.

IPv6 has a total address space of 128 bits—more addresses than can be expressed on a page; comparisons of the amount of addresses are inconceivable (think "enough addresses for every molecule in the solar system").

The basic structure of the IP address serving as network/host is unchanged. The routing protocols developed today are also expected to support IPv6 with modifications already in place. In fact, most of the structure to support IPv6 is in place in operating systems and network hardware. What has limited the adoption of IPv6 is need.

As IPv6 was developed to confront the imminent lack of IP addresses, other techniques under development allowed network administrators to conserve the existing number of IPv4 addresses already in use. These techniques have worked so well that even at present rates of allocation, the IPv4 address space is expected to last until 2010. This is especially true in the United States, which is blessed with an abundance of IP addresses.

In Europe and Pacific Rim nations, IPv4 address space is not in such great supply. Networks in these countries, including ISP backbones, have already deployed IPv6 as the primary Network layer protocol. IP used to enable personal consumer devices such as cell phones and personal game machines is also reintroducing the need for more IP addresses. As such, network professionals of the future need to be well versed in the capabilities of IPv6. Unlike the other protocols mentioned in this section, IPv6 will be in use more often as the global Internet matures.

IPX

Based on the Xerox XNS protocol, Internetwork Packet Exchange (IPX) was developed by Novell to enable MS-DOS clients with the NetWare client software loaded to communicate with NetWare servers. Despite users who seem determined to force IPX over slow WAN links to remote offices, IPX was developed solely for use on the LAN. Like AppleTalk, what IPX provides in ease of use is at the cost of increased network traffic. On most LANs, this traffic is not an issue, but over slower frame-relay or dedicated point-to-point T-1 lines, this extra traffic becomes a point of network congestion.

In the IPX view of the world, only two devices exist: the client, which makes a request, and the server, which fulfills the request. But clients cannot communicate with other clients using IPX. They can only make requests of servers—specifically NetWare servers.

Like the TCP/IP protocol suite, IPX is just the Network layer of an entire suite of protocols. Like IP, IPX is a connectionless protocol. It provides the means for getting higher layer protocols like Sequenced Packet Exchange (SPX), Service Advertising Protocol (SAP), and the NetWare Core Protocol (NCP) to share server resources with clients.

Also like IP, IPX has an address space organized around the network/node. The network portion of the address is 32 bits and the node portion is 48 bits. While this is an extensive network space, allowing more than 4.2 billion networks, there is no hierarchy that would allow IPX routers to aggregate the billions of networks into manageable routing tables. The host portion of the address space is generally the 48-bit globally unique MAC address assigned to each network interface card at the time of manufacturing. This mapping of the host ID to an existing, unique identifier on each host makes assigning IPX addresses trivial. Only the network portion need be defined.

To ensure reachability in complex networks, IPX routers share information using some very IP-like routing protocols. Routing Information Protocol (RIP) for IPX operates in the same manner as the distance vector routing protocol for IP, with the exception of the calculation of the metric. In the event that the IPX network is too complicated for RIP, a link state routing protocol named the NetWare Link State Protocol (NLSP) operates much like the IP link state protocol, Open Shortest Path First (OSPF).

When Novell was the dominant server product for Windows operating systems, IPX was an important protocol to understand. With the recent emergence of the Internet and the loss of market share to Microsoft NT and 2000, the use of IPX has dwindled. Even Novell now recommends that existing users of IPX migrate to the TCP/IP suite, and recent versions of Novell's operating system defaults to TCP/IP installations. Nevertheless, many corporate LANs with IPX packets being passed continue to provide mission-critical operations in the midst of an IP network.

NetBEUI

The earliest implementations of the NetBEUI protocol were developed for use with IBM's LAN manager software. The network basic input output system (NetBIOS) is a set of commands that computers issue to receive or send information on a LAN. Microsoft implemented the protocol stack for use as its LAN protocol of choice and renamed it the NetBIOS Enhanced User Interface (NetBEUI). NetBEUI is easy to configure for novice users because the numeric indicators of *network* and *host* become the *domain* and *computer name* of NetBEUI, enabling users to enter any naming structure they wish.

Like IPX and AppleTalk, NetBEUI was designed with the LAN in mind and ease of use paramount. Clearly, the naming structure for networks and hosts is not scaleable and furthermore is not routable. Ease of use is enabled through extensive use of broadcast traffic on the network as well, so that users can readily find network resources.

All these elements make NetBEUI ideal for simple networks where network expertise is in short supply and scaleability is not an issue. With Microsoft as an important network operating system and with users' familiarity with the "network neighborhood," Microsoft has transferred the NetBEUI interface and capabilities to NetBEUI over TCP (NBT) that allows users the functionality of NetBEUI over a TCP/IP-enabled computer. This uses the IP addressing and routing structure to enable NetBEUI to be used across an enterprise network.

CHAPTER 8

The Transport Layer

The primary job of the Transport layer is to provide end-to-end, error-free delivery of protocol data units on behalf of the upper layers (Session, Presentation, and Application layers). The Transport layer must perform this function regardless of the particular service offered to it by the lower layers (Network, Data Link, and Physical layers). The protocol data units handled at the Transport layer are usually called either *segments* or *messages*. In this chapter, we will use messages.

Figure 8-1 visually demonstrates the role of the Transport layer in supporting consistent Quality of Service (QoS) in the face of varying degrees of lower layer reliability. The rightmost bar in the figure shows that the service offered by the lower layers is essentially unreliable (a connectionless packet network, for example). In this case, the Transport layer must be rather complex to offset the unreliability of the lower layer service and achieve the level of service expected by the upper layers. The middle bar shows an instance in which the reliability of the lower layers is fairly good (for a connection-oriented packet network, for example). In this case, the Transport layer can be less complicated and still achieve the service level desired by the upper layers. Finally, the leftmost bar shows that, if the lower layer service is highly reliable (such as a circuit-switched connection with a reliable Data Link layer protocol), little needs to be done at the Transport layer to achieve a consistent level of service. The figure demonstrates that *the complexity of the Transport layer is inversely proportional to the reliability of its underlying layers.*

There are a few notable exceptions to the aforementioned functions performed by the Transport layer, including a class of Transport layer protocols that provide unreliable, connectionless service to their upper layers. For example, the User Datagram Protocol (UDP) from the Internet Protocol (IP) suite offers little more than addressing to its upper layers. However, with the proliferation of applications that don't require guaranteed delivery (such as nonmission-critical applications) or applications that cannot tolerate retransmission as a method of error control (such as real-time applications like streaming video or IP telephony), these nonguaranteed Transport layer protocols are gaining in popularity. We will analyze both connection-oriented and connectionless classes of Transport layer protocols in this chapter, with an emphasis on the manners in which they perform their unique functions within the OSI protocol stack.

TRANSPORT LAYER OPERATIONS

To perform its task of end-to-end, error-free message delivery, the Transport layer supports a number of subfunctions, which are described in detail in the following sections of this chapter.

▼ Addressing

■ Connection establishment and release

▲ Data transfer
  ■ Data prioritization
  ■ Error detection and correction
  ■ Flow control
  ■ Failure recovery
  ■ Multiplexing

Transport Layer Addressing

In previous chapters, you learned about the requirements for addressing at the Data Link layer (to identify a station on a multipoint data link) and at the Network layer (to identify a port on a machine). At the Transport layer, another unique address is required to identify a process executing within a machine. Most computers are capable of executing multiple processes at any one time. Often, however, a computer has only a single identifier (such as a port address) at the Network layer. When a packet arrives at the correct Network layer port, the port must know for which internal process it is destined. The Transport layer address provides this information.

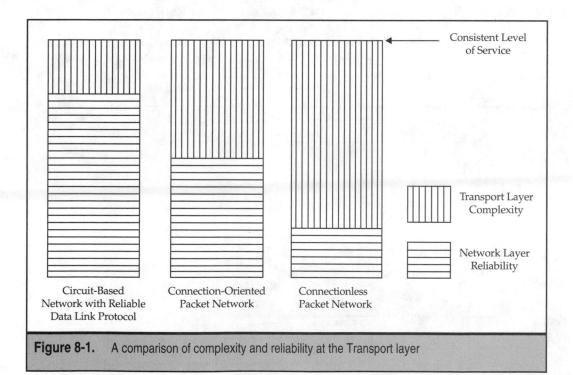

Figure 8-1.   A comparison of complexity and reliability at the Transport layer

Figure 8-2 shows the relationship between the Network layer address and the Transport layer addresses. In the figure, packets arriving for any of the four processes executing in the machine (Processes A, B, C, and D) all contain the same Network layer address—namely address XYZ. To deliver the contents of the packets to the proper processes, the Transport layer address is interrogated. For example, a message destined for Process B will have a transport address of 002. Likewise, a message destined for Process D will carry a transport address of 004.

A useful analogy in this regard involves the process of obtaining directions to visit a friend who lives in a large apartment building. To visit your friend, you'll need a street name (a data link address), a street number (a network address), and an *apartment number* (a transport address). Without each of these pieces of the address, you will not be able to find your destination.

Note that the Transport layer address is a *logical* address. It does not correspond to a hardware port, like a Data Link or Network layer address; rather, it corresponds to a software port that in turn corresponds to a particular process (an application) within the machine. Often, Network and Transport layer addresses are discussed as a single, unique entity. When this perspective is taken, the resultant addressed entity is called a *socket*.

Transport layer addressing is, however, fundamentally different from addressing either at the Data Link or Network layer. At the Transport layer, one is addressing a *process*, not a machine. To conserve computer cycles, processes are generally not active until they are required for some specific use (such as for connecting a user to an application). Therefore, when we attempt to connect to a peer Transport layer process in a remote machine, how do we guarantee that the process is available at a specific time? More importantly, how do we ensure that the process is ready to connect at a given address?

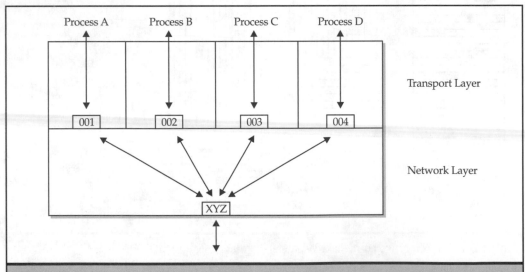

Figure 8-2.   The relationship between addresses in the Network layer and Transport layer

To draw upon another analogy, I can always dial your telephone number to ring your telephone. In other words, I can get to your network address, because it's published and easily available. When I ring your phone, however, is there any guarantee that you'll be there listening and ready for the call? The phone is a machine (with a fixed address), and you are a process (and not likely to be listening at all times).

Solutions to this problem have varied from the simple (such as the *well-known ports* approach) to the elegant (such as the *name server/process server* approach). The well-known ports approach is tantamount to saying, "I'll be home at my phone every day between noon and 2:00 P.M. Call then, and you'll be sure to get me." The name server/ process server approach is more complex and is analogous to this: "Call my friend (who knows me well), and let him know that you wish to talk to me. He will tell you where and when I can be reached. My friend will then call me with the information so that I'll be sure to be at the phone when your call is expected." To summarize, when attempting to establish a Transport level connection to a peer transport entity in a remote machine, the originating entity must know what Transport layer address to use to ensure accurate connection. One of several mechanisms to determine that address will be supported in any given system.

Connection Establishment and Release

Transport layer protocols that guarantee end-to-end, error-free message delivery are connection-oriented protocols. Therefore, a connection between peer transport entities must be established prior to data transfer. That connection must be managed during data transfer and released when data transfer is complete.

The establishment of a Transport layer connection is typically accomplished via a procedure called a *three-way handshake*. Figure 8-3 illustrates this method for establishing a Transport layer connection. The three-way handshake seeks to minimize the probability that a false connection will be accidentally accepted by requiring two acknowledgements to a connection request. As shown in the figure, Transport Entity A initiates the process by requesting a Transport layer connection with the Transport Entity B. Upon receiving the connect request from Entity A, Entity B will, under normal circumstances, acknowledge the connect request. Upon receiving that acknowledgement, Entity A (the original requester) will acknowledge the acknowledgement. When Entity B receives this second acknowledgement, the connection is established and moves into the data transfer phase.

This view of connection establishment considers the network a hostile environment in which packets (which may contain requests or acknowledgements) may be lost or destroyed at any time. By implementing the three-way handshake, the Transport layer minimizes the chance that a problem in the network will cause an erroneous connection. For example, if an acknowledgement for a connect request is not received within a specified period of time, the originator will issue a second request. This time value must be chosen carefully to ensure that an overly long network delay is not responsible for the "loss" of the original request. If timer values are too short, duplicate connect requests may arrive at the receiver. Moreover, if the second acknowledgement is lost in the network, the receiver will reject the connection after a time, presuming for some reason that the original

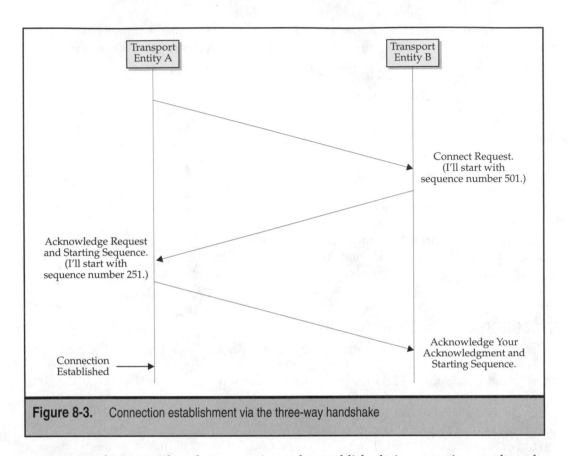

Figure 8-3. Connection establishment via the three-way handshake

requester no longer wishes the connection to be established. A connection results only when each of the three events (request-acknowledgment-acknowledgment) occurs within the appropriate time constraint. Thus, the three-way handshake ensures that both transport entities are fully committed to the connection. Aside from actually causing the establishment of a Transport level connection, the messages exchanged during the three-way handshake may serve to synchronize various parameters associated with the connection (such as beginning sequence numbers).

The release of a Transport layer connection is also controlled by a three-way hand-shake. In this fashion, connections can be released gracefully, without loss of user data. Moreover, connection release generally must occur twice for each established connection—once for each direction of transmission. It is not unusual for one side of the connection to be finished sending while the other side still has additional data to transmit. In this case, the connection in one direction is released and the connection in the other direction remains active until all data are transferred.

Data Transfer

During the data transfer phase of a transport connection, the Transport layer protocol must manage the connection to ensure end-to-end, error-free delivery of messages.

A number of functions must work together during data transfer to ensure that message delivery is guaranteed (sequencing and flow control) and that the delivered content is error-free (sequencing and acknowledgement). We will discuss each of these connection management functions in some detail in the following sections.

Data Prioritization

Many Transport layer protocols support two transmission priorities during the data transfer phase of a connection. The priorities are generally labeled *normal* and *expedited*. Messages labeled as expedited are typically given priority through the use of a special buffer management algorithm. For example, separate buffer pools can be established for normal traffic and expedited traffic. A simple algorithm like "empty the expedited buffers completely before removing anything from the normal buffers" would suffice to distinguish the two traffic types.

Some Transport layer implementations differentiate between normal and expedited data by using separate messages to transport each traffic type. An indicator of some kind in the Transport layer overhead signals which type of message, normal or expedited, is being received. Another approach is to support the partitioning of a single message into normal and expedited fields. When this approach is taken, the Transport layer overhead must indicate the break point between the two types of data within the message, so that the receiver can accurately recover the normal and expedited portions of the message.

Remember, however, that a given expedited message is recognized as such by the peer Transport layer entity only. The Network layer will remain unaware of any given message's priority and may, in fact, treat all messages (packets) equally. When coupled with a Network layer service that permits expedited packet flows (such as INTERRUPT service in ITU-T Recommendation X.25), expedited data at the Transport layer is a powerful mechanism for differentiating critical message flows from those of a more mundane nature.

Error Detection and Correction

As with many Data Link layer protocols, the function of the Transport layer is error detection and correction. The reader would do well at this point to ask "Why? If the Data Link layer is responsible for error detection and correction, why do we need to duplicate this function at the Transport layer?" The answer is that, unlike the Data Link layer, which is concerned only with errors committed by the Physical layer (bit errors), the Transport layer is concerned with errors committed by the Network layer (lost packets, out-of-sequence packets, and so on). This leads to an analogy between the Data Link and Transport layers. One can think of it this way:

Data Link:Physical as Transport:Network

Furthermore, in cases where the Data Link layer is not responsible for bit error detection and correction (or is absent altogether), this task may be performed by the Transport layer.

Detection of errors by the Transport layer is usually accomplished via a message sequencing function. Each message to be transmitted is given a unique sequence number by the transmitting entity. When a particular message is transmitted, a timer is started. The timer is stopped only upon receipt of an acknowledgement (positive or negative) for that particular message. If multiple messages are outstanding at any point in time, each is associated with its own timer function.

In some Transport layer implementations, acknowledgement of the last message in a sequence implies correct receipt of all messages up to that point. Receipt of a negative acknowledgement for any message requires that the transmitter return to the point of the error and retransmit all outstanding messages from that point on (called *go-back-N* retransmission). In other implementations, each message must be explicitly acknowledged by the receiver. If a negative acknowledgement is received in such a system, it pertains only to the specified message and requires that the transmitter retransmit *only* that particular message (called *selective* retransmission). Whereas go-back-N retransmission is most often used at the Data Link layer, selective retransmission is generally used at the Transport layer.

If a message timer expires, error correction is initiated in the form of retransmission of the message in question (in systems that use selective retransmission). Timer values must therefore be selected carefully to preclude the possibility of duplicate messages arriving at the receiver as a result of inordinately long network delays.

Flow Control

Each Transport layer entity has a limited amount of buffer (memory) allocated to it. Steps must be taken to ensure that the available buffer space is not exhausted under normal circumstances. This is a nontrivial issue when one stops to consider that messages are arriving unpredictably from the Network layer and can be passed to the upper layers only when they are ready to receive them. As a result of this requirement for buffer elasticity, the Transport layer typically uses some sort of explicit flow-control scheme.

The flow-control schemes implemented at the Data Link layer typically operate in the *sliding window* mode. In this mode, the acknowledgement of a frame carries with it an implicit invitation to open the window and send more frames. Thus, acknowledgement and permission to transmit are tightly coupled. At the Transport layer, we must *decouple* acknowledgement of messages from permission to send additional messages. Thus, an explicit flow-control scheme must be in place.

Many Transport layer protocols negotiate flow-control parameters during the three-way handshake process that establishes the connection. They may agree upon a maximum message size to be used over the connection. They may group some number of messages into a flow control group and use it as a maximum-transmit variable. They may simply require that the receiver indicate with every acknowledgement the exact amount of buffer space available at any given time. In every case, however, the receiver explicitly drives the behavior of the transmitter.

Acknowledgements not only serve as indicators of correctly received data, they also indicate the amount of new data that can be sent. In response to a series of correctly received messages, a transmitter might be told "acknowledge previously sent messages; send *n* more messages now." Another possible response, however, would be "acknowledge previously sent messages; send no more messages at this time." A subsequent acknowledgement would restart the message flow when enough buffer became available at the receiving entity.

Failure Recovery

Failures in networked systems can take two forms: network failures and computer failures. By far the easiest for the Transport layer to deal with are failures that occur in the network (such as node and link failures). If a network connection is lost (in virtual circuit environments) or if packets are lost due to a node failure (in connectionless environments), the Transport layer can use its sequence numbers to determine precisely what has already arrived at the receiver correctly and, therefore, what may need to be retransmitted. If the network failure is temporary (a new virtual circuit is established or a new route is found around a failed node), the Transport layer may continue to use the same network after service is restored. If the network failure persists for a long time, as it would for a complete network crash, the Transport layer, if it has been given the means, may use the services of a backup network to restore the transport connection (dial backup). Either way, network failures are easily dealt with by the Transport layer.

Failures of computers attached to the network are another matter altogether. As an end-to-end layer, the Transport layer executes under the operating system of the machine in which it resides. If the machine crashes, Transport layer execution is halted, and data in state tables may be lost. Thus, when the machine comes up and begins once again to execute, the status of Transport layer connections that existed prior to the crash is, at best, questionable. An approach that has been suggested is for the machine, once it is back up and running, to send broadcast messages to all other machines on the network to determine which of them had active transport connections to the failed machine at the time of the failure. If such information can be obtained (the amount of broadcast overhead is enormous in a large system), the restored machine may be able to recover some of the lost connections by relying on state information preserved by the machines that did not fail.

But even if this approach were workable, other problems exist in this scenario. As sequential state machines, computers must decide whether to acknowledge incoming messages either before or after writing those messages to nonvolatile memory. As it turns out, difficulties persist with either scenario. A detailed discussion of such difficulties is beyond the scope of this text. Suffice it to say, however, that, in the face of a machine crash, it is not likely that every existing Transport layer connection can be recovered or that data loss can be avoided even in cases where a given transport connection is restored.

Multiplexing

The Transport layer is capable of performing two types of logical multiplexing, which have been referred to as *upward* and *downward* multiplexing.

Referring back to Figure 8-2, you can see the case of upward multiplexing, which exists when multiple Transport layer connections share a single network connection. As noted previously, the transport address associated with each message permits the Transport layer to differentiate messages destined for one process from those destined for other processes. The advantage of upward multiplexing is generally cost savings. Network connections are conserved by the use of upward multiplexing. Where there is a cost associated with each network connection (fixed as with frame-relay PVCs or variable as with dial-up connections), upward multiplexing reduces cost by reducing the number of network connections required to support a given number of communicating processes.

Figure 8-4 illustrates downward multiplexing, which occurs when a single transport process uses multiple network connections. The principal advantage to downward multiplexing is enhanced performance. In cases where the network connection constrains throughput (such as small window sizes coupled with small maximum packet lengths), the use of downward multiplexing improves performance by aggregating multiple network connections into a single, higher throughput connection. Downward multiplexing at the Transport layer is therefore analogous to inverse multiplexing at the Physical layer.

Note that downward multiplexing has significance only if the underlying network is connection oriented. In a connectionless networking environment, limitations to throughput have nothing to do with constraints associated with network connections, as there are no connections to which constraints may apply. Also noteworthy is the fact that, even if sequential packet delivery is guaranteed across each individual network connection at the Network layer, the use of downward multiplexing at the Transport layer will likely introduce sequencing problems. To overcome such problems, Transport layers that perform downward multiplexing must also perform sequencing (at the transmitter) and sequence checking (at the receiver). At the receiver, a mechanism must be in place to resequence any out-of-sequence messages that arrive across the aggregated network connections.

A SPECIAL CASE: THE CONNECTIONLESS TRANSPORT LAYER

Thus far, we have concentrated on Transport layer implementations that operate in connection-oriented mode. After all, if the primary responsibility of the Transport layer is to guarantee error-free, end-to-end message delivery, a connection-oriented approach is clearly the best one to apply.

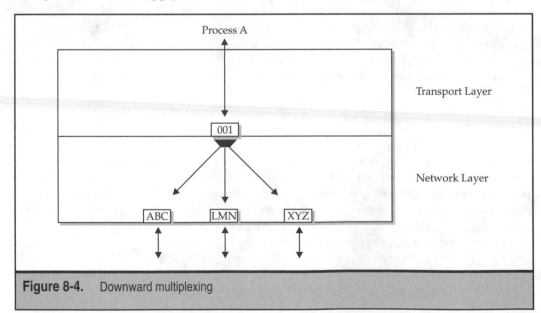

Figure 8-4.   Downward multiplexing

There are instances, however, in which guaranteed end-to-end message delivery is not required. The transport of nonmission-critical data is an example. Another example is any environment in which a human being is the ultimate requester of a given service. If a requested service is not delivered in a timeframe considered "appropriate" to the user, he or she will simply issue another request. Finally, processes that exchange real-time data (such as audio and video processes) do not have the luxury of achieving perfection through the use of message retransmission. In these cases, getting messages to their destinations without delay far outweighs the requirement that their contents be perfect.

In each of these instances, it is appropriate, even beneficial, to employ a connectionless Transport layer. The major advantage of a connectionless Transport layer over a connection-oriented one is efficiency. Because they perform far fewer functions than their connection-oriented counterparts (such as sequencing, acknowledgement, flow control, and so on), connectionless Transport layers are efficient in terms of transport overhead and throughput. This is just the ticket for noncritical and real-time message flows.

The astute reader may be asking at this point, "If error-free, end-to-end message delivery is not a requirement for these applications, why bother with a Transport layer at all?" The answer lies in the addressing function performed at the Transport layer. Even if guaranteed correct message delivery is not a requirement, the requirement to support multiple processes in a single machine is still present (see Figure 8-2). For this reason alone, we employ the services of a connectionless Transport layer.

TRANSPORT LAYER EXAMPLES

The Transport layer implementation that is arguably the most widely used on a global basis is the Transmission Control Protocol (TCP)/UDP combination from the IP suite (TCP/IP). From a theoretical standpoint, however, another Transport layer implementation, this one from the International Organization for Standardization (the ISO, at ITU-T Recommendation X.224) bears mention. To conclude this chapter, we will briefly explore each of these Transport layer implementations.

TCP and UDP

Figure 8-5 illustrates the position of both TCP and UDP within the IP suite. These Transport layer protocols were written expressly to function on top of the IP. IP supports connectionless Network layer service and, as such, is assumed to be unreliable in its delivery of packets across the network. Not surprisingly, TCP is a connection-oriented Transport layer protocol that supports error-free, end-to-end delivery of messages (called *segments* in TCP) in the face of IP's lack of reliability. TCP uses a minimum of 20 octets of overhead to deliver a segment reliably from one end user to another.

TCP addressing is accomplished by the use of well-known ports where various processes are listening for a connection at all times. For example, when a Web browser

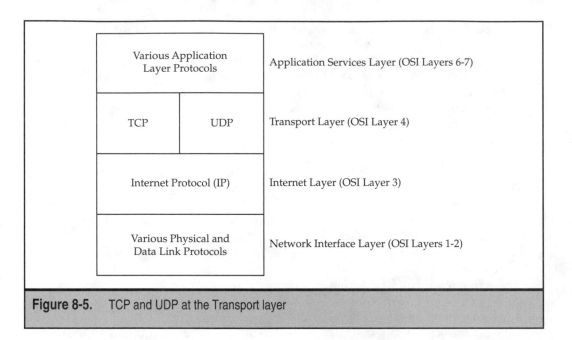

Figure 8-5. TCP and UDP at the Transport layer

wishes to obtain a Web page from a Web server, the Transport layer of the browser will use TCP port number 80 to establish the connection. For access to file-transfer capabilities, TCP ports 20 and 21 are used. Connection establishment and release are accomplished by the three-way handshake process.

TCP uses sequence numbers and acknowledgements to ensure accurate, sequential delivery of a stream of bytes from one machine to another. Flow control in TCP is explicit; each acknowledgement contains an indication of the amount of available buffer at the receiver at the time the acknowledgement was sent. TCP permits the placement of normal and expedited (called *urgent*) data in a single segment; overhead information permits the receiver to separate the two types of data upon processing a segment.

Interestingly, TCP was designed to work in environments that use the Serial Line Interface Protocol (SLIP) at the Data Link layer. SLIP does no bit-error control. As a result, TCP can perform bit error detection and correction (through retransmission) in cases for which that function is a requirement.

Also shown in Figure 8-5 is UDP, which offers a connectionless alternative to TCP for applications that do not require reliability (such as voice streams in IP telephony environments). In contrast to TCP's minimum 20 octet-per-segment overhead requirement, UDP requires only 8 octets of overhead on each segment to perform its functions. It has been speculated that, as the amount of real-time traffic on the Internet increases, UDP may someday surpass TCP as the most widely used transport protocol in the world.

The ISO Transport Layer Protocols

The ISO, in collaboration with the International Telecommunications Union (ITU), has constructed a standard for use at the Transport layer in systems that conform to the OSI Reference Model for communication. This transport standard is codified in the ITU-T Recommendation X.224.

Recommendation X.224 specifies a set of five Transport layer implementations, labeled Classes 0 through 4. The implementation chosen in any given networking environment is a function of the reliability of the underlying Network layer service. Table 8-1 shows how a particular Network layer service is mapped to one of the five Transport layer implementations. Network layer services can be classified as either A, B, or C. A type A network represents flawless Network layer services. A type B network is a connection-oriented packet network that makes errors only on a sporadic basis (and then informs users that errors have occurred). A type C network is a connectionless packet network and is assumed to be totally unreliable in its packet delivery characteristics.

As can be seen from the table, the ISO Class 0 Transport layer is designed to function with a type A network in which downward multiplexing does not occur. In a type A network where downward multiplexing is permitted, the ISO Class 2 implementation should be used, as it has the capability to perform resequencing operations when required. If the underlying network service conforms to type B, ISO Class 1 or 3 transport implementations are recommended, depending on whether or not downward multiplexing is (Class 3) or is not (Class 1) permitted. If a type C network is to be used under the Transport layer, the ISO Class 4 implementation is the protocol of choice. Like TCP, ISO TP Class 4 assumes total unreliability from its underlying network service.

The advantage of defining a number of Transport layer implementations, each suited to a different Network layer service, is efficiency, both in terms of overhead and throughput performance. As one might assume, each ISO transport protocol class is associated with more and more overhead as one moves sequentially from Class 0 to Class 4. Thus, the ISO transport protocol family permits a user to match the complexity of the transport protocol to the level of reliability expected from the Network layer service. As an aside, ISO has also defined a connectionless Transport layer protocol that is similar in orientation to the UDP protocol discussed previously.

Network Type	TP Class	Nomenclature
A	0	Simple Class
B	1	Basic Error Recovery Class
A	2	Multiplexing Class
B	3	Error Recovery and Multiplexing Class
C	4	Error Detection and Recovery Class

Table 8-1.   ISO Transport Protocol Classes

SUMMARY

The fact that the Transport layer occupies the exact center of the OSI protocol stack is somewhat misleading, because this layer represents both an end and a beginning. From a raw communications point of view, the Transport layer is an end. It is here that a consistent level of raw communications capability must be delivered unfailingly. If not for the Transport layer, upper layers could not rely on the predictably consistent performance of the communications medium. Through the use of a myriad of techniques, each discussed at some length in this chapter, the Transport layer does, in fact, offer the upper layers a reliable, consistent communications platform upon which to build.

From the point of view of the upper layers, the Transport layer is a beginning. Because the Transport layer takes care of everything communications oriented, the upper layers are free to concentrate on the provision of value-added services to their users (such as application programs). Chapter 9 will focus on such value-added enhancements to the basic networking functions that are made possible by the presence and proper functioning of the Transport layer.

CHAPTER 9

The Higher Layers

This chapter discusses the functions of the higher layers from the context of both the OSI Reference Model (the Model) and the Internet Protocol (IP) suite. The fundamental purpose of the higher layers, to enable interoperability, is discussed. OSI's three higher layers—Session, Presentation, and Application—are contrasted with TCP/IP's topmost layer, Application Services. Finally, a brief overview of example OSI and Internet applications is provided.

OSI AND TCP/IP: THREE LAYERS VS. ONE

While the responsibility of the lower layers is just to interconnect computer systems, higher layer services provide interoperability. Without the higher layers, messages can pass between application programs, but they are not necessarily understood at the receiving end. While OSI and TCP/IP both define higher layer services, each protocol stack has its own unique organization.

Figure 9-1 shows the relationship between the OSI and TCP/IP approaches. Layers 5 through 7 comprise the higher OSI services: Session, Presentation, and Application. TCP/IP, which preceded the OSI, has only one higher layer, called the Application Services layer.

Although they are covered by separate standards, some layers of the OSI Model are said to be *tightly coupled*—in other words, these layers work together very closely. The

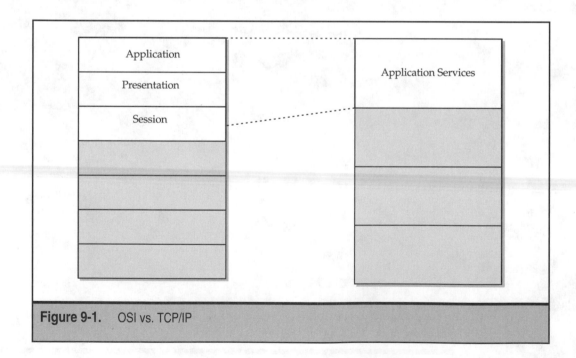

Figure 9-1.   OSI vs. TCP/IP

Presentation and Application layers are the most tightly coupled of the OSI layers and are often bundled together for implementation purposes. In fact, the functions of these two layers are so intertwined that the designers of TCP/IP (which predates OSI) simply lumped all the services they provide into the Application Services layer.

THE SESSION LAYER

The OSI Session layer provides value-added services (sessions) to the Transport layer's end-to-end message-oriented service. These services are useful in certain applications only; for many applications, the Session layer may be of limited utility. The Session layer's services are optional; if none are to be used, the entire layer can be omitted. Next is a list of several functions specified for inclusion in the Session layer, most of which are useful only in connection-oriented transport connections:

▼   Session establishment and release

■   Dialog management

■   Synchronization

■   Activity management

▲   Exception reporting

In connection-oriented services, connection establishment and release requests, as well as data transfer requests, are passed to the Transport layer. When the session is finished, the Session layer usually makes a graceful termination as opposed to an abrupt one. By using a handshake procedure, no data is lost when one side wants to end the conversation but the other does not. In a handshake, a series of signals acknowledges that the information transfer can take place between two hosts, or that the host at one end of a conversation has no more data to send and wishes to terminate the connection.

> **NOTE:**   Many of the Session layer functions are absent in connectionless networks in general, and TCP/IP in particular. This is not to say that the Session layer functions are unimportant. Indeed, connection-oriented TCP incorporates several session services, such as dialog management. Other session functions, such as synchronization, are provided by the TCP/IP applications.

An important general rule to remember about the Session layer is that without connections, no sessions are possible. However, exceptions to the rule exist, and even OSI defines a "connectionless" Session layer. The connectionless Session layer has few useful features for managing dialogs and the like. Even without sessions being mapped to connections, connectionless networks support file transfers and even some financial transactions without full OSI session support. Sometimes, connectionless sessions, and perhaps even connection-oriented sessions, add delay in the form of layer processing and complexity to an already overworked network protocol.

Sessions are quite useful whenever connections exist between clients and servers across a network. Even in the IP suite, the connection-oriented TCP protocol (running at layer 4, the Transport layer) enhances the underlying connectionless IP protocol (running at layer 3, the Network layer) with several session-like functions.

THE NEED FOR THE PRESENTATION
AND APPLICATION LAYERS

If connectivity were the only issue, any layer of protocol above the Transport layer wouldn't be necessary. The Network layer provides paths from one data terminal equipment (DTE) device to another, and the Transport layer ensures that the path is correct. Although the Session layer acts as a "window" to the communication-oriented layers, it is not required for communication.

When data arrives at the destination application, it must be interpreted correctly and processed according to your request. The correct interpretation and processing is referred to as the *interoperability* issue. To achieve interoperability, agreements about the form and meaning of the data transmitted must have been established. These agreements, or protocols, transcend the communication network—in other words, they are independent of the physical means by which data is transmitted. The agreements concerning the syntax and semantics are provided by the upper layer protocols.

Consider the following example, illustrated in Figure 9-2. At each of the two hosts, a user is running an application program. Five of the seven layers have been implemented in each host. The Session layer establishes a connection between the two application programs so they can pass messages between them, and sets the dialog rules: two-way alternate, two-way simultaneous, or one-way communications. The Session layer calls on the Transport layer to ensure that the messages arrive at the receiving host without error. The Transport layer may packetize the messages; it then invokes Network layer services to route the packets through network nodes to the remote host. The Data Link layer makes sure the frames arrive correctly at each node. Finally, the Physical layer sends the bits by representing 0s and 1s on the physical medium. The applications on the two hosts can now send messages to one another. Connectivity between two applications has been achieved.

All that this elaborate scheme assures is that the content of the message sent by application program A (such as client software) is identical to the content of the message received by application program B (such as server software). Networks do not automatically ensure that the content of the message make sense and are understood by the two systems involved. It is the role of the Presentation and Application layers to provide this meaning. Using the lower layers, connectivity, but not necessarily interoperability, has been achieved.

Imagine the following scenario in the world of human communications. An American wishes to communicate with a Spaniard via telephone, though neither person speaks the other's language. The American picks up the phone and dials the correct

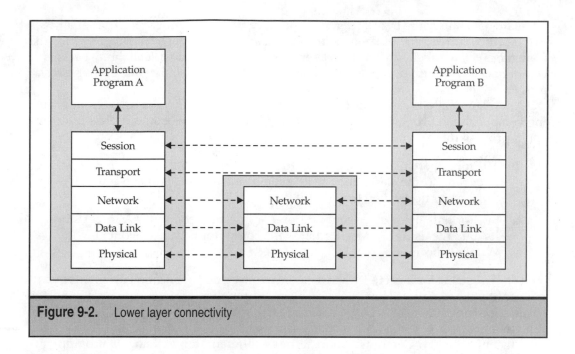

Figure 9-2.   Lower layer connectivity

number. The Spaniard picks up the phone and greets the caller, saying "Buenos dias." The American, who speaks only English, hears the Spanish greeting but does not understand what it means.

When the Spaniard picked up the phone, the connection was established between the two people. They could speak to each other because there was connectivity. However, they could not understand each other because they spoke different languages. Therefore, they did not achieve interoperability.

In our analogy, the Spaniard and the American need language translators, equivalent to the upper layer protocols, to understand one another. One option is to place a Spanish-English translator between the two people to translate in both directions. While this certainly works for this particular communication session, problems arise if either the Spaniard or the American wishes to speak to someone who does not speak Spanish or English. If they want to establish a conference call to a German speaker, they need two more translators—one for Spanish to German and another for English to German. If a fourth language entered the conversation, three more translators would be required, resulting in a total of six translators required for four people to speak and understand each other!

Suppose, however, that each speaker could talk to anyone without knowing all languages. Let each person have one translator, who translates that person's language into a standard language, such as Esperanto (the "universal" language that was invented by Russian physician Dr. Zamenhov in 1887). If Esperanto were used, the Spaniard and American could communicate using two translators: one to translate Spanish to Esperanto for the

Spaniard, and another to translate English to Esperanto for the American. The advantages to this setup soon become apparent as more languages are added. The German entering the conference call requires only one translator (German to Esperanto) for all three speakers to communicate. A fourth person adds another translator, bringing the total to four translators, and so on.

A similar approach can apply to electronic communications. Take two computers using an application program such as electronic mail to send a document. Two problems could prohibit these programs from exchanging messages:

▼   The format of the mail message header might not be the same on both systems. For example, the Destination field could be the first field on one computer, but the second field on the other.

▲   File formats themselves could differ. For instance, computer A might store its data file using the Mac format while computer B stores its data file in Windows format.

We can establish connectivity between the two systems' application programs by implementing the same set of protocols for layers 1 through 5. After this has been done, the application programs can exchange data files—computer A can send a file to computer B. But what happens when the application program receives the data file and tries to decipher the information? Computer B is expecting a different format than what computer A sent, and is thus looking for fields in the wrong order. In other words, it has received the file, but it has no idea what's in it. Even if the file could be deciphered, the format is incompatible. To accommodate this problem, some other processing must occur between the Session layer and the application program.

The Application and Presentation layers take care of this problem, and here's how it's done. First, computer A's e-mail application hands the e-mail message to the Application layer, which changes the format to some standard agreed upon by the two systems prior to communication. This format is generally known as a *canonical form*.

Since computer A is a Macintosh in our example, the data file still represents the information in Mac form. Computer B, however, expects the file in Windows form. To accommodate this, the Application layer hands the data file to the Presentation layer, which converts the Mac coding to some standard representation (like Esperanto). The Presentation layer then hands the data file, in standard format and standard code representation, to the Session layer, and the file is transferred to computer B.

The Session layer on computer B passes the standard data file to the Presentation layer, which converts the standard code to Windows form. It passes the Windows data file up to the Application layer, where the application program can pull information out into its application-specific format.

The problem has been solved. Connectivity and interoperability have been achieved by implementing the seven layers of the OSI Model in both systems. Agreement of the protocols used at each layer is required for communication to occur.

The Presentation Layer

The Presentation layer is responsible for specifying the syntax of the transmitted information. Presentation layer specifications and services are general in nature, so that many different applications can invoke a single Presentation layer service.

An excellent example of a Presentation layer service is encryption, which provides security. E-commerce customers often feel uncomfortable sending credit card numbers in plain text over the World Wide Web (WWW). Therefore, Web sites must employ encryption protocols that protect their customers' financial information. But the Web is not the only application that uses the Internet for transport. E-mail is another popular Internet application that can use encryption to send private information. The Web and e-mail are two distinct applications, but both can require the security services encryption provides.

Another example, compression, is a Presentation layer service that reduces the amount of information sent between two computers, reducing the amount of bandwidth required for transmission and/or the actual connect time between the two systems. Compression removes information from the transmission that could be computed at the receiving end. Compression techniques send only enough information for the receiver to allow it to re-create the original transmission correctly.

The more predictable the information, the better the capability for compression. If the transmission consists of a string of random information, for example, compression will be impossible. By random information, we mean that no subset of the transmission can be used to predict the values of the rest of the transmitted data.

Compression has taken on great importance as providers have attempted to squeeze more information into band-limited channels. For instance, many virtual private network (VPN) implementations use *tunneling* protocols (when a packet from one protocol is wrapped in another) to send non-IP and/or encrypted data over IP networks such as the Internet. Tunneling protocols add substantial overhead to transmissions. Most VPN hardware and software, therefore, will compress the data before transmission to improve overall efficiency of the communication.

Compression simply trades bandwidth for processing power. Because compression requires processing, as does decompression, sending compressed signals conserves bandwidth at the expense of computing resources. As the compression ratio increases (as fewer and fewer bits are sent to represent an entire transmission), the requirement for processing increases. These days, while the price/performance curve of computing equipment has made computing power readily available, this is less true for bandwidth availability.

Code conversion is somewhat more straightforward than compression and encryption. For example, if one machine uses ASCII and the other uses EBCDIC, tables can be used to convert one another's text. For example, an *A* in ASCII is represented decimally by the number *65*, while it is representd as *193* in EBCDIC. However, some codes don't translate at all, or at least not quite correctly. EBCDIC sends *DEL* (for delete) as *7*, but the closest ASCII counterpart is *BS* (for backspace), which is *8*.

Another problem can occur while sending numbers. To send numbers that are more than 1 byte long, some machines transmit (and receive) the high order byte first, while others transmit the low order byte first. For example, an Intel microprocessor would send a bill for $1 as 00000001. It would transmit the lowest byte, *01*, first, followed by three *00* bytes. If a Motorola microprocessor received the data, it would think that the first byte was the highest. The result would be a bill for $1,000,000!

NOTE: These two methods of ordering bytes have come to be called *big endian* and *little endian*. A computer scientist compared the problem to one of the voyages in Jonathan Swift's *Gulliver's Travels*, in which Big-Endians fought a war over which end of a soft boiled egg should be broken open.

Although Intel, Motorola, and other chip makers have now built the necessary conversions into their hardware, other problems can arise. The International Organization for Standardization (ISO) created a standard notation for data so that programs on different machines, written in different languages, can communicate. It is called the Abstract Syntax Notation 1 (ASN.1).

The Application Layer

The Application layer is the seventh (and top) layer of the OSI Model. This layer provides services directly to user application programs. To avoid incompatibilities between programs and services, the Application layer defines standard ways in which these services are provided. This standardization also relieves programmers from having to re-create the same functions in each and every network application program they write.

The result is interoperability of services between programs. For instance, many applications require file transfer services to exchange files between their computers. If both applications agree to use the FTAM (File Transfer, Access and Management) Application layer service, for example, they can send and receive files via a commonly defined interface. The Application layer defines this common interface to various application services.

Many different types of Application Layer services exist, and each service may invoke other Application layer services. This is a unique feature of the Application layer. For instance, before a file can be sent, an association between the sender's and receiver's Application layers must exist. Both file transfer and association establishment are Application layer services, and both services are required for successful transfer of information.

These Application layer services come in two forms, called *common* and *specific*. Common services are useful to many different types of applications, such as association establishment and termination. Specific services are typically associated with a single or small group of applications, such as a message handling service.

Given this architecture, different application programs can call these services via a published interface, which obviates the need for a programmer to write a message handling system, for example. It also means that these services can interoperate between different applications.

The Application layer provides services to user programs running in a computer. These services are not the applications themselves. The Application layer provides a set of open and standard application program interfaces (APIs) for programmers to use to accomplish common network application tasks such as file transfer, remote login, and the like. Obviously, not many programs do all of these things, so each API set is given its own Application layer "compartment" to run in. That way, programmers can use only the API set they need (such as the file transfer set) and keep the code modules smaller, requiring less memory, as a result.

The implementation of the Application layer can be quite complex because it provides services to a wide range of applications. Most other protocol stacks take a more relaxed approach and just "bundle" all of the higher layers (Session, Presentation, and Application) into a loose collection of code modules (such as a dynamic-link library in Windows or a *lib* module in UNIX). Programmers link to whichever set of APIs the application needs. This "application services" approach is used in the IP suite (TCP/IP) and results in an emphasis on "vertical" compartments rather than "horizontal" layers. This Application layer compartment structure is shown in Figure 9-3.

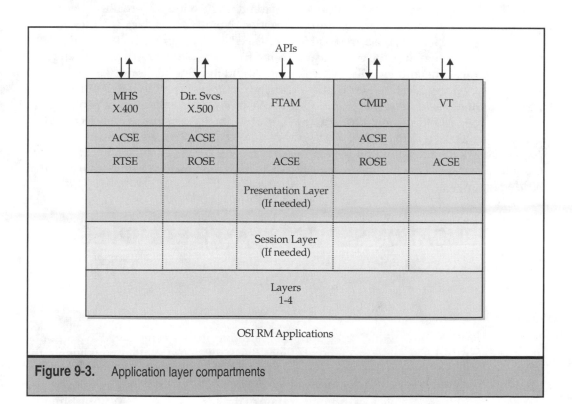

Figure 9-3.   Application layer compartments

Each basic API set can share an underlying Presentation layer and Session layer. Since a wide variety of needs exist for the services provided at layers 6 and 5 (such as encryption and session logs), it is common to bundle these implementations as well; however, the compartments are looser at these layers than in TCP/IP implementations. Some applications forego true Presentation and Session layer services and rely on the Association Control Service Elements (ACSEs), Remote Transfer Service Elements (RTSEs), and Remote Operation Service Elements (ROSEs) to do the job. Figure 9-3 shows dotted lines for bundling at layers 6 and 5, which indicates their use is on an as-needed basis.

Application layer implementations are the most visible part of the OSI for most users. Network technical support specialists may be familiar with Ethernet frames (layer 2) or routing protocols (layer 3), but users tend to focus on "What can the network do for me?" (layers 5, 6, and especially 7). So the way in which the Application layer services are packaged for use is important in understanding how users perceive Application layer services.

THE APPLICATION SERVICES LAYER

TCP/IP's Application Services layer provides the end user a window into the network and offers many useful functions. While the organization of the TCP/IP stack is quite different from that of the OSI stack, the same basic jobs still need to be accomplished. In TCP/IP, the uppermost layer combines the previously covered functions of OSI's Application and Presentation layers, and also includes some features found in OSI's Session layer.

The TCP/IP Application Services layer, as the name suggests, contains the standard protocols that provide services to end-user applications. This collection of protocols includes Simple Mail Transfer Protocol (SMTP) used for Internet e-mail and Hypertext Transfer Protocol (HTTP) used by the World Wide Web. Now that we have moved up the protocol stack to understand the services each OSI layer provides, let's move back down the stack from applications that are well known to most readers to examples that are often hidden from view.

TCP/IP APPLICATION SERVICES LAYER EXAMPLES

Following are some examples of TCP/IP Application Services layer applications. In this section, we'll take a look at e-mail and the Web.

E-mail

E-mail is one of the most widely used applications today; it has been viewed by corporate America as an aid to worker productivity since the early 1980s. While many lament the passing of the handwritten note, e-mail is the communication king of the 1990s. E-mail requires central network administration for the local server (post office and mailboxes) and gateway, as well as possible local administration of the client software. Administration is an essential service for the end user and thus an important issue to deal with.

The X.400 protocol is a standard that many anticipated would become a dominant player in the e-mail arena. The original version contains basic functions; an update includes secure messaging and interaction with the X.500 Directory Services. While it does contain more features than the Internet's Simple Mail Transfer Protocol (SMTP) standard, X.400 is scarcely used today because SMTP is simpler to use, is in place, and has worked well for a long time—plus users are familiar with it. TCP/IP does not support X.400 because TCP/IP is not an OSI protocol.

The SMTP is a popular messaging program that is widely used on all Internet-attached systems for passing messages over the Internet. The Department of Defense (DOD) developed SMTP as part of the ARPANet research project, and this popularized e-mail. SMTP relies on TCP/IP for transporting the messages.

Beyond simply handling SMTP text messages, other functions and protocols have been defined for Internet e-mail. Mapping between the ITU-T X.400 Message Handling System (MHS) and SMTP has been described. Several extensions to SMTP have also been defined, including the Multipurpose Internet Mail Extensions (MIME).

SMTP defines a protocol for the exchange of mail messages between e-mail servers. It does not, however, provide much in the way of a user interface—it does not provide a clean mechanism for users accessing and managing their e-mail. For example, SMTP provides no commands so that a user can reply to a message, forward a message to another user, send carbon copies (CC) of messages, or perform other basic e-mail functions.

To fill this gap, the Post Office Protocol (POP) was developed. It provides a simple user interface to SMTP. POP transfers e-mail from an e-mail server to the user's computer system. POP also supports enhanced capabilities such as filters for message size or to screen e-mail by sender. POP is intended to be used across a LAN. The POP server maintains users' mailboxes; each user's system has POP client software, which provides a rudimentary interface so that users can access and manage their mail.

The World Wide Web

For many people, the Web *is* the Internet. In fact, it is a collection of servers that use the Internet as a transport facility. These servers provide access to documents formatted in HTML (Hypertext Markup Language), as well as other forms of information such as image and sound files. Although the Web has been around for only a few years, relatively speaking, it has become a mainstream communication medium faster than did the telephone, radio, or television industries.

Like all Internet applications, the Web relies on an Application Services protocol. HTTP (Hypertext Transfer Protocol) is the native WWW protocol that transports all the HTML-formatted pages and other Web content from a Web server to the user's browser. The protocol itself is fairly simple, describing a few basic commands and some header information that indicates things such as where a visitor may have been referred from (to track advertising effectiveness), how old a Web page is (to determine whether a cached, or locally stored, page can be reused), and which browser program is being used (to possibly display content optimized for a particular version).

Web interactions are *stateless* because HTTP provides no mechanism for keeping track of sessions between a browser and a server. With the introduction of *cookies*, returning users can now be recognized by sending a stored identification number back to a server, which can look up customer profile information in a database. This is how e-commerce sites can not only greet you by name, but also provide you with product recommendations that will presumably encourage you to part with more of your money!

HTTP has no inherent security mechanisms, with all data sent in the clear at the Application Services layer. The Secure Sockets Layer (SSL, described in the "Encryption" section later in the chapter) provides HTTP with data privacy by intercepting and encrypting the data before it is handed to the Transport layer.

OSI APPLICATION LAYER EXAMPLES

Following are some examples of OSI Application layer applications. In this section, we'll take a look at directory services and remote procedure calls.

Directory Services

ITU-T Recommendation X.500 describes an Application layer protocol that provides on-line, distributed directory service. In its most basic form, X.500 allows users on one e-mail network to look up the names and e-mail addresses of users on another, interconnected network. As such, X.500 constitutes the equivalent of an electronic "white pages" for users of X.400 services.

The X.500 Recommendation is global in scope and also incorporates database features, making it possible to make such inquiries as, "Show me the names and addresses of all manufacturers of statistical multiplexing equipment in the Federal Republic of Germany." X.500 directories have been extended to include other types of information. These extensions make X.500 suitable for almost any kind of directory, e-mail or otherwise.

A "yellow pages" function can be supported without duplicating entries by the use of alias names and pointers to locate one source of data with multiple listings. X.500 databases are often used in conjunction with AAA (authentication, authorization, and accounting) servers in VPN and dial-up remote access systems.

The development of the Lightweight Directory Access Protocol (LDAP) began in the early 1990s at the University of Michigan. Like all major universities and many organizations, the university had many huge directories with constantly changing users (the students). Directory maintenance had become a nightmare of remote logins and hand-made changes. Inconsistency was common. On the other hand, the full X.500 directory services were seen as overkill. The university wanted the convenience of one access method and maintenance procedure for directories, but not all of the vast capabilities that X.500 had come to include. So they invented a scaled-down version of X.500 called LDAP, a "lightweight" version of the full X.500 DAP.

LDAP was and is better suited for the Internet and use with TCP/IP (technically, the IP suite) than X.500. LDAP is a nice match for Web applications, because Web browsers combine e-mail, file transfer, and other capabilities all in one client, which may make it necessary for the browser to access a number of directories when accessing a server.

LDAP has attracted a lot of industry attention as directories have become more critical for networked applications. For example:

▼ IBM includes LDAP support in a variety of platforms, including OS/400 (as part of Directory Services), AIX, OS/390, NT, and Windows.

■ Netscape supports LDAP in its Web browser and Netscape Directory Server products.

■ Novell includes LDAP in its NetWare Directory Service (NDS).

▲ Microsoft supports LDAP in the Microsoft Exchange messaging product, as well as in their Active Directory service in Windows 2000.

LDAP is widely supported by all the major players in the WWW and network applications arena, because LDAP offers all of them a way to access each other's information, allowing users to employ their applications software without worrying about changing directory structures or maintenance procedures. Users are no longer confined to using application software because of a certain directory structure. In short, LDAP offers all of the advantages of any standard way of doing things.

The LDAP was developed because manual directory maintenance was becoming physically impossible in large networks. LDAP offers a universal mechanism for locating and managing users and network resources. Users are not only located for e-mail and messaging purposes, but resources such as remote printers and databases can have entries in a network directory—all need to be accessed and maintained.

LDAP works regardless of the structure of the directory in which it is used to access. A whole range and variety of otherwise proprietary network directory structures can be accessed without needing a special program and procedures. LDAP greatly simplifies the X.500 Directory Access Protocol (DAP). LDAP can support X.509 public key encryption (which came out of the X.500 specifications), making it very popular for authentication mechanisms.

Remote Procedure Call

Remote Procedure Call (RPC) was developed to create a better and more efficient way to distribute computing power between a client and server. Many servers on a network basically download executable code over a network to a client computer. The ability to execute the code depends on the client, not the server. All of the software code, the applications, and any subprograms (the *procedures* to RPC) must be in local memory.

For example, consider a database application running on a LAN. The client does not need to have the database stored on its hard drive, nor the full database application.

However, the software needed to access the database must be installed on the client. The user interacts with the database by making a request for information using the client software interface; then the data is downloaded over the LAN, placed in local memory, and displayed. If the user wishes to manipulate the data (such as to filter out or add different data elements), requests are sent to the server, which returns the new data in the requested form.

RPC is a bundling of layers 5, 6, and 7 functions. RPC allows procedures (or subprograms, subroutines, or whatever these subordinate code modules are called in a given architecture) to execute off the remote server and return the results over the network to the client issuing the RPC message (call). Naturally, this allows for applications to execute faster (less code to download), requires less resource use on the client (memory), and is more efficient for the servers and software maintenance (one procedure/ many users).

Unfortunately, many implementations of RPC technology exist, and most are proprietary (some are low-cost licenses) and vary by operating system—that is, for example, UNIX server RPCs are built and run differently than Windows RPCs. Usually the client portion of the RPC can be run on a different platform than the server portion of the RPC (such as Windows client, UNIX server). This is due to the inclusion of several Presentation layer features in RPC.

Today's object-oriented approaches perform much the same purpose as RPC. CORBA (Common Object Request Broker Architecture), developed by the Object Management Group, allows pieces of software to communicate no matter what language they were written in—very much a canonical form. Microsoft's DCOM (Distributed Component Object Model) is a competing method.

The Open Software Foundation (OSF) was a collection of UNIX software and hardware (workstation) vendors seeking to standardize all aspects of UNIX—the group merged with X/Open Company Ltd. to form the Open Group in 1996. After the use of RPC became common on UNIX platforms, the OSF sought to standardize RPC implementation to minimize differences between software vendors' RPC implementations. Its plan was called the Distributed Computing Environment (DCE), since RPCs were always seen in client/server environments and not technically restricted to one client to one server activities.

DCE includes its own directory service specification to find the remote RPC servers and the server stubs themselves on the network. DCE allows for multithreaded (that is, the client may continue processing while waiting for an RPC response) and concurrent (that is, many clients can simultaneously access a single RPC server) operation. DCE also includes security, time, and distributed file services.

OSI PRESENTATION LAYER EXAMPLES

Following are examples of OSI Presentation layer applications. In this section, we'll take a look at compression and encryption.

Compression

Compression comes in many forms, some geared toward particular types of files. For example:

▼ **JPEG (Joint Photographic Experts Group)**   Used for compressing color images. The technique is considered "lossy" because it can dramatically reduce file sizes, though at the cost of losing some detail and, hence, picture quality. JPEG is a popular format for images used in Web pages because the graphical files can be downloaded and displayed quickly. JPEG is the most common form of storing pictures taken with digital cameras.

■ **MP3**   Part of the MPEG (Moving Picture Experts Group) digital video compression standard. MP3 is one of three encoding schemes (called *layers*) that compress audio signals. By removing extraneous parts of a sound signal that the human ear can't discern, it can reduce a CD sound file by a factor of 12 with no perceptible loss in quality. The extremely small size of MP3 files has created great controversy in the recording world, as copyrighted music can be illegally distributed over the Internet with ease.

▲ **LZ78**   Named for the developers Abraham Lempel and Jakob Ziv, who developed the scheme in the late '70s, is a common compression algorithm found in many forms. The LZC variant is used in the UNIX "compress" program, LZW forms the basis of the popular GIF (Graphics Interchange Format) file format, and LZS can be found in VPN client software.

Encryption

Many examples of encryption are in use today. Some cryptographic schemes are more generally useful, while others were developed for specific environments. Example implementations include these:

▼ **PGP (Pretty Good Privacy)**   Developed by Phil Zimmerman to provide a security mechanism that everybody could use. It's popular for encrypting e-mail messages and files stored on hard drives, and is now part of an entire suite of security products offered by Network Associates. At the heart of PGP is a form of cryptography known as *public key* cryptography, which makes it possible to exchange secret cryptographic keys across an untrusted medium such as the Internet. A freeware version of the PGP application is available at **http://www.pgp.com**.

■ **S/MIME (Secure Multipurpose Internet Mail Extensions)**   Another e-mail protection scheme. Like PGP, the standard uses public key cryptography to provide e-mail security. S/MIME support is a built-in feature of many e-mail applications, such as Netscape Messenger and Microsoft Outlook. In contrast to PGP, this protocol requires something called a digital certificate that provides additional security and trust. The X.509 standard describes the format of certificates.

▲   **SSL**   The security protocol used to protect transactions across the Internet. While it can be used with other applications, SSL is primarily used in the WWW for electronic commerce. It, too, relies on public key cryptography to secure data transmissions. SSL also uses X.509 certificates so a Web browser can verify the identity of the Web server that a user is visiting.

Details on how secret key cryptography, public key cryptography, and digital certificates work are found in Chapter 23.

OSI SESSION LAYER EXAMPLES

Following is an example of an OSI Session layer application. In this section, we'll take a look at NetBIOS.

NetBIOS

The network basic input/output system (NetBIOS) was jointly developed by Microsoft and IBM in 1984 for the first IBM LAN offering, Token Ring. Many network software implementors know NetBIOS as a "Session layer for LANs." Indeed, NetBIOS was intended as a Session layer interface for network application programmers that would relieve them of much of the details of network coding, including session history logs and the like. IBM needed session logs to run SNA effectively on LANs. NetBIOS still lives and is supported in Microsoft's Windows product family.

NetBIOS can be used for both connection-oriented and connectionless services, but the power of NetBIOS is clearly in its LAN client/server session capabilities. NetBIOS is a rather complex architecture, with its own protocol for name management and other general purpose services. However, NetBIOS often runs on top of the IP suite today.

SUMMARY

The Session, Presentation, and Application layers form the higher layers of OSI. The functions of these three layers are bundled together in the TCP/IP world in the Application Services layer. The Session layer is really the last layer that is concerned with simply getting data delivered between two hosts. Above that, the upper layers are necessary because connectivity is only one piece of the communications puzzle. Without interoperability, the act of communications is literally nonsense. To provide meaning to data is the whole point of data communications, and to that end many applications have been created. While these applications are generally Internet-based, with TCP/IP having achieved dominance over ISO's model, there are vestiges of OSI in much of what we use today.

PART III

The PSTN

CHAPTER 10

Transmission

When the telephone companies were building their infrastructures during the last century, it was natural for them to divide the network into functional groupings along the lines of technology—access, switching, and transmission components. The *access* component includes the equipment and facilities that connect the subscriber to the serving wire center, which in most instances is also a switching office. *Switching* comprises the equipment responsible for connecting one subscriber to another across the network. When two subscribers are connected to the same switch, that switch can make the end-to-end connection between them. If, however, subscribers wishing to converse are connected to different switches, a path between switches must be provided to complete the call. *Transmission* technology has traditionally played the role of switch interconnection and, therefore, includes the equipment and facilities required for the support of *interoffice* connectivity.

As might be expected in a large network, the number of active connections between switches may be quite large at any one point in time. As a result, transmission technology almost always involves some sort of *multiplexing* at the Physical layer. The particular multiplexing technology used in transmission systems varies with the characteristics of the signals to be carried (such as baseband or broadband) as well as with the historical point in time when one examines the network.

We will take a chronological view of the evolution of transmission technology in this chapter. We will examine four major transmission technologies in the order listed here:

▼   Analog metallic transmission systems (such as N-Carrier systems)

■   Digital metallic transmission systems (such as T-Carrier systems)

■   Digital optical transmission systems (such as SONET)

▲   Analog optical transmission systems (such as WDM/DWDM)

The first two systems are called "metallic" because they operate over predominately copper physical facilities. The remaining systems are referred to as "optical" in reference to their use of fiber-optic physical facilities.

NOTE: This chapter covers only transmission systems characterized as *wireline* systems. Wireless transmission systems will be covered in Chapter 14.

Examining the preceding list, you will note that, although originally developed in the context of interoffice transmission systems, some of these technologies (such as T-Carrier and SONET) are commonly used in access scenarios. In fact, a general phenomenon associated with the deployment of high-bandwidth technologies in the network is that these technologies are first applied in *trunking* (transmission) scenarios and then they migrate outward as customers demand faster access services.

ANALOG METALLIC TRANSMISSION SYSTEMS

In the early telephone network, analog voice information was carried in broadband form on an end-to-end basis across all elements of the network. Access lines carried broadband electrical representations of voice signals. Switching nodes supported dedicated space division paths through the fabric, again for broadband signal transfer. Transmission systems were thus optimized for the transport of these broadband signals between telephone company central offices. To be both cost-effective and efficient, many narrowband (e.g., 4000Hz) signals, each representing a single voice conversation, were multiplexed onto the transmission system for transport between central offices. An example of such a broadband multiplexer is shown in Chapter 4, Figure 4-14, which illustrates how many narrowband channels are multiplexed into a wideband facility. Each 4000Hz channel is capable of carrying a single voice conversation from one central office to another. An elaborate multiplexing hierarchy grew out of this basic concept of frequency division multiplexing (FDM). That hierarchy served well for many years, but it was not without its drawbacks.

Perhaps the most significant disadvantage of FDM transmission technology is the problem of signal quality. As a broadband signal traverses a transmission facility, it loses energy to the medium in a process called *attenuation*. If not boosted in some way, the signal would die out completely over some distance. In the broadband world, the solution to the attenuation problem is *amplification*. Unfortunately, within a given frequency range, amplifiers are nonspecific devices—that is, they amplify not only the signal but also any noise component that may have been introduced on the facility. This effect is cumulative, so after passing through a number of amplifiers, the signal becomes so distorted as to be unintelligible.

An additional drawback associated with FDM technology is the cost involved in maintaining high signal quality. To preserve signal quality over distance, expensive and highly linear electronics are required. Such a high degree of linearity in such circuits as amplifiers is impractical due to cost, because of the sheer number of amplifiers required in a large network. Even with the most expensive amplifiers, however, noise is still introduced into the system and will accumulate on an end-to-end basis. For these reasons, the use of FDM technology in the terrestrial telephone network has been mostly abandoned in favor of digital techniques.

This is not to say that FDM does not play an important role in modern telecommunications systems. Wireless telephone and data networks depend on FDM technology, as do many cable TV (CATV) and satellite delivery systems. In fact, as we shall see later in this chapter, the most leading edge optical transmission technology, *wavelength division multiplexing* (WDM), uses FDM in the optical domain to increase the information carrying capacity of fiber-optic strands. Whereas FDM is far from a dead technology, it has given way to digital technology in most contemporary wireline telephone and data networks.

DIGITAL METALLIC TRANSMISSION SYSTEMS

In 1962, AT&T commissioned the first digital metallic transmission system between a central office in downtown Chicago and another in Skokie, Illinois. Since that time, the takeover of the network by digital technology has been swift and complete. With the exception of access facilities (local loops), today's telephone network operates entirely in the digital realm (for switching and transmission). Digital transmission systems are optimized to transport baseband signals.

Why Digital?

The move from broadband signaling to baseband signaling in the telephone network may seem odd at first glance. After all, the human voice is broadband—that is, the sound pressure waves that are emitted by the human vocal tract vary continuously between some minimum and maximum values of intensity (see Chapter 4). Why, then, would we wish to go to the trouble of converting a broadband signal to a baseband signal for purposes of transmission through a network?

The answer is quite simple. Baseband signals, like their broadband counterparts, lose energy as they traverse a physical medium. In the broadband world, this attenuation phenomenon is dealt with through the use of amplifiers. As discussed previously, amplification is problematic because of cumulative noise build-up over distance. Although baseband signals are subject to the same attenuation as broadband signals, the way that the attenuation problem is approached is quite different in the baseband realm. Rather than amplify a baseband signal when its energy drops below some predetermined level, digital transmission systems use a technique called *regeneration* to recover an attenuated signal. Through a device commonly called a *repeater* (or *regenerator*), an attenuated baseband signal is literally reconstructed from scratch at each repeater location.

Figure 10-1 shows the repeater regeneration process graphically. The input function of the repeater reads the 1s and 0s, which, although attenuated (and perhaps corrupted by noise), are still recognizable as either 1s or 0s. At its output, the repeater reconstructs

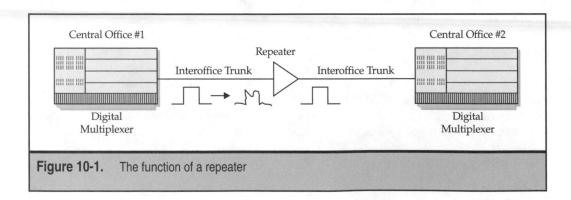

Figure 10-1. The function of a repeater

the incoming bit pattern with perfect fidelity. In the process, any noise that may have been introduced into the signal in its trip down the facility is "left behind" at the repeater. Thus, using a digital transmission system, the signal that arrives at the receiver is an exact copy of the signal sent by the transmitter, even if thousands of miles separate the two.

This advantage of digital technology alone was enough to provide impetus for the conversion of the entire terrestrial network to digital technology. As an added benefit, digital electronics tend to be inexpensive when compared with their analog counterparts. Not only could the telephone companies offer vastly superior service to their users by converting to digital systems, but they also saved money in the process.

Analog-to-Digital Conversion: The Codec

An issue that arises when contemplating the move from analog transmission to digital transmission is the fact that the human vocal tract emits broadband signals, which must be converted into baseband form so that they may be carried by a digital transmission system. This process, often called *analog-to-digital conversion*, is performed by a device called a *codec*, which is a contraction of the terms "coder" and "decoder." A codec designed to convert voice signals into baseband form does so by stepping through a sequence of four processes. Operation of the codec is illustrated in Figure 10-2 and is explained in the following paragraphs.

The operations of filtering and sampling are closely interrelated and will be described together. For a receiver to reconstruct a broadband signal from its digitized representation, the original signal must be sampled at a rate that is twice as high as the highest frequency contained in that signal. This axiom was first elucidated by Harry Nyquist at Bell Laboratories in the middle of the last century and is therefore called the "Nyquist sampling rate." Elsewhere in this book (see Chapter 1) it has been established that a frequency range between 300 and 3300Hz is critical in the transmission of voice signals. If we capture and transmit only the energy in this band, the resultant voice signal will be both intelligible (I know what you said) and recognizable (I know who you are).

Because analog metallic transmission systems used guard bands to lessen crosstalk between channels, the frequency band typically associated with a broadband voice signal is 0–4000Hz. According to Nyquist, then, to capture perfectly all of the information contained in this signal, the sampling rate must be 8000 samples per second. We must, however, ensure that no spurious high-frequency components remain in the voice signal prior to the sampling operation. The presence of such high-frequency components leads to a phenomenon called *aliasing*. A detailed discussion of aliasing is beyond the scope of this text—suffice it to say that when aliasing occurs, it becomes impossible for the receiver to reconstruct the original broadband signal accurately from the set of samples that it receives. Thus, we band-limit the broadband voice signal to 4000Hz prior to sampling. When applied to this band-limited signal, a sampling rate of 8000 samples per second will guarantee perfect fidelity in the analog-to-digital conversion process. The end result of the sampling operation is a set of samples that can be used by the receiver to reconstruct the original broadband voice signal. This process is called *pulse amplitude modulation* (PAM).

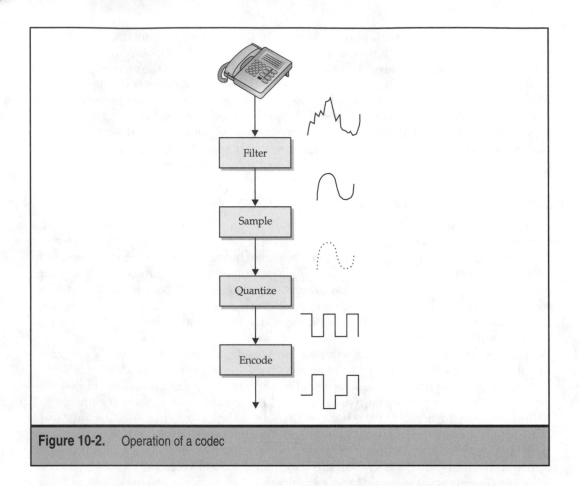

Figure 10-2.   Operation of a codec

The PAM samples must now be assigned discrete values in a process called *quantization*. Quantization results in the output of a digital representation of the original broadband signal. The astute reader will recognize that whenever a continuously varying signal is converted to a signal in which only discrete values are allowed, the introduction of noise is unavoidable. This "rounding-off error" is called *quantizing* noise. In general, the more bits used to represent each sample of the broadband signal, the less quantizing noise. Research performed 50 or so years ago shows that to provide "toll-quality" voice, each sample needs to be represented by at least an 8-bit discrete binary value. In that there is little advantage to using more than 8 bits per sample, the number 8 was settled upon as optimal for use by voice codecs.

The discrete binary representations produced as a result of the quanitzation operation are then encoded using a specific code set. For voice networks in North America, an encoding technique called "μ-255" (pronounced "mu-two fifty five") encoding is used (most of the rest of the world uses another approach called "a-law" encoding). The choice of this particular encoding technique was driven by the requirement for "ones density" in the digital transmission network. The code set avoids long strings of zeros in the data stream, because such events can cause loss of synchronization in the transmission system (such as repeaters and multiplexers).

To summarize the operation of a voice codec, the broadband voice signal is band-limited to 4000Hz and is then sampled 8000 times per second. Each sample results in an 8-bit discrete binary value that is then encoded using μ-255 (or a-law) encoding. The resultant real-time data stream from an operating voice codec is 64,000bps, or 64Kbps (8000 samples per second × 8 bits per sample). This signal rate is called "digital signal level zero," or simply DS-0. (Readers desiring a highly detailed description of the voice digitization process are directed to ITU-T Recommendation G.711.)

An interesting aside is that, as the cost of electronics has plummeted during the past 40 years, the location of the voice codec in the telephone network has shifted from the core to the edge, and finally, to the user. When first implemented, voice digitization was performed only on transmission (trunk) facilities. Switches still handled broadband signals, as did access lines. In the early 1980s, the advent of digital switching necessitated that codecs move from trunk-side locations to line-side locations. The "channel bank" was born, and every line had its own dedicated codec in the central office. In the era of Integrated Services Digital Networks (ISDN), the analog-to-digital conversion process moves out to the handset.

T-1 Multiplexing

The concept of T-1 multiplexing is simple. A T-1 multiplexer uses time-division multiplexing (for a review of TDM, see Chapter 4) to combine 24 DS-0 channels for transmission over a four-wire, full-duplex facility. In addition to the bandwidth occupied by the DS-0s (24 × 64Kbps = 1,536,000bps) is additional T-1 framing overhead (8000bps) that must be accounted for. The data rate of a T-1 facility is therefore 1,544,000bps, or 1.544Mbps.

Figure 10-3 shows the format of a T-1 frame. The frame is divided into 24 time slots, each of which is 8 bits wide—for a total of 192 bits per frame. At the beginning of each frame, however, is a single framing bit, making each frame 193 bits long. To accommodate the Nyquist sampling rate, frames must repeat at a rate of 8000 per second (i.e., 125 μsec frame duration). Once again, we see that the T-1 data rate is 1.544Mbps (193 bits per frame × 8000 frames per second = 1,544,000bps).

Looking closely at Figure 10-3, you can see that each time slot on the T-1 transmission system belongs to a separate pair of users (that is, a separate voice conversation). When a call is connected across the transmission network between two switching offices, a vacant time slot on a T-1 facility is found and the call is assigned to that time slot for its duration. A T-1 facility, therefore, represents 24 simultaneous voice conversations between two points.

Telephone Signaling in the T-1 Environment

The signaling information required to establish, manage, and release calls in the telephone network must often traverse transmission links between switches. Looking again at Figure 10-3, each of the 24 channels that make up the T-1 frame is occupied by an 8-bit

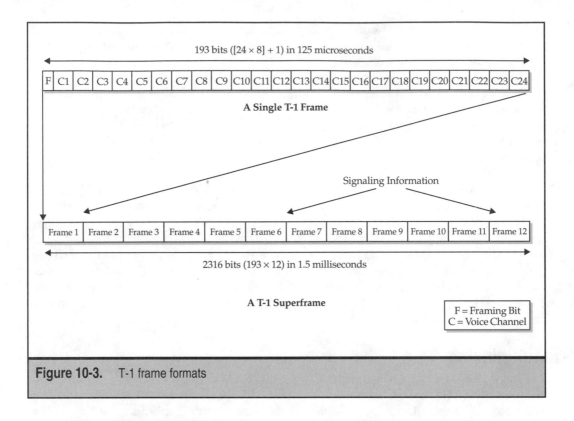

Figure 10-3.   T-1 frame formats

voice sample from a user. The only additional bandwidth available in the T-1 frame is the framing bit, which is, of course, used for framing and synchronization. Where, then, can telephone signaling information be accommodated in the T-1 frame? ×

The technique used to carry signaling information in the T-1 frame is quite elegant. It involves the grouping of T-1 frames into structures called *superframes*. As shown in Figure 10-3, a T-1 superframe is composed of 12 T-1 frames that are concatenated together. Each of the 12 frames in the superframe carries its own framing bit. Thus, a superframe contains 12 framing bits; 6 of these (those from frames 1, 3, 5, 7, 9, and 11) are used for T-1 synchronization, leaving 6 (those from frames 2, 4, 6, 8, 10, and 12) to be used for other purposes. These 6 "leftover" framing bits are used to indicate frames within the superframe that carry signaling information. Specifically, frames 6 and 12 carry signaling information. In frames 6 and 12, the least significant bit of every voice channel is "robbed" and is used for telephone signaling. Thus, in frames 6 and 12 of the superframe, the voice samples are truncated to 7 bits to allow for the transport of the signaling information. This "bit-robbed signaling" technique does not cause perceivable degradation of the voice signal, presumably because it happens so infrequently that it cannot be detected by the human ear.

To elaborate on the preceding discussion of signaling, a newer T-1 frame format, called *extended superframe* (ESF) has been defined. An ESF consists of two superframes concatenated together; therefore, an ESF contains 24 framing bits. Some of these are still used for frame synchronization and as indicators for frames that carry bit-robbed signaling (now found in frames 6, 12, 18, and 24 of the ESF). The remainder, however, now support enhanced functions such as error monitoring (through a CRC-6 calculation) and non-intrusive testing (through a 4Kbps embedded operations channel).

The Anatomy of a T-1 Span

Figure 10-4 illustrates the components and topology of a T-1 circuit. At either end of the span are T-1 terminal multiplexers. A terminal multiplexer terminates the T-1 signal and is typically capable of delivering the information contained in each T-1 channel (each DS-0) to some user device. If the user device were, for example, an analog telephone set, the terminal multiplexer would also be responsible for performing a digital-to-analog conversion through a built-in codec function. On the other hand, if the user device were some sort of data terminal (such as a router), the terminal multiplexer would be responsible for converting the T-1 signaling format (such as bipolar return to zero, or BPRZ) to a format compatible with the data terminal through a built-in channel service unit (CSU) function.

Depending on the particular application, *drop/insert multiplexers* may be deployed along the T-1 span. A drop/insert multiplexer permits the manipulation of the individual DS-0s at a given location. For example, a drop/insert multiplexer located in an office building might drop off 6 of the 24 channels in the T-1 for use at that location. Traffic from that same location might then be inserted into the now empty channels for delivery to the next multiplexer down the span. The use of drop/insert multiplexers along a T-1 span thus supports bandwidth management functions on the span.

Regardless of the presence or absence of drop/insert multiplexers on a T-1 span, regenerators are required at intervals of approximately 6000 feet. As described previously, these regenerators receive an attenuated signal, discover the 0s and 1s in that signal, and send out a perfect replica of the signal on the T-1 span. As long as inter-regenerator spacing rules are observed, no distance limitation is associated with a T-1 span.

Finally, the physical circuits that support a T-1 span consist of two twisted pairs of copper cable—one for transmission in each direction. As discussed in Chapter 4 (see Figure 4-11),

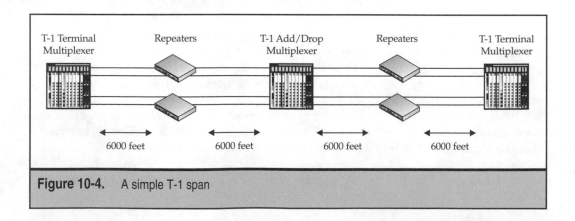

Figure 10-4. A simple T-1 span

signaling on a T-1 span is via the BPRZ technique. This signaling scheme is also often referred to as alternate mark inversion (AMI).

T-1 Technology for Data Applications

Thus far, our discussions of T-1 multiplexing technology have focused on the role of T-1 as an interoffice transmission mechanism for telephone networks. Another way of looking at T-1 technology, however, is that a T-1 circuit offers 1.536Mbps worth of bandwidth that can be useful in data communications environments. In fact, the use of T-1 in data environments is commonplace today. Having been originally designed to transport μ-255 encoded voice streams, the use of T-1 for data is associated with several problems.

As used by telephone companies for voice, T-1 is *channelized*. That is, every T-1 circuit is divided into 24 channels of 8 bits each for the transport of G.711-encoded voice. Such channelization may or may not be appropriate for data applications. Certainly, data applications have no inherent requirement to obey the 64Kbps boundaries imposed by T-1 channelization. For data applications, an unchannelized (or variably-channelized) T-1 circuit might be preferable. The good news is that many T-1 equipment vendors support unchannelized T-1 technology. They do so with devices called *intelligent multiplexers*. Using an intelligent multiplexer, a router that requires 256Kbps of throughput on a T-1 can be assigned four DS-0 channels to achieve that throughput level. Likewise, a videoconferencing system that operates at 384Kbps can be assigned six DS-0s on the T-1 for its use. Intelligent multiplexers permit the mixing and matching of different signals (such as voice, video, and data) on a single T-1 span. They also support dynamic bandwidth reconfiguration, both on-demand and on an automatic (time-of-day/day-of-week) basis. Unfortunately, the techniques used by intelligent multiplexer vendors to achieve this flexibility in the use of T-1 bandwidth are proprietary. The establishment of an intelligent T-1 network, therefore, presupposes that all multiplexers will be obtained from the same equipment vendor.

Another important consideration in the use of T-1 in data environments involves the maintenance of network synchronization. If you recall the discussion of Physical layer synchronization in Chapter 4, you'll remember that long strings of zeros in the data stream cause synchronization problems when using the BPRZ signaling scheme. In BPRZ, the periodic transmission of 1 bits is responsible for providing the line transitions critical to the maintenance of synchronization. This "ones density" requirement states that strings of no more than 15 zeros in a row are permitted to cross a T-1 span. In voice environments, this requirement poses no problem. The μ-255 voice encoding scheme guarantees that the longest string of zeros possible in digital voice scenario is 13 (the actual bit pattern being 1000000000000010). In data environments, long strings of zeros are encountered fairly frequently. How, then, can we use T-1 for data without risking loss of network synchronization?

The classical solution to this so-called ones-density problem in T-1 is an approach called *zero code suppression* (ZCS). In ZCS, the provider of the T-1 circuit simply inserts a 1 bit in the least significant bit position of every T-1 channel in the frame. Note that the insertion of this bit does not in any way interfere with the user data being transported. However, the bandwidth of each channel is reduced from 64Kbps to 56Kbps. The use of

ZCS explains why, when a data customer orders a DS-0 channel, a 56Kbps facility is delivered. Although effective, the use of ZCS is clearly wasteful of T-1 bandwidth.

A newer approach to the ones-density problem is called *binary eight zero substitution* (B8ZS). In a B8ZS environment, the transmitter examines the data stream continuously. Whenever it sees a string of eight consecutive zeros, it removes them from the data stream and substitutes a special code that can be unambiguously recognized by the receiver. Obviously, the special code must have two characteristics: it must contain 1 bits, and it must be unique. These characteristics are met by using a code that not only contains ones, but that also contains a pair of "bipolar violations" (a bipolar violation is an instance in BPRZ where two consecutive ones *do not* alternate with respect to polarity). At the receiver, when the special code is encountered, it is removed from the data stream and the eight zeros are replaced in their proper location. In this fashion, B8ZS supports "clear channel" operation of T-1 for data applications.

T-1 Applications

T-1 technology has been deployed to meet the requirements of various applications over the years. Some of these applications, like interoffice trunking and service to digital loop carrier systems, are situated within the carrier network and are not seen by end users. Others, like connections to PBX equipment and access to other networks are access applications for T-1 and, as such, are highly visible to end users.

Interoffice Trunking in Telephone Networks

T-1 technology was originally developed as a multiplexing and transmission system to interconnect telephone switches in different central offices. In that capacity, it served well for many years. With the growth of the public switched telephone network, the need for more bandwidth between central offices was greater than T-1 could provide. T-1 then became the foundation for a digital multiplexing hierarchy that supported rates greater than T-1 could provide. Today, links between telephone company central offices are most often supported using fiber-optic transmission technology at very high data rates. At the base level, however, even these high-speed optical connections can be broken down into their T-1 components.

Digital Loop Carrier Systems

T-1 technology has been used to extend the digital reach of the central office into communities served by telephone companies. The use of digital loop carrier (DLC) systems makes this possible. In a DLC environment, a terminal multiplexer called a *central office terminal* (COT) attaches to a telephone switch. One or more T-1 spans connect the COT to another terminal multiplexer called a *remote terminal* (RT). The RT is located close to a group of telephone customers, such as those in a residential neighborhood or a business park. Broadband voice signals are carried in analog format only as far as the RT. At that point, they are digitized (by codecs built into the RT) and transported back to the central office (24 at a time) over T-1 circuits.

The use of DLC technology in this fashion is sometimes called *pair gain*. Deployment of DLCs is beneficial both for the carrier and for the customer. For carriers, it reduces the number of physical facilities required to serve some number of customers. One of the more popular DLC systems is the SLC-96 (pronounced "slick ninety-six") system from AT&T (now Lucent Technologies). SLC-96 uses four T-1 circuits (plus one for backup) between the RT and the COT to deliver 96 voice channels to a remote location (such as a residential neighborhood). The savings on physical facilities is substantial, going from a requirement for 96 pairs without the SCL system to 10 pairs with. For customers, the use of DLC technology tends to improve service quality by shortening the length of access loops considerably.

Complex Bandwidth Management

The use of drop/insert multiplexers in the T-1 environment permits carriers to manipulate signals in the DS-0 domain. DS-0s can be dropped at locations that require bandwidth and can be inserted into unused slots in the T-1 span to make most efficient use of the capacity available between locations. DS-0s can be extracted from one T-1 span and placed onto another span in a process called *cross-connection*. The manipulation of channels (drop/insert and cross-connect) in a T-1 stream to support the most efficient use of available bandwidth is sometimes referred to as *circuit grooming*.

T-1 for Private Line Connectivity

During the 1970s and 1980s, many business customers constructed their own private networks for voice transport. At the heart of such networks is a device called a private branch exchange (PBX)—essentially a small telephone switch that resides on the customer premises. To interconnect PBXs in different locations, point-to-point private T-1 circuits proved to be economical when compared to analog "tie-line" facilities. Today, PBXs are commonly equipped with T-1 interfaces that support not only connections between PBXs, but also connection of the PBX to the public switched telephone network.

In a similar fashion, point-to-point private T-1 circuits are used to interconnect nodes (such as routers) in private data networks. Once again, the major impetus for the use of T-1 in this environment is cost savings. Compared to an equivalent number of 56Kbps data circuits, a T-1 circuit is a considerably less expensive alternative.

T-1 for Voice Access

As mentioned, it is common practice for telephone companies to support T-1 connections to their networks for PBX customers. Doing so is beneficial to the carrier in that it saves physical facilities and renders the PBX customer easier to manage than if similar connectivity were to be provided using analog tie-lines. For the PBX owner, T-1 reduces the number of facilities that must be purchased and maintained. Moreover, the numbers of line cards and their associated expenses (power, floor space, etc.) decrease when T-1 is substituted for tie-lines in these settings.

T-1 for Data Access

Many of today's high-speed data services such as frame relay and Asynchronous Transfer Mode (ATM) are accessed using T-1 circuits. In these cases, the T-1 span originates at the serving node in the central office (a frame relay or ATM switch) and terminates at the customer's premises, typically on a frame relay access device (FRAD) or a router. Once again, T-1 offers an economical alternative for access to these statistically multiplexed, efficient backbone networks.

T-1 has also become popular for access to the Internet (perhaps through a frame relay or ATM network). For organizations that can justify the expense of a T-1 connection, its always-on, guaranteed bandwidth characteristics make it ideally suited to the Internet access environment.

ISDN Primary Rate Interface (PRI)

T-1 circuits are used in the provisioning of ISDN PRI service. The ISDN PRI is a physical T-1 whose channels are used somewhat differently from those of a traditional T-1. Instead of allowing all 24 channels to carry user information (with bit robbing for signaling), a PRI uses only 23 of its 24 DS-0s for such purposes. Instead of employing bit-robbed signaling in each channel, the PRI reserves the twenty-fourth DS-0 channel for the transport of packetized signaling information—thus the terminology "23B plus D." Each of the B channels is a "clear channel" DS-0 that is available to the user. The D channel is the signaling channel. (For additional information on the PRI and on the signaling protocols used over the D channel, see Chapter 13.)

M1/3 Multiplexing

As the requirement for additional bandwidth between telephone company central offices increased, a multiplexing hierarchy was developed that extended beyond the T-1 rate. Table 10-1 shows the North American Digital Signal Hierarchy (the table also includes digital signal hierarchies used in Japan and in most of the world outside of North America; we will refer to these aspects of the table later in the section "Digital Optical Transmission Systems"). The hierarchy begins, of course, at the DS-0 level (64Kbps). The next level, DS-1, is achieved by multiplexing 24 DS-0 signals for a data rate (with framing) of 1.544Mbps. This is also the data rate of a T-1 circuit.

T-1 and DS-1

At this point, it might be instructive to differentiate between the terms *T-1* and *DS-1*. Although many people use the terms interchangeably, there is a subtle difference in their meaning. DS-1 refers to a service type; T-1 refers to the technology used to deliver that service. DS-1 service can be delivered using technology other than T-1. For example, it is common practice today to deliver DS-1 service using a technology called *high-speed digital subscriber line* (HDSL). The service provided by HDSL is DS-1 service, and to the user, it is indistinguishable from DS-1 service provided over a T-1 facility. In other ways, however (signaling, frame format, repeater spacing), HDSL is quite different from T-1.

Carrier Level	North America	Europe	Japan
DS-0/E-0/J-0	0.064Mbps	0.064Mbps	0.064Mbps
DS-1/J-1	1.544Mbps		1.544Mbps
E-1		2.048Mbps	
DS-1c/J-1c	3.152Mbps		3.152Mbps
DS-2/J-2	6.312Mbps		6.312Mbps
E-2		8.448Mbps	
E-3/J-3		34.386Mbps	32.064Mbps
DS-3	44.736Mbps		
DS-3c	91.053Mbps		
J-3c			97.728Mbps
E-4		139.264Mpbs	
DS-4	274.176Mbps		
J-4			397.2Mbps
E-5		565.148Mbps	

Table 10-1.   Global Digital Signal Hierarchies

In Table 10-1, you can see that two DS-1s can be multiplexed into a DS-1c at a rate of 3.152Mbps. Likewise, 2 DS-1cs (or 4 DS-1s) can be multiplexed into a DS-2 running at 6.312Mbps. Finally, 7 DS-2s (or 28 DS-1s) can be multiplexed into a single DS-3 running at 44.736Mbps. DS-1c and DS-2 are sometimes encountered within a telephone company

DS-3c and DS-4

DS-3c and DS-4 were never deployed in any real quantity due to the vagaries of the physical medium required to support them. These rates mandate the use of wave-guide media for transport. Although there was much interest in waveguide technology in the 1970s and 1980s, fiber optics proved to be a better solution for the support of very high data rates such as DS-4 at 274.176Mbps. The only widespread use of DS-4 technology was in the connection of microwave modems to microwave antennas located in the same building. The modem, for example, might be on a lower floor of the building, whereas the antenna was on the roof. The deployment of waveguide in the building's elevator shaft offered a convenient solution to the problem of connectivity in such settings.

central office, but they are rarely seen outside of that environment. DS-3, however, is used both as a common interoffice transmission mechanism as well as an access mechanism for customers requiring bandwidth in the 45Mbps range.

Figure 10-5 illustrates a technique called M1/3 (pronounced "M-one-three") multiplexing. As its name implies, an M1/3 multiplexer combines 28 T-1s into a single T-3 (the technology used to transport DS-3 service). M1/3 multiplexing is a two-stage process that involves an intermediate T-2 (DS-2) step. We will step through the M1/3 multiplexing process.

To begin, 24 DS-0s are multiplexed (in the fashion previously described) into a T-1 by use of a T-1 multiplexer. From a timing standpoint, this multiplexing operation is *isochronous*. That is, each of the DS-0s that make up the T-1 is driven from a single clock source associated with the T-1 multiplexer.

The second stage of the M1/3 process involves combining four T-1s into a T-2 by use of a T-2 multiplexer. If you multiply 4 times 1.544Mbps, you do not arrive at the T-2 rate of 6.312Mbps. The astute reader will recognize that, if there is framing overhead associated with the T-1 multiplexing operation, some additional overhead must be incurred in going from T-1 to T-2. Although that is the case, the T-2 framing overhead does not totally account for the difference between 6.176Mbps (i.e., 4 × 1.544Mbps) and the T-2 rate of 6.312Mbps. The remainder of the unaccounted-for bits are there for the purpose of "stuffing" the signal to account for timing differences between the subrate T-1s. Unlike the T-1 multiplexing stage, you cannot be certain that each of the T-1s that make up the T-2 is driven from the same clock source. Two clocks can meet a particular specification, but they may not be running at exactly the same rate. Such clocks are said to be *plesiochronous* with regard to one another. To multiplex plesiochronous data streams, bits must often be inserted (or "stuffed") into one data stream to accommodate these timing differences. As you will see later in this section, this stuffing operation becomes problematic when it comes to complex bandwidth management functions in metallic transmission networks.

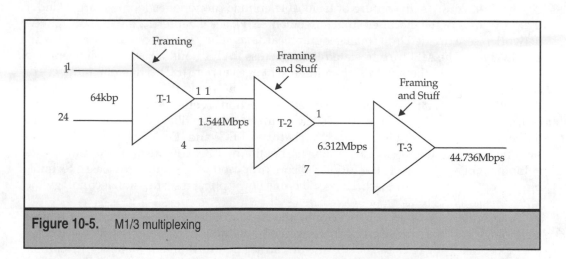

Figure 10-5.   M1/3 multiplexing

The third stage of the M1/3 process involves combining seven T-2s into a single T-3 by use of a T-3 multiplexer. Again, if you multiply 7 times 6.312Mbps, you do not arrive at the T-3 rate of 44.736Mbps. Once again, the unaccounted-for bits represent T-3 framing overhead and stuff bits.

The result of the stuffing operations highlighted in Figure 10-5 is a lack of visibility of T-1s within the T-3. To recover any individual T-1 from the T-3 data stream, the entire T-3 must be *demultiplexed* by a reversal of the process shown in the figure. In that the provision of complex bandwidth management functions at the 1/3 level requires that T-1s be visible for manipulation (such as drop/insert and cross-connection), the requirement to completely break the T-3 into its constituent T-2s and then further to the T-1 level stands as an obstacle to efficient, cost-effective enhanced T-1 services. As we will discuss in the section "SONET and Simplified Bandwidth Management," one of the goals in the development of standards-based optical TDM technology was to enhance the visibility of subrate signals within a multiplexed aggregate signal.

DIGITAL OPTICAL TRANSMISSION SYSTEMS

Although in the 1970s T-3 circuits were sometimes delivered using coaxial media, these proved problematic in terms of installation and maintenance. By the early 1980s, fiber-optic technology was being used almost exclusively to deliver DS-3 service. The fiber systems employed for this purpose were asynchronous in nature. They used TDM over single-mode fiber cable to transport multiple DS-3s, typically in interoffice settings. The data rates of these systems varied, but 565Mbps was fairly common. The biggest problem with these fiber-based transmission systems was that they were proprietary. The equipment on each end of a connection had to come from the same vendor. Furthermore, if end-to-end management of a large network was desired, all equipment in that network needed to come from a single vendor.

By 1984, AT&T's divestiture of its local telephone monopolies loomed large. Under the provisions of the modified final judgement (MFJ; see Chapter 1), the local telephone networks were divided into small geographical areas called local access and transport areas (LATAs). The MFJ further stated that, to cross LATA boundaries, a local telephone company required the services of an interexchange carrier (IEC). Thus, the numbers of connection points between local exchange carriers (LECs) and IECs was about to increase dramatically. In that these "mid-span meet" connections required a high-speed transmission system to be effective, the proprietary nature of existing fiber systems was a substantial roadblock to efficient interconnection between LECs and IECs. The involved carriers saw the trouble brewing on the horizon and petitioned Bellcore (now Telcordia) for development of a solution. The solution Bellcore proposed was a standards-based technology for optical mid-span meet. They called this new optical transmission technology the *synchronous optical network*, or *SONET*.

A secondary problem to be solved by the standardization of optical mid-span meet technology was the lack of compatibility among metallic transmission systems around the globe. Referring back to Table 10-1, you can see that three major metallic carrier hierarchies are in use in different parts of the world. North America uses the T-Carrier system. Europe and much of the rest of the world uses a system referred to as E-Carrier.

Japan, due predominately to post-war reconstruction by the United States, uses a hybrid carrier system that combines elements of T-Carrier with unique elements of its own. That system is commonly called J-Carrier. The designers of the SONET standards crafted those standards in such a way as to encompass the transmission requirements of the entire world into a single system. Thus, the global counterpart to the SONET standards in North America came to be known as the *synchronous digital hierarchy*, or SDH.

SONET/SDH Signal Hierarchy

Table 10-2 shows the SONET/SDH signal hierarchy. In the SONET realm, the signals are referred to as Synchronous Transport Signal (STS), whereas in the SDH world, they are called Synchronous Transport Module (STM). As seen in Table 10-2, the base rate for the SONET hierarchy is STS-1 at 51.84Mbps. Note that this rate coincides with the practical upper end of the North American digital signal hierarchy, which is DS-3 at 44.736Mbps (see Table 10-1). In fact, a single DS-3 service can be mapped to the STS-1 SONET signal with some room left over for network and user overhead.

The next step in the SONET hierarchy is STS-3. Note that unlike in the metallic transmission hierarchy, high-rate SONET signals are exact multiples of lower-rate signals. Thus, STS-3 at 155.52Mbps is exactly 3 times the STS-1 rate of 51.84Mbps. Likewise, the 622.08Mbps STS-12 signal is exactly 4 times the 155.52Mbps signal at STS-3 (or 12 times the STS-1 rate of 51.84Mbps). The STS-3 data rate in SONET is also equivalent to the base rate of the SDH hierarchy (i.e., STM-1 at 155.52Mbps). Because the E-Carrier metallic hierarchy has no rate that is close to DS-3 (E-3 is substantially slower at 34.386Mbps), STM-1 was chosen as the SDH base rate by virtue of its ability to accommodate the E-4 metallic signal at 139.264Mbps.

Table 10-2 demonstrates clearly that, unlike the potpourri of incompatible data rates that characterize metallic transmission networks around the globe, SONET and SDH were designed from the ground up to support interoperability among global optical transmission networks.

SONET Level	Data Rate	SDH Level
STS-1	51.84Mbps	
STS-3	155.52Mbps	STM-1
STS-12	622.08Mbps	STM-4
STS-48	2.48832Gbps	STM-16
STS-192	9.95328Gbps	STM-64
STS-768	39.81312Gbps	STM-256

Table 10-2. The SONET/SDH Signal Hierarchies

The Anatomy of a SONET Span

Figure 10-6 illustrates the components of a simple linear SONET span. The span originates and terminates on end equipment that uses the transmission capability of SONET to move information across the span (such as digital circuit switches with integrated SONET interfaces). Such end devices typically attach to a SONET multiplexer, which may integrate traffic from other pieces of end equipment. One or more drop/insert multiplexers might be present along the span, as shown in the figure. As with any digital transmission scheme, attenuated signals are, at intervals, restored by the action of repeaters (or regenerators) as shown. Unlike metallic transmission systems, whose repeaters are spaced about a mile apart, SONET repeater spacing is more on the order of 24 to 72 miles. SONET is designed for use with single-mode optical fiber. Any one strand of fiber in a SONET environment operates in a simplex fashion. Duplex communications are thus supported over two separate strands, one for transmission in each direction.

Figure 10-6 also illustrates some terminology unique to the SONET environment. At the lowest level of abstraction, the fiber connection that attaches any one piece of SONET gear to any other is called a *section*. That is, the point-to-point connections between repeaters, between a repeater and a multiplexer, and between a multiplexer and an end user are all SONET sections. The spans between SONET multiplexers (which may include any number of repeaters) are called *lines*. Thus, the spans between the drop/insert multiplexer and the other two multiplexers in the figure are SONET lines. Finally, the span between the devices that use SONET for transport (the end devices, from a SONET perspective) is called a *path*. A SONET path is shown between the two circuit switches in Figure 10-6.

These designations are important because, in the tradition of a layered architecture, SONET protocol functions are segregated according to those that occur at the section layer, those that occur at the line layer, and those that occur at the path layer (see the following section).

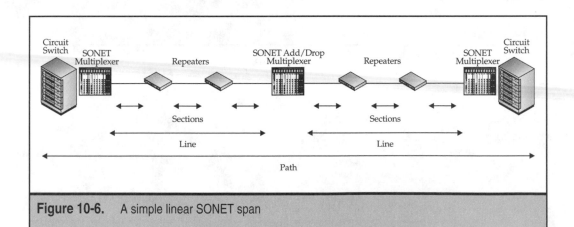

Figure 10-6.   A simple linear SONET span

SONET as a Layered Protocol Architecture

Figure 10-7 shows the four layers of the SONET architecture. (Although the nomenclature is slightly different, these layers are the same as those defined for the SDH.) Note that the four layers of the SONET protocol stack correspond to the Physical layer of the OSI Reference Model. (You can read more about how information moves along these layers later in the chapter, in the section "The SONET Layers Revisited.")

At the top is the path layer, whose main function is to map user services into a structure that is compatible with the SONET standards. In this context, the term *user service* refers to circuit-based (metallic) transmission streams (such as T-1, T-3, and E-4) as well as commonly-encountered packet-based schemes (such as ATM cells and IP packets). The path layer uses a designated area of the SONET frame, called a *synchronous payload envelope* (SPE), to structure the bits from the user into a format that can be carried by SONET equipment.

The next layer down is the SONET line layer, which is responsible for multiplexing functions in SONET. Whenever low-bit-rate SONET signals require multiplexing into a higher bit-rate aggregate, the line layer equipment performs its functions. The line layer can multiplex three STS-1s into an STS-3. Likewise, the line layer can multiplex four STS-12s into an STS-48.

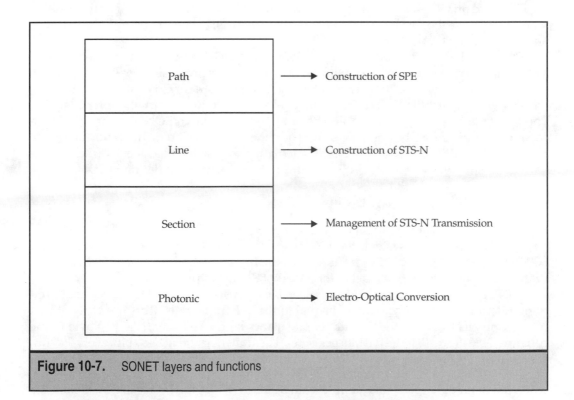

Figure 10-7.   SONET layers and functions

The next layer down, the section layer, is responsible for managing the transmission of the STS-N signal across the SONET fiber facility. Section layer equipment is concerned, therefore, with SONET simple framing operations that support the maintenance of basic line synchronization, among other things.

Finally, the photonic layer (not recognized by many SONET texts) is responsible for electro-optical and opto-electrical conversion. Later, we will establish that SONET/SDH is really not an optical networking technology, but rather an electrical networking technology that uses fiber optics for transport. Whenever a SONET signal must be manipulated (regenerated, dropped or inserted, or terminated) in a SONET network, it must be converted from an optical signal to an electrical signal. After manipulation, it must again be converted from the electrical domain to the optical domain for further transport over optical fiber. This process, termed *O-E-O* (for *optical-to-electrical-to-optical*) is the task of the photonic layer in the SONET/SDH architecture. By default, O-E-O must occur in every node in a SONET/SDH network.

STS vs. OC: What's the Difference?

As mentioned previously, STS is an acronym for "Synchronous Transport Signal." Another commonly used term in SONET is *optical carrier* (OC). Thus, we can refer to an STS-1 or an OC-1. From a data rate point of view, the two designations are equivalent—that is, an STS-1 and an OC-1 are both associated with a data rate of 51.84Mbps. For this reason, many people use the two terms interchangeably. There is, however, a difference between an STS-N and an OC-N.

STS should be used when referring either to a SONET frame format or when SONET signals are carried electrically (sometimes over short distances using coaxial cable). OC, on the other hand, is used only when referring to the SONET signal on a fiber-optic facility. If you compare the STS (electrical) signal with its equivalent OC (optical) signal, you'll note differences. For example, the ones density issue in SONET is handled by a scrambling function at the photonic layer. Thus, the bits delivered to the section layer (the STS signal) are different from the bits that actually traverse the fiber facility (the OC signal). Interestingly, there is no corollary to the STS vs. OC differentiation in SDH.

The SONET STS-1 Frame Format

Figure 10-8 illustrates the format of a SONET STS-1 frame. Because SONET is a byte-interleaved multiplexing system, the lowest level of granularity used when discussing SONET is almost always the byte. As seen in the figure, an STS-1 frame is 90 bytes (columns) wide by 9 bytes (rows) deep, for a total of 810 bytes per frame.

SONET frames are transmitted at a rate of 8000 per second (equal to the Nyquist sampling rate for voice digitization, and equal to the T-1 frame rate). This frame repetition rate permits us to derive the STS-1 data rate using the frame format. An 810-byte frame contains 6480 bits (810 bytes x 8 bits per byte). The transmission of 6480 bits every 125 microseconds (i.e., 8000 times per second) yields the STS-1 data rate of 51,840,000bps (or 51.84Mbps). In that SONET frames are always transmitted at a rate of 8000 per second, when the data rate triples, for example from STS-1 to STS-3, the frame size must increase

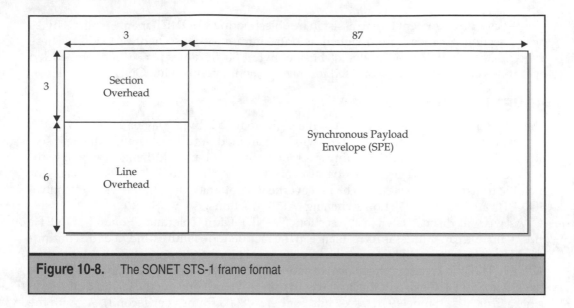

Figure 10-8.   The SONET STS-1 frame format

accordingly. It comes as no surprise, then, that an STS-3 frame is three times as large as an STS-1 frame (270 bytes wide by 9 bytes deep for a total of 2430 bytes).

SONET frames are sent in serial fashion down the fiber facility. The use of the rectangular documentation format shown in Figure 10-8 is purely for simplicity's sake. When a SONET frame is transmitted, the sender begins in the upper left corner of the frame and reads across the entire first row of bytes. When it reaches the end of the first row, it returns to the left side of the frame and begins sending the second row of bytes. In this way (much like how you read a book in English), the sender continues to send until it finishes with the last byte in the lower right corner of the frame.

Examine Figure 10-8 again, and you can see that the STS-1 frame is subdivided into various fields. The SONET transport overhead comprises two fields: the section overhead and the line overhead. Section overhead accounts for 9 bytes in the upper left corner of the frame (3 columns by 3 rows). Line overhead accounts for 18 bytes in the lower left corner of the frame (3 columns by 6 rows). Although a byte-by-byte description of each byte of the SONET transport overhead is beyond the scope of this text, we shall examine each overhead component and its function briefly in the following sections.

SONET Section Overhead

The first 2 bytes of the section overhead are used for frame synchronization. The framing pattern X'F628 is protected in the scrambling operation, so it never appears anywhere but in the first two section overhead byte positions. One byte of the section overhead is used for error detection at that layer. The mechanism employed for error checking is a bit-interleaved parity (BIP) check. The value of the BIP-8 count in any given frame is derived from the *previous* frame. Thus, no attempt is made to correct BIP-8 errors; they should be viewed as a management tool to assist in troubleshooting a SONET span. Finally, 3 bytes of the section overhead are assigned as datacommunications channels (DCC). These bytes make up a 192Kbps (3 bytes x 8 bits per byte x 8000 frames per second) embedded operations channel (EOC) to be used for management purposes in SONET networks.

SONET Line Overhead

The first 3 bytes of the SONET line overhead make up a 2-byte "payload pointer" and a 1-byte pointer justification field. These will be examined in further detail in the next section, as they underlie SONET's ability to support complex bandwidth management without the need for full demultiplexing at every node in the network (as is the case for metallic transmission systems). The line overhead contains a byte that is used to control SONET's automatic protection switching (APS) functions.

As we will discuss later in the section "SONET Fault Tolerance," one of SONET's most valuable capabilities is its resilience in the face of node failures and fiber discontinuities. The APS byte in the line overhead plays a role in the support of fault tolerance in SONET. The line layer overhead also contains a BIP-8 byte for error checking at the line layer. Another DCC exists at the line layer. It uses 9 bytes of line overhead and thus represents a 576Kbps channel (9 bytes x 8 bits per byte x 8000 frames per second). As with the DCC bytes at the section layer, the line layer DCC bytes are designated as an EOC to be used for management purposes in SONET networks.

The SONET STS-1 Synchronous Payload Envelope (SPE)

Examining Figure 10-8 one last time, you can see that the part of the STS-1 frame that is not accounted for by transport overhead (section and line overhead) is called the *synchronous payload envelope* (SPE). You can think of the SPE as that portion of the SONET frame that is available for the mapping of user services (such as T-1s and ATM cells). The STS-1 SPE is 87 bytes (columns) long by 9 bytes (rows) deep for a total of 783 bytes. The capacity of the SPE is therefore 50.112Mbps.

Figure 10-9 shows the SONET STS-1 SPE in detail. Note that the SPE contains a single column of overhead, called *path overhead*. This overhead controls important functions at the path layer, but it also reduces the SPE service-carrying capacity to 49.536Mbps (which turns out to be just fine for a single DS-3).

SONET Path Overhead

SONET path overhead is used by the path layer to provide various control functions. One byte continuously verifies signal continuity at that layer; another byte indicates the internal composition of the payload contained within the SPE. A BIP-8 byte in the path overhead supports error checking at the path layer. A final byte of note in the path overhead provides continuous path status and performance information to other SONET nodes (if they incorporate a path layer), other SONET end equipment, and SONET managers.

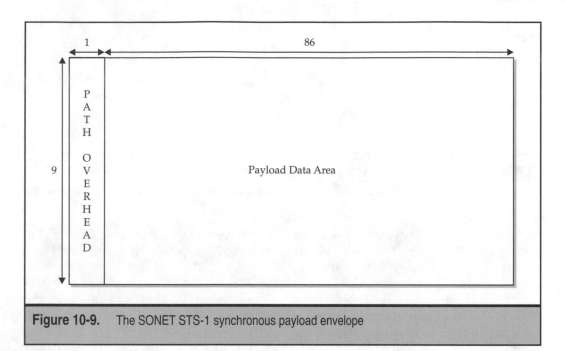

Figure 10-9.   The SONET STS-1 synchronous payload envelope

NOTE: SONET is an *opaque* set of protocols. That is, a mapping standard must be in existence for a particular user service for that service to be accommodated by SONET. As you will see, mapping mechanisms for most popular signal types are part of the SONET standards.

The Internal Composition of the SPE

As mentioned previously, SONET can carry many metallic transmission services (such as DS-1, DS-2, and DS-3) as well as several packet-based protocol types (such as ATM cells and framed IP packets). It is the path layer's responsibility to map such services into the SONET SPE in a fashion prescribed by the SONET standards. Next, we will look at three distinct service types and describe how they are mapped by the path layer into the SPE.

Virtual Tributaries and Transport of Subrate Metallic Services   *Subrate metallic services* are those levels of the North American digital hierarchy (see Table 10-1) that run at data rates less than the DS-3 rate of 44.736Mbps. These include DS-1, DS-1c, and DS-2. SONET transports these services within the SPE by using *virtual tributaries* (VTs). A VT is a column-wise grouping of bytes within the SPE such that the number of bytes allotted to a given VT is sufficient to carry the metallic service intended. A total of four VT types have been defined for SONET. They are listed in Table 10-3.

The VT-1.5 is defined for the transport of DS-1 services over SONET. A VT-1.5 is a block of three columns and nine rows of the SPE. A block of 27 bytes repeating every 125 microseconds represents a bandwidth channel of 1.728Mbps. A DS-1 signal at 1.544Mbps

VT Type	Columns	Rows	# of Bytes	Capacity	Service
VT-1.5	3	9	27	1.728Mbps	DS-1
VT-2	4	9	36	2.304Mbps	E-1
VT-3	6	9	54	3.456Mbps	DS-1c
VT-6	12	9	108	6.912Mbps	DS-2

Table 10-3.   SONET Virtual Tributaries

fits quite nicely into that amount of VT bandwidth, with some left over for overhead to control VT-1.5 mapping functions. As shown in Table 10-3, each VT type is defined specifically for the service type it accommodates.

Unfortunately, the mapping of VTs into the SPE is not quite as simple as that. VTs must exist within a structure called a *virtual tributary group* (VTG) to be mapped into the SPE. All VTGs are 12 columns wide and 9 rows deep, for a total of 108 bytes. Looking back to Figure 10-9, you can see that the useable number of columns in the SPE is 86. Dividing that number by 12 (the number of columns in a VTG), results in 7, with 2 columns left over (these happen to be columns 30 and 59, which are never used when mapping VTGs into an SPE). Thus, a single SPE can accommodate 7 VTGs. A VTG-1.5, therefore, is composed of 4 VT-1.5s, each consisting of 3 columns each, for a total of 12 columns. Likewise, a VTG-2 is composed of 3 VT-2s consisting of 4 columns each, again for a total of 12.

All of the foregoing discussion is important because of two rules involved in mapping VTGs into SPEs: Rule one says that you cannot mix VTs within a VTG. For example, you cannot combine DS-1s and E-1s in the same VTG. The second rule, however, states that you *can* mix VTGs within an SPE. Thus, it is perfectly legal to have a VTG-1.5 and a VTG-2 in the same SPE. This aspect of the SONET standards can lead to waste. For example, if you have only a single DS-1 to transport, an entire VTG-1.5 will need to be allotted within the SPE (and the bandwidth represented by the three remaining VT1.5s is wasted).

Asynchronous DS-3 Mapping   We have previously established that, with a capacity of 49.536Mbps (exclusive of path overhead), a single STS-1 SPE is ideal for the transport of DS-3 service at 44.736Mbps. The mechanism whereby DS-3 services are mapped into SONET is called *asynchronous DS-3 mapping*. In this process, the incoming DS-3 signal is treated as a generic bit stream. The DS-3 bits are converted into SONET bits and carried transparently through the SONET network. Note that, when using asynchronous DS-3 mapping, the contents of the DS-3 (perhaps derived from the M1/3 multiplexing process) are not visible to SONET. SONET provides nothing but an invisible transport service.

Super-Rate and Packet-Oriented Services—Concatenated Frames   There are instances during which the incoming service operates at a data rate that is higher than STS-1. For example, a Gigabit Ethernet (GbE) stream would overrun the capacity of a single STS-1 under high-use conditions. There are also times when the STS-1 frame format becomes an artificial constraint to throughput. If, for example, you want to send a group of IP packets, why should your maximum payload capacity be limited to 49.536Mbps? The issue of *super-rate* payloads is dealt with in SONET by the use of *concatenated* frames. Figure 10-10 illustrates the difference between a super-rate frame (STS-3) that has been constructed from constituent STS-1s and a concatenated frame (STS-3c) that is intended for use by a super-rate service.

The diagrams shown at the top of Figure 10-10 illustrate the multiplexing of three STS-1 frames into a single STS-3 frame. Note that the frame becomes three times as large (to account for the increased data rate when the frame repetition rate must be constant at 8000 frames per second). The STS-3 frame, however, is clearly composed of its constituent STS-1 frames. Looking closely, you can see that three columns of path overhead are associated with an STS-3 frame. (The knowledgeable reader will recognize the stylization of this diagram. The three columns of path overhead would not appear exactly as shown if they were part of a real STS-3 frame.) STS-3, therefore, must be thought of as *channelized* STS-3. Likewise, STS-12 may be composed of 12 STS-1s or 4 STS-3s. Either way, the subcomponents of the super-rate frame are visible in that frame.

The diagram at the bottom of Figure 10-10 shows an STS-3c frame that has been constructed by the process of concatenation. The fact that the frame is concatenated is indicated by the placement of a lowercase *c* after the STS designation (STS-3*c*, STS-12*c*, etc.). Note that only one column of path overhead is associated with a concatenated frame, regardless of its size. Whereas it is true that the STS-3 frame and the STS-3c frame are exactly the same overall size (270 columns by 9 rows), it is in their internal SPE format that they differ. STS-3c can therefore be thought of as *unchannelized* STS-3. The payload capacity of an STS-3c frame, for example, is 149.76Mbps.

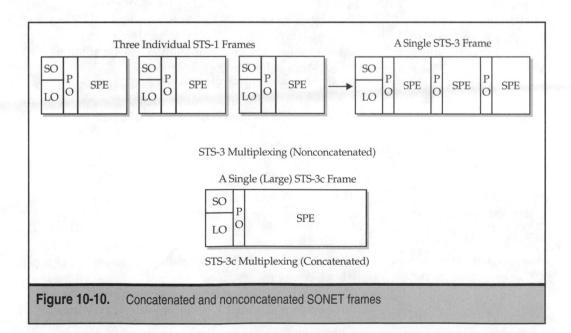

Figure 10-10.   Concatenated and nonconcatenated SONET frames

> **NOTE:**   It is often stated mistakenly that the equivalent to the SDH STM-1 frame is the SONET STS-3 frame (see Table 10-2). However, the SONET STS-3c frame is actually most like the SDH STM-1 format.

Concatenated frames are most often used to carry packet-oriented services such as ATM cells or framed IP packets. Mapping such services directly into a concatenated SPE is efficient and avoids having to recover some subordinate signal (such as a VT-1.5) on the way to recovering the original data units. When data units such as cells and packets are mapped directly into SONET, the process is row, rather than column, oriented. We have established that with subrate metallic services, SONET transport is column oriented (VT oriented). Once again, because it is efficient and direct to transport cells and packets in the order in which they arrive at a SONET node, row multiplexing is used for these service types. Also note that if some sort of interframe synchronization is required at the OSI Data Link layer by these protocols, the protocols, not SONET, are responsible for providing such timing information.

The SONET Layers Revisited

With the preceding discussion of SONET topology and layer overhead behind us, it's a good idea to look again at the SONET layers and functions shown back in Figure 10-7. Let's build a scenario to track the progress of a subrate metallic transmission service (such as DS-1) as it is mapped and multiplexed onto a SONET stream (such as OC-12).

Services come into the path layer. To make the arithmetic simple, we will assume that 336 DS-1s need to be transported over a SONET OC-12 link. As the DS-1s arrive, they are first mapped into VT-1.5s at the path layer. Each VT-1.5 carries a single DS-1. The path layer then combines the VT-1.5s into VTG-1.5s at a rate of four VT-1.5s per VTG-1.5. Now, the path layer populates the SPE by arranging seven VTG-1.5s into a single SPE (leaving columns 30 and 59 blank). Finally, the path layer appends its overhead to indicate, among other things, the internal composition of the SPE. The resultant SPEs, each containing 28 VT-1.5s (or 28 DS-1s) are handed down to the line layer.

The line layer is responsible for multiplexing the SPEs from the path layer into the required STS signal level (in our example, STS-12). In our example, each SPE is placed in an STS-1 frame. A 2-byte "payload pointer" value (additional detail to follow) is computed and placed into the line overhead of each STS-1 frame. The resultant STS-1 frames (minus section overhead) are multiplexed into an STS-12 frame by the process of *byte interleaving*. The result is then passed to the section layer.

The section layer appends basic framing information (X'F628), among other things, to the STS-12 frame and passes it to the photonic layer. The photonic layer is responsible for electro-optical conversion (at the transmitter, the reverse at the receiver) and management of the physical fiber facility.

The result of this layered "division of labor" is an optical carrier system that is standards-based, easy to provision and manage, capable of very high data rates (such as 40Gbps), and that exhibits a high degree of robustness (fault tolerance). Furthermore, the use of a hierarchy of pointers permits complex bandwidth management functions (such as drop/insert and cross-connection) to be performed without a full demultiplex of the aggregate signal. Let's conclude our discussions of SONET with a look at the benefits of this optical transmission technology.

Why Do Carriers (and Customers) Love SONET?

Prior to the development of SONET and SDH, carriers and customers lived in a world of globally incompatible, low-speed metallic transmission systems. If fiber was being used for transport, the protocols that operated over those fiber links were proprietary. Metallic transmissions systems had a number of flaws that the SONET designers sought to overcome in the design of standards-based optical transmission. In the following discussion, we will list each of the benefits of SONET/SDH.

SONET/SDH for Standards-Based Optical Mid-Span Meet

The impetus for the development of SONET standards was the lack of a standards-based optical mid-span meet technology that could be used to interconnect carriers. AT&T's divestiture of its local operating monopolies in 1984 heightened awareness of the seriousness of the coming interconnect problem between the newly created Regional Bell Operating companies (RBOCs) and the traditional interexchange carriers (such as Sprint, AT&T, and WorldCom, to use their current names).

SONET has provided standards-based optical mid-span meet on a ubiquitous basis throughout North America. SDH has done the same in many other developed (and developing) nations around the world. Even if there were no other benefit for carriers, this one saves enough in expenses to justify SONET's deployment.

SONET/SDH for Standards at Rates Above 45Mbps

Prior to the development of optical technology, data rates on metallic transmission systems were limited to about 50Mbps. These services (such as DS-3 and E-3) could be delivered over coaxial cable, but coaxial cable proved a costly medium to deploy and maintain. The advent of fiber-optic technology permitted many 50Mbps streams to be time-division multiplexed onto a single strand of fiber. Unfortunately, these optical solutions were vendor-specific.

SONET and SDH have fulfilled their promise of the standards-based optical transport of high-bit-rate baseband signals. The rate at which R&D laboratories and equipment vendors are extending SONET data rates is breathtaking. In 1987, OC-3 systems (155.52Mbps) were just available from equipment vendors. Today, OC-768 (39.81812Gbps) systems are becoming available. Paradoxically, some believe (this author included) that SONET may give way eventually to another technology, *wavelength division multiplexing* (WDM). WDM promises exponential decreases in the cost of bandwidth and will be discussed in its own section later in the chapter.

SONET and Simplified Bandwidth Management

Because of the bit-oriented, plesiochronous nature of metallic transmission systems, manipulation of a subrate within an aggregate is problematic. Discussions in this chapter have established that, because of the bit-stuffing operations undertaken in metallic carrier systems, a full demultiplex is required to gain visibility (and therefore usability) of a particular subrate signal (such as a T-1 riding on a T-3). This approach is equipment-intensive, failure-prone, and cumbersome.

SONET is a system of position multiplexing. For example, when the line layer maps the SPE into the STS frame structure, it sets 2 bytes of overhead, the payload pointer that indicates where that SPE begins relative to the STS-1 frame. Figure 10-11 illustrates this function of the payload pointer.

Figures 10-8 and 10-9 might have left you with the impression that the SPE is in some way always perfectly aligned with the STS-1 frame. That is, one could assume that the position of the first byte of path overhead would always follow the third byte of section overhead in every frame. In SONET parlance, this is called *locked mode* operation. Locked mode was popular in the early days of SONET, but it has been superceded in practical use by an alternative called *floating mode* operation. In floating mode, the function of the payload pointer is of critical significance.

Figure 10-11 illustrates that the position of the SPE relative to the STS-1 frame is not necessarily locked. In fact, in floating mode, the SPE can begin anywhere within the structure of the STS frame. The benefit, therefore, of floating mode operation is that buffer requirements at each node are lessened considerably. As data rates increase, this becomes increasingly significant.

When floating mode is employed, the line layer "points" to the beginning of the SPE within the STS frame. Once the location of the beginning of the SPE is known, additional overhead at the path layer can provide indicators as to the internal contents of the SPE (for example, the SPE shown in the figure might contain 28 DS-1s organized as 7 VTG-1.5s of 4 VT-1.5s each). It should also be mentioned that if timing discrepancies exist between the clock that generates the STS frame format and the clock that maps the SPE into the STS frame, the payload pointer can be moved (or *justified*) 1 byte at a time to accommodate

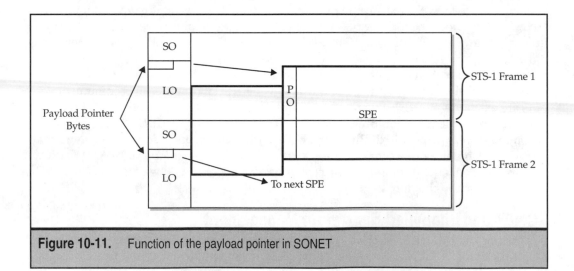

Figure 10-11.   Function of the payload pointer in SONET

these differences. In this rather elegant fashion, SONET overcomes the timing problems associated with plesiochronous clocking environments without overly complicating the performance of complex bandwidth management functions.

The upshot of this discussion is that SONET supports visibility of subrate payloads (all the way down to the DS-0 level, if desired). As such, SONET can manipulate those payloads individually, without necessitating a complete demultiplex of the aggregate signal. This not only reduces the cost of complex bandwidth management services (and hopefully, the price), but it permits the development of new services (such as drop-and-repeat) that were not previously associated with metallic transmission environments.

SONET Fault Tolerance

Metallic transmission systems were intended for use between telephone company central offices, where there would be plenty of interoffice redundancy. As such, they were not designed as fault-tolerant systems. However, whenever the capacity of a transmission system increases, the potential for catastrophic failure increases proportionately. It is one thing to appease the 24 users of a failed voice T-1 circuit; it is quite another thing to lose 21,504 DS-1 services because of a single fiber cut. To this end, the designers of SONET/SDH ensured that adequate protection mechanisms were a part of the standard. Although using redundant, diverse-routed fiber facilities can offer fault tolerance in point-to-point and linear SONET topologies, the optimum topology for exploiting SONET's fault tolerance is the *ring*.

SONET ring topologies fall into two basic categories, unidirectional rings and bidirectional rings. Generally, unidirectional rings protect at the path layer, whereas bidirectional rings protect at the line layer. Unfortunately, a technical discussion of SONET protection ring operations is well beyond the scope of this text. The reader is referred to any one of the excellent technical texts on the subject or directly to the standards themselves (see Telcordia TA-496 or TR-1230 or their ANSI equivalents).

In essence, a SONET fault-tolerant ring is a circularly closed collection of point-to-point links between drop/insert multiplexers. Under normal operations, traffic flows in one direction around the ring (or is duplicated and flows in both directions around the ring). If a node failure or a fiber discontinuity is detected, traffic is routed around the ring in the other direction to its destination (or the remaining duplicate signal is accepted at the receiver). In either case, the ring can survive a catastrophic node or facility failure. The recovery time for any ring topology is generally less than 50 milliseconds.

To be fair, one must recognize that the drawback to a SONET/SDH protection ring architecture is inefficiency. Under all circumstances, 50 percent of the available bandwidth on the ring must be reserved in case of failure. In the case of bidirectional rings, the capacity held in reserve can be offered (at a low price) to customers who can live with absolutely no service guarantees. Interestingly, organizations like public school systems have been able to take advantage of these "best-effort" SONET services. (If someone were to offer you nonguaranteed (but in reality nearly flawless) DS-3 service at dial-up rates, I suspect you would take it and be happy.)

SONET Operations, Administration, Maintenance, and Provisioning (OAM&P)

Generally, a failure in a metallic transmission circuit (such as a T-1) required that the circuit be taken out of service for test and repair. Remember that these networks were originally designed to be deployed in interoffice trunking settings where it was assumed that plenty of redundancy would be present to offset the effects of a single circuit failure between central offices. As a consequence, there was little support for nonintrusive testing and maintenance in early metallic transmission networks. In fact, until the introduction of the extended superframe format into T-1 technology, little if any bandwidth was available for the interchange of management information between network nodes and management systems. Such embedded operations channels (EOCs) are critical for the support of remote service provisioning, real-time service management, and rapid rectification of network failures.

Once again, the designers of SONET understood that to reduce costs for carriers and to increase service satisfaction for customers, the new optical protocols would require robust OAM&P capabilities—thus, the presence of the data communications channels at the section and line layers in SONET. Adding together the DCC at the section layer and the DCC at the line layer, we see that no less than 768Kbps (a full half of a T-1) has been designated for use as an EOC for management information flow. Moreover, a standard message set is associated with management operations over the DCCs, theoretically permitting end-to-end management and provisioning in a large, multivendor SONET network.

Other OAM&P features of SONET include the BIP-8 counts embedded within the section, line, and path layers. The presence of the BIP-8 function at each of the SONET layers assists craftspeople in pinpointing the exact location of signal degradation along a particular SONET span. Finally, other SONET overhead bytes (such as path continuity and path status) assist in the management of a SONET infrastructure.

In summary, the designers of SONET/SDH were able to overcome most of the drawbacks associated with older metallic transmission systems (such as global incompatibilities, relatively low transmission rates, complex bandwidth management, little or no OAM&P capabilities, and little or no fault tolerance). In addition, SONET/SDH overcomes the deficiencies associated with older asynchronous fiber-based transmission systems (such as lack of a standards-based optical mid-span meet).

ANALOG OPTICAL TRANSMISSION SYSTEMS

The demand for bandwidth in backbone transmission networks has been growing explosively for the past decade. The causes of this growth are manifold, but most center on the uptake of TCP/IP in the Internet, in enterprise intranets, and in business extranets. The demand for bandwidth is growing so fast, in fact, that SONET cannot scale fast enough to keep up. It has been suggested by many recent sources that the demand for bandwidth is currently doubling every nine months!

One solution to the problem of "bandwidth exhaustion" is simply to deploy new fiber cable. But even in instances where rights-of-way can be obtained, this is an expensive proposition. It has been estimated that the cost of deploying a new buried fiber route runs between $75,000 and $250,000 per linear mile, depending on the particular geography.

Another solution to the requirement for additional capacity in backbone networks is to deploy a technology that would pack more information carrying capacity on existing fiber routes. This is precisely what wavelength division multiplexing (WDM) does. It is not surprising, therefore, that WDM technology has grown in popularity in recent times.

Wavelength Division Multiplexing

WDM is a form of frequency division multiplexing. Unlike older FDM systems that operated over metallic media at the low end of the electromagnetic spectrum, WDM operates over fiber-optic media using the infrared portion of the electromagnetic spectrum (see Chapter 4, Figure 4-6). Figure 10-12 illustrates the process of WDM.

At the input of the WDM process, each signal to be transported (these are typically baseband signals, such as SONET) is assigned for modulation to a separate laser. Each laser emits its signal at a different wavelength (λ, pronounced "lambda") between the frequencies of 191.1THz and 196.5THz (in the 1550 nanometer range). The lasers are chosen so that the spacing between frequencies is regular (50GHz spacing is common in today's WDM systems). The output of each laser is combined, using FDM, on a single strand of fiber. Thus, a 4-channel WDM system can multiplex, for example four OC-48 streams on a single fiber, therefore increasing system capacity by a factor of four. Likewise, a 16-channel WDM system can support a sixteenfold increase in capacity without resorting to the deployment of new fiber cable.

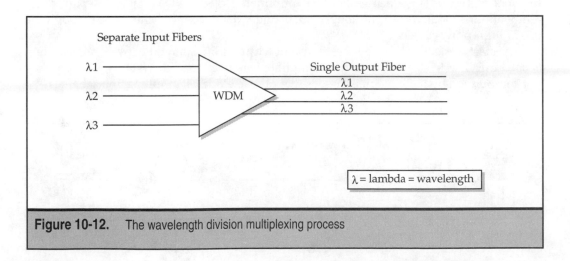

Figure 10-12.   The wavelength division multiplexing process

Current state-of-the-art WDM systems can carry about 100 channels. Such systems are referred to as *dense wavelength division multiplexing* (DWDM) systems because of their high channel densities. Although there is no general industry agreement on the terminology, this author considers systems that support 16 channels (wavelengths) or less as WDM systems, whereas those that support more than 16 channels as DWDM systems. As channel densities inevitably increase in the future, we can expect additional nomenclature distinctions. (How about UDWDM for "ultra-dense wavelength division multiplexing" for systems that support in excess of 100 channels?)

A simple comparison of SONET with and without DWDM illustrates the compelling power of DWDM technology. In a pure SONET OC-48 environment, one strand of fiber can transport 2.48832Gbps. With the deployment of a 100-channel DWDM system in this environment, that same strand of fiber can now carry 248.832Gbps! Moreover, no additional fiber cable is required to realize this capacity increase. Conservative estimates of the cost of deploying DWDM (using an 80-channel system as a benchmark) indicate that the cost of DWDM is about $6000 per mile—compared to deploying new fiber at up to $250,000 per mile.

The Anatomy of a WDM/DWDM Span

Figure 10-13 shows the components of a WDM/DWDM span. At each end of the span is a DWDM terminal multiplexer. These multiplexers are responsible for mapping input services into specific wavelengths (λ) for transport through the system. In the middle of the span is an optical add/drop multiplexer (OADM). Using technology associated with optical diffraction gratings, the OADM supports drop/insert functions on a lambda basis. At intervals along the span are optical amplifiers. These are responsible for boosting the level of the optical signal, which, like any other signal, becomes attenuated as it traverses the fiber medium. The spacing of optical amplifiers along a WDM/DWDM span is currently about 120 kilometers, although much longer distances are achievable through the use of high-power lasers.

The alert reader might wonder at this point why amplification rather than regeneration is used in WDM systems. Moreover, if amplifiers are used in such systems, why are they not confounded by the same noise issues that plague electrical amplification schemes? To

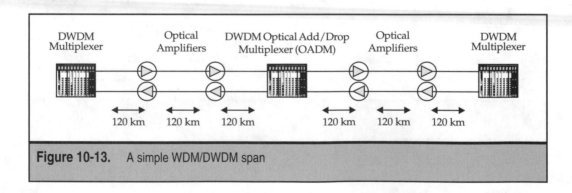

Figure 10-13. A simple WDM/DWDM span

the first issue, WDM is a form of FDM. As such, the signals transported in a WDM system are broadband in nature. Amplification therefore must be employed to counteract the effects of attenuation in a broadband signaling environment.

Earlier in this chapter, we established that as a broadband signal is amplified many times over a long distance, noise tends to accumulate to the point that the signal is no longer discernible amidst the noise. In that discussion, however, we were focused on metallic FDM systems. Metallic systems, by their nature, are highly susceptible to noise in the form of electromagnetic interference. Fiber-based systems, on the other hand, are not nearly as affected by electromagnetic disturbances, so noise problems are nowhere near as pronounced in the optical domain. In fact, not only is amplification desirable in optical FDM systems, it turns out to be the basis for the practical deployment of such systems.

EDFA: The Basis for Cost-Effective WDM/DWDM

The development of optical amplification was critical to the commercial success of WDM. In a baseband optical multiplexing system (such as SONET), attenuated signals require regeneration at regular intervals. At the repeater location, optical signals are converted to electrical signals for the regeneration process to proceed. After regeneration, the electrical signals are reconverted to optical signals and transmitted on the output fiber. We have previously referred to this double conversion as O-E-O, for optical-to-electrical-to-optical.

The implication here is that at every node location (repeaters and multiplexers), a laser is required at the output of the O-E-O process. Such lasers typically cost thousands of dollars. In a single wavelength system such as SONET, the need for a laser at each node location is not a significant cost obstacle to deployment. Now, however, imagine the use of regeneration in a 100-channel DWDM system. At each node location (including regenerators), 100 lasers would be required. Furthermore, since inter-repeater distances are not that great (about 50 km or so), thousands of lasers (at a cost of millions of dollars) would be required over a fairly long point-to-point span. Clearly, the regeneration function needs to be eliminated, or greatly reduced, for WDM and DWDM to be commercially viable technologies.

When multiple wavelengths are combined onto a single strand of fiber, a perplexing effect can be observed. Energy from short wavelengths "bleeds" into longer wavelengths, providing an amplification function for the longer wavelengths. First regarded as an impairment in multi-wavelength transmission systems, this "stimulated Raman scattering" effect became the basis for modern optical amplification technology. In the early 1960s, Elias Snitzer, while working at American Optical, Inc., discovered that if a section of fiber was treated ("doped") with a rare earth element (Neodymium), the Raman scattering effect was enhanced. In the mid-1980s, David Payne, then working at the University of Southampton in the U.K., found that doping fiber with the rare earth element Erbuim specifically enhanced the Raman scattering effect at about 1500 nanometers (nm). Interestingly, this is the frequency band in which WDM/DWDM systems operate. Thus the Erbuim-doped fiber amplifier (EDFA) was born.

Figure 10-14 shows a schematic diagram of an EDFA. Signals enter the EDFA in an attenuated state at a frequency of 1550nm. As the signal passes through the Erbuim-doped

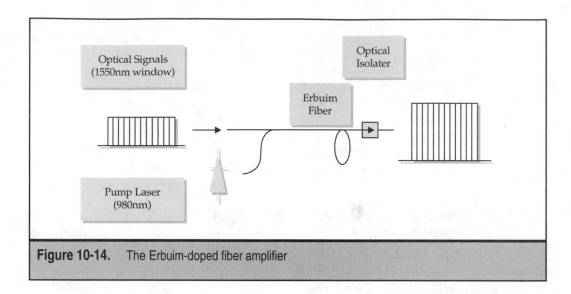

Figure 10-14.   The Erbuim-doped fiber amplifier

section of fiber, the energy from a high-frequency (980nm) pump laser is combined with the signal to amplify it. Prior to leaving the EDFA, the pumping frequency is removed from the signal by an optical isolator to leave an amplified representation of the original input signal. Note that the EDFA amplification effect is a wideband effect. That is, a single pump laser can amplify a wide range of signal frequencies. As such, the EDFA represents a *single laser solution* to the problem of attenuation in WDM/DWDM networks.

The EDFA had two immediate beneficial effects on the development of WDM/DWDM technology. First, it eliminated the requirement for regenerators at short intervals along a WDM span. The inter-EDFA distances in modern WDM networks are on the order of 120 km. About 8 such amplifiers can be traversed before signal regeneration becomes necessary. Thus, inter-repeater distances are extended upward to 1000 km by the use of EDFAs. Work in optical laboratories today suggests that with a combination of high-power lasers and enhanced optical amplifiers, inter-repeater distances of 5000 km are achievable! Thus, it is not far-fetched to propose undersea optical routes that will require no signal regeneration whatsoever. In addition, the EDFA represents an *all-optical* solution to the problem of attenuation. Although electrical power is required to run the pump laser in an EDFA, at no time does O-E-O conversion take place in the amplification process. The quest to extend the all-optical properties of the EDFA to other forms of WDM/DWDM transmission equipment (such as drop/insert multiplexers and cross-connects) is driving the industry toward a new golden age of networking.

The Dawn of the Optical Network

When the acronym is expanded, SONET yields "synchronous optical network." Whereas this author fully agrees with the classification of SONET as synchronous, he would contend vociferously that SONET is *not* an optical network. As established previously, signal

Telecommunications: A Beginner's Guide Blueprints

Table of Contents

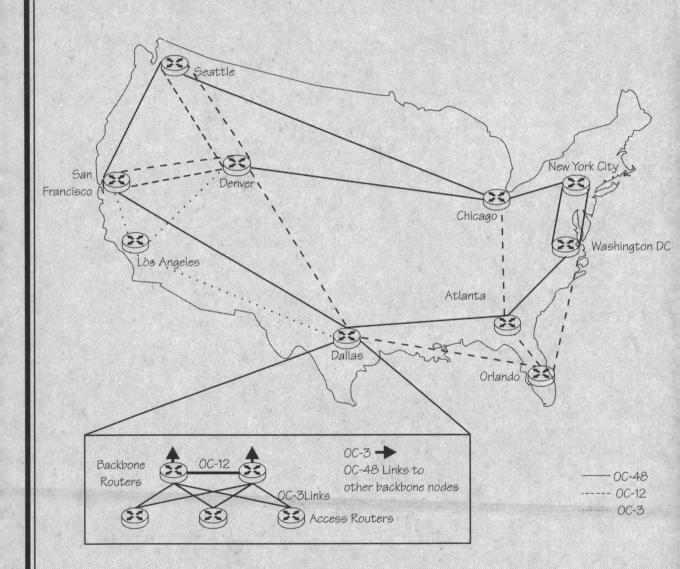

Backbone Routers

OC-12

OC-3 ➤
OC-48 Links to other backbone nodes

OC-3 Links

Access Routers

———— OC-48
------- OC-12
·········· OC-3

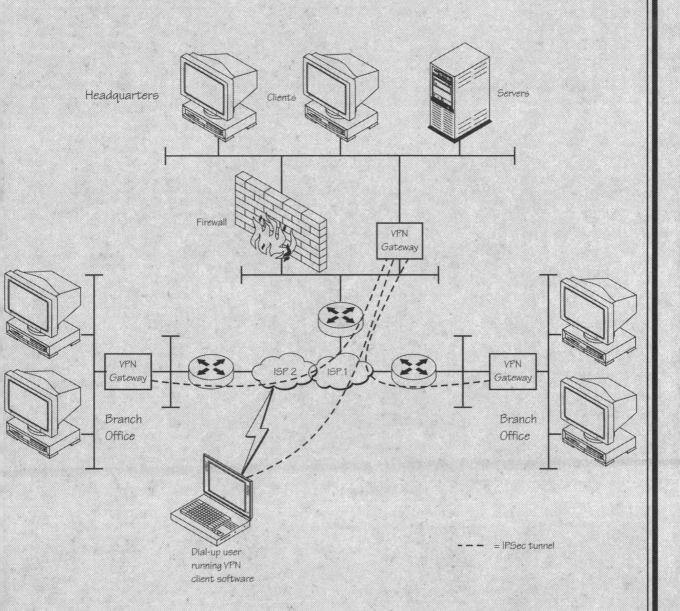

Headquarters

Clients

Servers

Firewall

VPN
Gateway

VPN
Gateway

ISP 2   ISP 1

VPN
Gateway

Branch
Office

Branch
Office

Dial-up user
running VPN
client software

– – – = IPSec tunnel

Circuit Switching

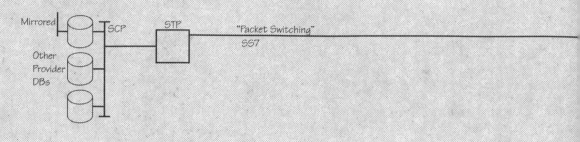

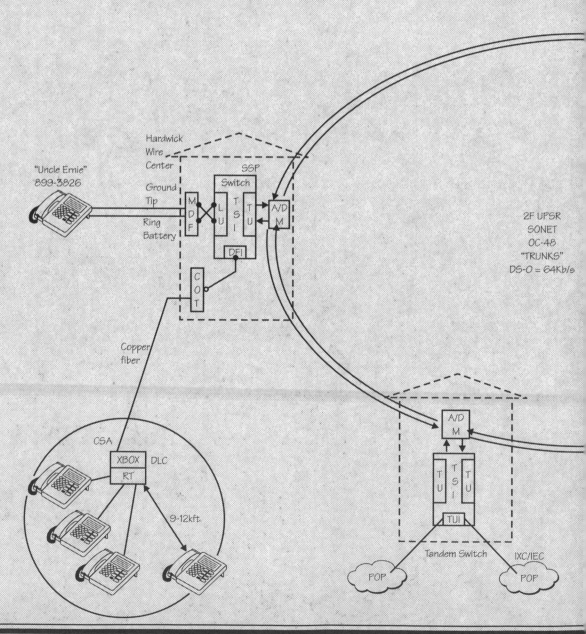

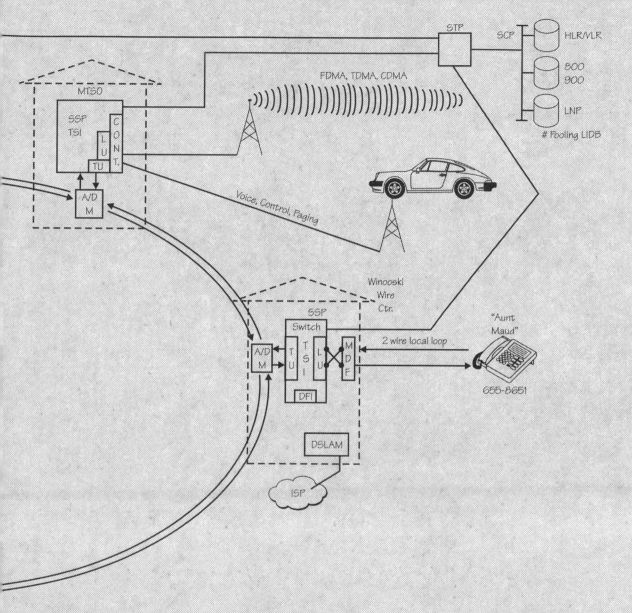

The Internet Exposed: From Your Home to the Server

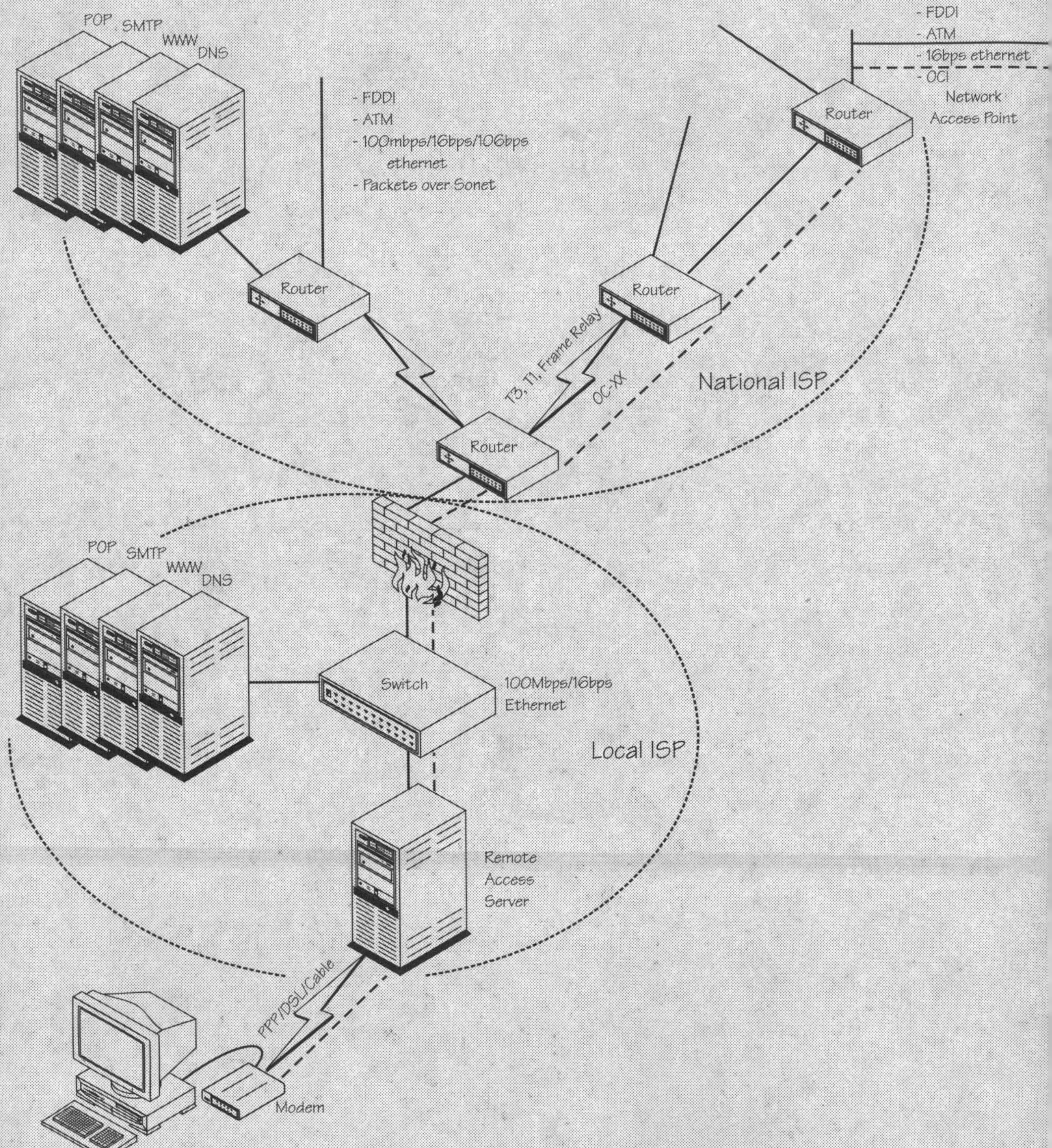

POP SMTP WWW DNS

- FDDI
- ATM
- 100mbps/16bps/106bps ethernet
- Packets over Sonet

Router

Router

- FDDI
- ATM
- 16bps ethernet
- OC1

Network Access Point

Router

T3, T1, Frame Relay

OC-xx

National ISP

Router

POP SMTP WWW DNS

Switch

100Mbps/16bps Ethernet

Local ISP

Remote Access Server

PPP/DSL/Cable

Modem

6

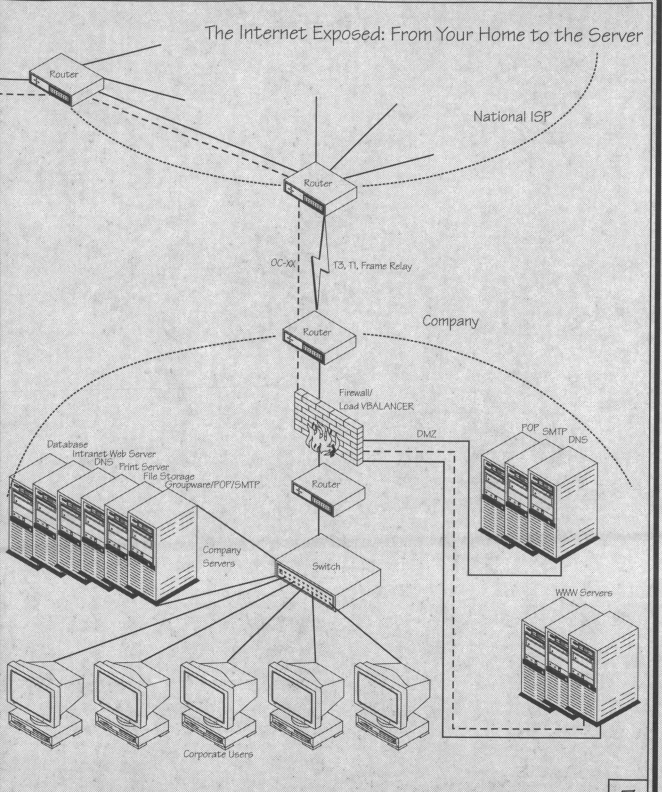

The Internet Exposed: From Your Home to the Server

Router

National ISP

Router

OC-XX   T3, T1, Frame Relay

Router   Company

Firewall/
Load VBALANCER

DMZ   POP SMTP
                         DNS

Database
Intranet Web Server
DNS   Print Server
     File Storage
     Groupware/POP/SMTP

Router

Company
Servers

Switch

WWW Servers

Corporate Users

7

Corporate Data Network

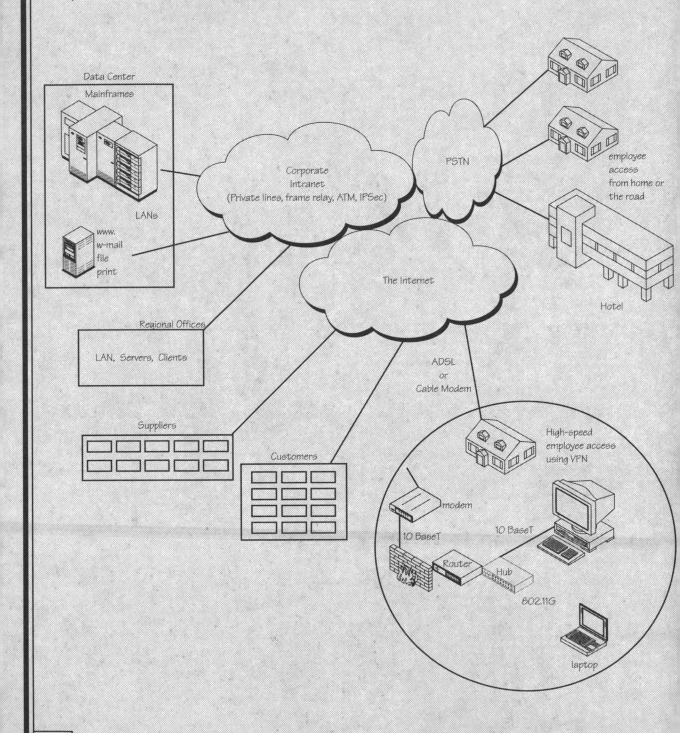

Data Center
Mainframes
LANs

www.
w-mail
file
print

Corporate
Intranet
(Private lines, frame relay, ATM, IPSec)

PSTN

employee
access
from home or
the road

Hotel

The Internet

Regional Offices

LAN, Servers, Clients

ADSL
or
Cable Modem

High-speed
employee access
using VPN

Suppliers

Customers

modem

10 BaseT

10 BaseT

Router

Hub

802.11G

laptop

manipulation at SONET nodes is performed in the electrical domain. As such, SONET is better thought of as an electrical network that uses fiber media for transmission.

The development of the EDFA may well be considered the first step toward the deployment of a true optical network. In this context, "optical" means *all optical*. By eliminating the requirement for O-E-O conversion at intermediate nodes (amplifiers), the EDFA forms the basis for a *true* optical network.

Currently, OADM can be performed without the need for O-E-O conversion. Using diffraction grating technology, in which the grating is actually etched into the fiber strand during manufacture, optical add/drop multiplexers can manipulate (drop or insert) signals at the wavelength level without the need for O-E-O conversion.

Next on the horizon is all-optical switching, in which wavelengths are cross-connected using one of a number of competing technologies. Promising in this regard is micro-electro-mechanical systems (MEMS) technology. In one MEMS-based approach, tiny moveable mirrors are used to direct individual wavelengths through a cross-connect matrix. Another approach uses liquid crystal display (LCD) bubble technology to switch individual wavelengths between input ports and output ports. The maturation of these technologies will lead to instant provisioning of large amounts of bandwidth in future optical networks. This capability, in turn, will usher in a variety of new, optically-based service offerings (such as optical virtual private networks, bandwidth trading, and optical dial tone) that will meet the capacity needs of users into the far future. It is truly an exciting time to be a part of the industry of transmission technology.

SUMMARY

Figure 10-15 summarizes the contents of this chapter. The figure illustrates how the transmission technologies discussed in this chapter are combined to form a modern telecommunications network. It also highlights the area of the modern network in which each technology is appropriate and commercially effective.

Currently, metallic transmission technologies are more often used for network access than for any other purpose. Thus, a customer with a PBX might use a DS-1 service (delivered over a T-1 circuit or by other means such as HDSL) to gain access to the switched telephone network or to interconnect with other PBXs using point-to-point transmission facilities.

When the customer's DS-1 service arrives at a carrier central office in a metropolitan area, it is likely to be multiplexed onto a SONET transmission vehicle. This process is covered in detail in this chapter.

For long-haul transmission between metropolitan areas, the carrier will likely multiplex the SONET stream (containing the customer's DS-1) onto a wavelength using WDM technology. As established in this chapter, WDM technology permits the carrier to expand capacity easily and cost-effectively without the deployment of new fiber cable.

This chapter has covered an extensive amount of information. We began with a discussion of FDM technology and ended with a discussion of, you got it, FDM technology! This chapter once again proves that, in this industry, there is nothing truly new under the sun.

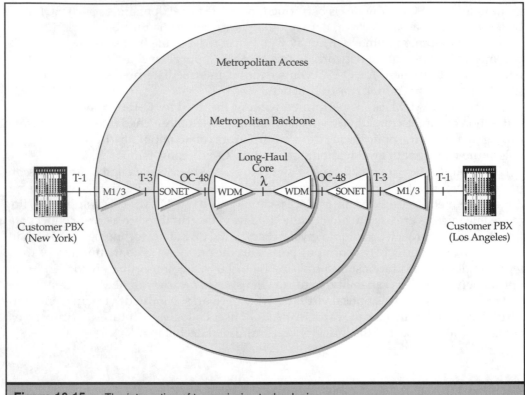

Figure 10-15.   The integration of transmission technologies

CHAPTER 11

Switching in the PSTN

Prior to the advent of the Internet, the only truly international communications network was the Public Switched Telephony Network (PSTN). The PSTN routes and carries voice calls using circuit switches, which have their roots back in the late 1800s when Almon B. Strowger, an undertaker of Kansas City, MO, filed his patent application on March 12, 1889 for the first commercially viable telephone switch. Since that time, switching systems have spawned two "sciences:" the science behind the advancement of the hardware and software technologies that form the architecture of the switching systems themselves, and the engineering science that is used to estimate the sizing of these switching systems. This traffic estimation, or engineering, allows all users equal access to the facilities of the switching systems. In this chapter, we explore these sciences.

TRAFFIC ENGINEERING

This section includes a discussion of the models used for traffic engineering and the problems they were designed to solve. Included in these models are the concepts for sizing queuing (waiting) systems and blocking (busy tone) systems.

Many systems that we encounter each day are sensitive to the amount of traffic traveling through them. The sensitivity is reflected in the type of service that we receive. Although these systems are designed to handle a certain amount of traffic, no system is prepared to deal with any and all traffic simultaneously. Rather, the systems assume that not everyone wants service at the same time. Traffic engineering is the practice of designing such systems to handle an appropriate load.

Consider highways, for example, which are very sensitive to traffic levels. If the traffic load is light, everyone can proceed quickly. If traffic is heavy, we encounter delays. Similarly, supermarkets are sensitive to the amount of shoppers at the store. As aisles and checkout lines become crowded, we suffer delays as customers join a line (or queue) for service. The longer the line, the greater the delay in reaching the goal of bringing home the bacon. Such systems are called *queuing* systems.

Telephone systems, however, behave in a fundamentally different manner. In the telephone network, if a customer encounters an "all circuits busy" condition, the call is cleared and the caller is connected to an announcement. No matter how long the caller waits, he will never be connected to the desired party unless he hangs up and dials again. Such a system is called a *blocking* system, because if no facility is available to handle a customer's call, the call "traffic" is blocked.

Fixing Traffic Problems

The answer to traffic problems is simple: provide more of the resource that everyone wants. In highway systems that means more roads, or more lanes on existing roads. In supermarkets it means more open checkout lines. In telephone systems it means more switching equipment and trunks.

However, providing the needed relief may not be easy. We all know how much time and money is required to fix a highway traffic problem. The supermarket problem may not be as difficult because it could simply mean scheduling additional personnel at peak times. But the telephone network is more like the highway system than the supermarket in that relief takes more time and money as equipment is acquired, installed, and configured. Moreover, a balance must be maintained between good service and the cost of providing that service. Just as excess supermarket checkers or unused lanes are a waste of money, so is excessive telephone equipment.

In short, traffic engineering is the practice of providing just enough equipment to afford good service while minimizing the capital cost so that a profit can be earned.

Measuring the Load

To perform the traffic engineering function, an engineer requires data about loads. Telephone equipment is extensively instrumented so that load measures can be taken for this purpose.

One measure of load is a count of *events* that occur during a time interval, such as an hour. Events of interest might be the number of calls initiated on a switch, the number of attempts to find an idle trunk in a trunk group to another switch, or the number of calls failing to find a trunk and thus being routed to an announcement. These events are called *peg counts*; in the case of calls failing to find an idle trunk, they are called *overflow*.

The other essential measure of traffic is *usage*, the amount of time that a trunk or other piece of equipment is busy during an hour. Usage measurements are taken on a sample basis. For example, all trunks in a group may be sampled once every 100 seconds. If a trunk is found to be busy, it is assumed to have been busy for the full 100 seconds. This is a reasonable assumption for equipment such as trunks and switching networks that stay busy for periods of several minutes while a conversation is in progress. However, common control equipment that can take only a fraction of a second to a few seconds to handle a call must be sampled much more frequently. In each case, the usage is expressed in *hundred call seconds per hour*, or CCS. Because each hour contains 3600 seconds, equipment that is busy for the full hour registers as 36 CCS.

By dividing usage data by peg count data, an engineer can derive a figure for average *holding* time, or the average length of time a piece of equipment is busy for each event or call.

Traffic Data Collection

Traffic engineers seek data from what is called the *time consistent busy hour*. That is, if the peg counts identify three peak hours during the day, the hour having the highest average load over time will be chosen as the hour from which engineering data will be collected. If equipment is sufficient to provide the objective grade of service during this busy hour, service should be even better in all other hours.

Having established the busy hour, data is then collected over some period to furnish the load information. Some equipment is engineered to handle the load over the average busy season, which is the three months of the year having the highest traffic volumes in the busy hour (data in these months is taken on weekdays only, with weekend traffic being less). Much of the control equipment, as well as equipment in toll offices, is engineered to handle the load on the 10 highest traffic days or on the highest single traffic day. If high-day engineering is used, the objective grade of service could be much higher (such as blocking 5 or 10 percent of the calls rather than just 1 percent).

Extreme value engineering is sometimes used to determine the required equipment for handling loads. In this case, the single highest hourly usage value for each day is sought. That is, engineers use what is called the *bouncing* busy hour rather than a *time consistent* busy hour.

Traffic Forecasting

Once traffic data has been collected over a period of years, engineers can develop normalized numbers that change in a predictable way. One such number is CCS/line value, which can be determined for the switching network and the control equipment of an office. This value remains stable over time and can be plotted. Typically, it will grow in a linear fashion, and linear regression techniques can be used to forecast its value in the future.

The ever-increasing levels of Internet traffic have upset this process. In some cases, telephone companies have installed separate switches expressly for use by Internet service providers (ISPs) to avoid adding ISPs to switches that serve business and residence customers because of the service problems that Internet traffic can cause. The usage numbers that justify these installations are discussed later in this chapter in the sidebar "Internet Service Providers and Internet Traffic Issues."

Equipment Provisioning

With a forecast CCS/line and a line growth forecast, the engineer can accurately predict the load on office equipment several years into the future—such as when the installed equipment will reach its capacity and how much additional equipment should be installed to handle future loads. With this information, scenario studies can determine a company's most economical relief plan. That is, for example, should one update be planned that will provide enough equipment for the next four years, or should two updates be planned, each providing for two years' growth?

Grade of Service

Because traffic-sensitive systems are not designed to handle the maximum load, a measure of the level of service must be provided to allow an engineer to assess how well the system is behaving. Enough of the proper equipment must be provided to handle some objective load level at a particular grade of service.

In telephone systems, the most important criterion is the *probability of blocking*, the probability that when a call is presented to a component of the network (such as a switch, trunk, etc.), the call is blocked because no server is available. The normal grade-of-service objective in the busy hour is to block no more than 1 percent of offered calls.

Some telephone equipment, such as digit receivers, allows customers to queue for service. For such equipment, the appropriate grade-of-service measure is the probability of a certain percentage of callers incurring delay that exceeds a threshold. For example, the ability of customers to receive dial tone is measured in this way. A typical service level for a digit receiver would be a .01 percent probability of waiting more than 3 seconds for a receiver (a dial-tone delay).

Telephone Traffic Characteristics

In engineering telephone equipment, it is important to understand traffic characteristics so that sufficient equipment is provided to handle the expected load. If, for example, we provide only enough equipment to handle the load offered between 3 A.M. and 4 A.M. at the objective grade of service, clearly service will be very poor at higher traffic periods during the day and evening.

We know empirically that traffic is unevenly distributed based on time of day, day of the week, and month of the year. For example, local calling is higher in the winter months than in summer months. Conversely, toll calling is higher in the summer. Traffic during the day peaks in the morning, afternoon, and evening with peak sizes depending on the type of area served by a switching office. For example, a switch in midtown Manhattan sees peaks between 10 and 11 A.M. and 2 to 3 P.M. but much less traffic occurs in the evening. Residential switches behave quite differently.

Telephone Traffic Assumptions

Statistical techniques are used to model telephone system behavior. In constructing the statistical models for telephone traffic, the engineer must make certain assumptions that make the problem tractable. One common assumption is that traffic is offered from an infinite number of sources that are independent of one another—that is, the load varies in a random fashion. However, in many instances in telephony, this condition does not hold true (for example, there is nothing random about the traffic stimulated by a radio disk jockey who proclaims "I'll give $95 to the ninety-fifth caller").

Another set of assumptions are the averages that we assign to telephone calling. We know from empirical studies that the average local call is about 3 to 5 minutes long and the average toll call is about 10 minutes long. We also know that the average residential line initiates about 3.5 calls per day. This knowledge is used to forecast future traffic loads. One can take the averages and multiply them by the forecasted lines to determine future load. However, if the averages abruptly change, service problems can result.

Increasing data traffic on the telephone network is one factor causing such a change. For example, increased use of fax machines has dramatically altered the usage averages. Internet access has also upset engineering models. Usage adjustments for these traffic sources are shown in Table 11-1.

Type of Connection	Calls Per Day	Minutes Per Call
Voice Calls	3.5	3–7 (local) 10 (long distance)
Fax Calls	10	1
Internet Calls	5	60

Table 11-1.   Modern Usage Assumptions

Internet Service Providers and Internet Traffic Issues

As millions of U.S. households and organizations connect to the Internet and World Wide Web, ISPs are appearing to satisfy this hunger for Web access. In most cases, these ISPs supply Internet access on switched local loops and trunks, and then onto leased trunks from carriers to ISP points of presence (POP). As a result, the ISPs are generating more and more traffic on the PSTN.

Unfortunately, most PSTN facilities were engineered for 3-minute voice calls, but Internet access and Web browsing sessions usually last much longer. Sixty minutes seems to be an average session time, and much longer average session times have been claimed. (One account speaks of a *two-week* Internet session; the local exchange carrier [LEC] manually terminated the connection, only to get a telephone call 5 minutes later: "What happened to my Internet access?") Low flat-rate monthly charges from ISPs give many people little incentive to hang up at all. As a result, other calls may be blocked on the PSTN. (There are tales of 9-1-1 calls being blocked—no switch capacity or outbound trunk—due to increased Internet usage in a neighborhood.)

When Internet connections are used across dedicated circuits built for short voice connections, the impact on PSTN resources can be considerable. Studies have shown that ISP use can be as high as 26 to 28 CCS, which approaches the maximum on the circuits (36 CCS). In contrast, voice use peaks at a modest 12 CCS and averages only 3 CCS. In addition, the system's busy hours have shifted from late afternoon (voice) to late evening (data). This has led to trunk growth numbers that are four times those forecasted for normal voice growth.

A key problem for LECs today is how to deal with the overload conditions imposed on local switches by customers accessing the Internet. The most likely solution to relieving the problem is to get Internet traffic off the circuit-switched network. Technologies such as asymmetric digital subscriber line (ADSL) offer the opportunity to address this issue (see Chapter 20 for a discussion of ADSL).

Applying the Information and Determining Blocking Probability

After realistic (or estimated or guessed) usage statistics have been determined, these numbers are applied to a model that is used to size the PSTN equipment. The primary statistical model used for traffic engineering telephone equipment is called the Blocked Calls Cleared (BCC) model. It is used for equipment such as switching networks, trunks, and most service circuits. In this model, a customer arriving for service and finding no available servers is cleared from the system. The grade of service measure for BCC is the probability that a call is blocked given a particular offered load. This model is also called the *Erlang B* model, so named for its creator, Agner Krarup Erlang of Denmark, who published *The Theory of Probabilities and Telephone Conversations* in 1909.

While one can perform the calculation to determine the probability of blocking given a certain offered load and number of servers, Erlang B tables have been developed to allow the engineer simply to look up the information. An engineer uses these tables by first determining the load expected on a piece of equipment. Then, knowing the allowable grade of service, the engineer refers to the table to find how many servers are required.

Efficiency Considerations

One interesting phenomenon in traffic-sensitive systems is the efficiency at which they operate. For example, having a single checkout line in a grocery store during light periods might result in times when the checker is idle and other times when several people are in line. The load that one checker can handle and still provide good customer service is quite low. As the number of checkers increases with load, the greater the amount of time that each stays busy while still affording good customer service.

The same is true on highways. A single four-lane road can operate more efficiently than two, two-lane roads. In other words, a single four-lane road can handle more traffic than two, two-lane roads at the same rate of delay. This principal, known as the *Law of Large Numbers*, is also true for telephone networks.

Trunk Groups and the Law of Large Numbers   Whenever things are grouped together for any purpose whatsoever, the Law of Large Numbers is in effect. Basically, this law states that large groups are more predictable than small groups when it comes to mathematical properties. To see why, consider the average height of people figured in groups of 10 versus groups of 100. You would notice a lot of variation in the average height by determining only 10 measurements at a time. One group's members could each be over 6 feet tall (1.8 m), while another's members might be under 5 feet 6 (1.7 m), and so on. Your results, then, would be skewed based on the heights of the small sample of subjects. But when taken in groups of 100 at a time, our measurements would show much less variation, producing results that are closer to the average height for the population as a whole.

In the PSTN, service providers often use the Erlang B traffic tables to predict the probability, in percentages, that a call will be blocked. A level of service known as P.05, for example, will block only 5 percent of the calls attempting to use this group, or 1 call in 20.

The P.1 level of service will block 10 percent of the calls attempted, or 1 in 10, and so on. These levels of service are built into tariffs and contracts, and they are closely watched by state regulators. Since larger trunk groups can support higher occupancies, the PSTN tends to aggregate traffic into large groups to more predictably achieve a level of service.

Trunk Group Occupancy   Since the Law of Large Numbers applies, large trunk groups can be utilized closer to 100 percent at a given level of service, such as P.05.

For instance, trunk groups of 25 can be 75.9 percent occupied and only block 1 call in 20, while trunk groups of 200 can be 94.3 percent occupied. Larger trunk groups can be pushed closer to the edge, while smaller trunk groups need lots of slack.

However, this higher loading factor on larger trunk groups comes with a price. Because they operate closer to the edge, large trunk groups are more sensitive to traffic increases than small trunk groups. For example, a traffic increase of 14.2 percent will double the blocking to 1 call in 10 (P.1) on a trunk group of 25, but only a 7.9 percent increase will produce the same effect on a trunk group of 200.

In some cases, traffic increases have doubled the loads on trunk groups. This lowers the level of service on a trunk group of 200 to P.5 service, which blocks 1 call out of every 2. Since these levels of service are built into tariffs and contracts—and watched by state regulators—there are hefty consequences to these degraded levels of service.

Level of service affects the users as well as the service providers. For example, just after the 1989 California earthquake, PSTN traffic increased dramatically. As a result, friends and families trying to telephone one another to or from San Francisco jammed the telephone system and many were faced with blocked lines. In larger cities, added traffic pressures on large trunk groups can quickly overload the trunking network. Yet in smaller towns with modest trunk groupings, service degraded much less under the same increases.

The higher efficiencies of large trunk groups have resulted in large tandem switching offices (tandems switch trunk to trunk) across the country. When one of these tandem offices burned in Hinsdale, Illinois, the large trunks could not handle the traffic increases on alternate routes and quickly blocked many of the calls attempted on them.

Managing the Network

The traffic engineer must attempt to strike a balance between good service and low cost. If congestion in one part of the network is left unchecked, it can spread to other areas far removed from the source. Such situations can bring the network to its knees. Active management of network resources and traffic is necessary to deal with these circumstances. To that end, two types of network management are used: protective and expansive.

Protective controls are those that restrict traffic in some way. For example, under congested conditions, a switch could automatically stop retrying if the first attempt to set up a talk path through the switch fails. Another type of control is used in a disaster situation. In this case, traffic into the affected area is limited in an attempt to keep resources free to allow calling out of the area. This is easily done in modern switches that can be instructed to route, for example, 90 percent of the calls into a numbering plan area (NPA) straight to an all-trunks-busy announcement. This has the effect of cutting off traffic close to its source to keep the destination switches from becoming congested.

Special Cases

Situations arise that can result in severe traffic overloads. Some of these, such as Christmas Day calling loads, can be anticipated. It is clearly not prudent to provide enough equipment to accommodate such an extreme case, since much of the equipment would remain idle for the remainder of the year. Therefore, network managers must cope with congestion on such heavy days. Other situations, such as natural disasters, cause focused calling. The resulting volume cannot be anticipated or accommodated and network congestion results.

Expansive controls can be taken to use available network resources better. For example, traffic between New York City and Miami is usually routed on trunks directly between the two cities. However, when this traffic load is heavy on Christmas morning, for example, network managers may route part of the load via trunks to switches in the western United States. Such switches will be lightly loaded at this time because of the three-hour time difference. The result of this type of active management is better service during overload periods.

TRAFFIC SWITCHING

So far in this chapter, we have discussed the engineering that is necessary to size, model, and manage a PSTN switching system comprising loops, trunks, and switches. In this section, we take a microscopic look inside a PSTN switch. To add some basis for the architecture of today's modern switching machines, a quick look at the history of switch architectures is also provided.

The purpose of switching is to permit the exchange of information without requiring a private line between two stations. Instead, we give each station a link to a switched network and ask the network to convey the information. In Chapter 6, we introduced circuit switching and packet switching. Remember that circuit switching works by establishing an end-to-end connection between two stations. The link is dedicated for the stations' use for the duration of the conversation and, once established, involves no delay in transmitting the information except for propagation delay on the facility. This type of switching was developed for voice telephony. In contrast, packet switching was developed to transport data by allocating bandwidth only when it is needed to transmit information. In this chapter, our attention is focused on the design of circuit switches for the PSTN.

Circuit Switching Requirements

Following are three main requirements that circuit switches must satisfy:

▼   To establish connections in the network between the two parties that want to communicate. This connection must be dedicated to the two parties and must last for the duration of the call to ensure that no delay in voice transmission occurs.

■ To receive, store, and transmit signaling information, such as dialed digits, ringing, or other audible tones and recorded announcements. Signaling information can be passed over the same path that the parties use for conversation, but problems arise when this method, known as *in-band signaling*, is used. It is commonplace now for signaling (at least between switches) to be exchanged over a separate path using *out-of-band signaling*.

▲ To perform routing and translation on the call based on signaling information received. That is, the switch must use the dialed digits to determine where the called subscriber is located and the most efficient route for reaching that location.

Switch Structure

Switches consist of three main components: the switching network, the control function, and the interfaces to the outside world (such as loops and trunks).

▼ The *switching network* is the component in which the circuit-switched connections are established. In other words, the network furnishes the talk path through the switch between the two stations.

■ The *control unit* is the portion of the switch that interprets signaling information, performs routing and translation functions, and tells the switching network how to establish a path between interface units.

▲ The *interfaces* perform the signaling functions between switches, and between the switch and the subscriber. They also terminate the wire-based or wireless facilities used for transmission.

If we look only at the switching network portion of the office, we can identify two types of networks: space-division switching and time-division switching.

Space-division switching provides a separate two- or four-wire path between stations. This separate path "in space" for each conversation gives the method its name. When manual switchboards were in use, operators provided space-division switching by manually connecting different pairs of cords, thereby furnishing a talk path. Later, electromechanical switches performed space-division switching.

The alternative way to furnish a connection through a switching network is via time-division switching, in which many pairs of stations share a transmission path at discrete time intervals. This type of switching requires that the analog voice signal be converted to digital and that many voice samples are interleaved during switching.

Digital Time-Division Switching: TSI, TMS, and TMT

An implementation of the time-division switch network is the *time slot interchanger* (TSI), a solid-state device that permits the switching of voice samples between inputs and outputs.

Manual Switchboards: A History

In the early 1990s, the last manual switchboard providing commercial services was finally retired. The same technology used in the manual switchboard has been in continuous service for more than 100 years now. The concept is simple: lines and trunks are terminated at the switchboard.

Each trunk or line is provided with a signaling ability to alert the operator for a request for service. Two popular signaling schemes were a common battery operation and a ringdown operation. In the common battery operation, the telephone's switchhook completed a circuit and drew power from a battery. This power activated an indicator that notified the human operator of an incoming call. When the telephone was placed on-hook, the circuit was broken and the indicator was extinguished. In a ringdown signaling system, a magneto used at the telephone created a ring current. The operator was notified when this ringing caused a bell to sound. (Note that this same system is still in use today in our current telephone system to indicate that an incoming call has arrived.)

Control and routing in these manual switchboards was performed by the human operator. Eventually, one of the first mechanical automatic switchboards was created, according to legend, by a small-town businessman who was upset by the bias the local operator showed while routing calls. It seemed that the operator was the wife of a competing businessman in this small town and sometimes misrouted her husband's competition's calls.

The businessman's invention allowed the caller, rather than the operator, to determine the destination of a call. Each phone was equipped with a dial that sent digits (on-hook/off-hook pulses on a common battery circuit). Each pulse of a digit advanced the connection one "step" on the switch. Each step was a relay connection to another telephone. This stepper switch design was the predominate switch architecture used through the 1960s. But as silicone technologies advanced, the mechanical stepper switching network was replaced with a solid-state network of a similar design. Instead of relays creating the connections for the voice path, a device called a solid-state crosspoint created the connection.

Figure 11-1 illustrates a single TSI that supports the same six lines on the input and output sides. Connecting the lines to the TSI are interfaces that convert the analog voice signal to a digital bit stream for switching, and then convert it back to analog on the output side. The TSI consists of two buffers that are merely memory chips: a controller that is a microprocessor, and a map that shows which inputs are to be connected to which outputs.

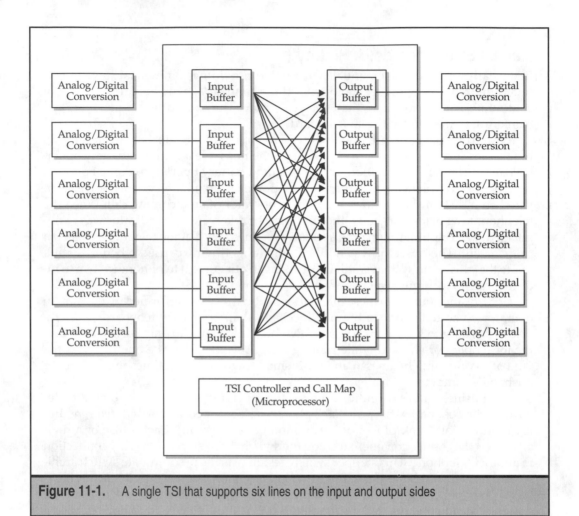

Figure 11-1.   A single TSI that supports six lines on the input and output sides

The TSI digitizes a sample of each input line and places it in the input buffer in each time slot. The controller then uses the map to determine where each voice sample goes in the output buffer. The output interface then puts each sample on its corresponding output line.

You can probably guess that there is a maximum size for every element of a switch, and this is certainly true of a TSI. The number of lines that can be supported is limited because of the limitation on the number of voice samples that can be interchanged in each time slot—about 256 lines per TSI. But because switches must be able to handle tens of thousands of lines, many TSIs must be used along with a mechanism for connecting them to increase the load. This function is served by a *time-multiplexed switch* (TMS). The TMS is a rapidly reconfigurable, solid-state, space-division switch. It permits voice samples to be sent from one TSI to another, and is capable of being reconfigured about 256 times in each time slot (1/8000 of a second) to allow pairs of TSIs to be connected and samples to be exchanged.

Modern digital switches use the TSI and TMS, or equivalent devices, as building blocks to construct their switching network portions. These devices can be combined in different arrangements to create a network with high speed and a low probability of blocking. A common arrangement is the *time-space-time* (TST) switch, which uses a TSI, followed by a TMS, followed by another TSI, to pass the voice samples. Lucent's 5ESS and Ericcson's AXE are both examples of TST switches. Nortel Networks' DMS is a TSTS switch and Seimens EWSD is rather unique. It uses the TST construct in its smallest configuration and grows space switches to form TSST and TSSST switches handling larger and larger line capacities. Some of these switches are discussed later in the section "Digital Switch Requirements and Examples."

Switch Control Section

The control section of a switch performs the routing and translation functions to identify which ports are to be connected and then to command the network to make the connection. It also supervises connections in progress and takes down connections after conversations are complete.

The control section also performs a surveillance function to ensure proper switch operation. For this purpose, it executes diagnostics during slow traffic periods and allows access by maintenance personnel for testing.

Wired Logic vs. Stored Program Control   The two types of common control switches are *wired logic* and *stored program control* (SPC). Wired logic common control was used on older switches such as the #5 Cross-bar (5XB). This type of control is hard to implement and even harder to change.

Modern switches use a digital computer as the control element, and a generic program is used to make it behave as a switch. Such a switch is said to have SPC. Two types of data are stored in memory for use by the program. *Translations*, the first type, is a database identifying all switch resources (for example, lines and trunks and their addresses—line equipment numbers and telephone numbers, class of service, etc.). *Transient* call store information, the second type, contains volatile information about each call in progress.

Software for special services such as Centrex, call waiting, and call forwarding can be installed separately from these generic services. Consequently, each switch needs to contain only the software essential to its own operation.

There are a number of obvious benefits to using stored program control in modern switches. First, call setup can proceed far quicker than in older electromechanical switches. Another is the obvious ease with which modern switches can be updated with new software to support network changes.

Switch Interfaces

The study of switch interfaces is a study of the history of telephony. Most modern switches have a repertory of interfaces that support all telephone equipment, from ringdown and manual equipment to state-of-the-art digital devices. In broad terms, these interfaces can be divided into two categories: lines and trunks.

A line unit is used to connect a local loop. This could be as simple as a circuit board in an electronic office. Functions of a line interface are

▼ Analog to digital conversion

■ Digit reception and analysis

■ Tone generation

■ Loop test and reporting

■ Ring generation

■ Battery

■ Overvoltage protection (lightning)

▲ Timing

A trunk unit performs important functions for a trunk. Specific functions include signaling, transmission, amplification or the repeating of the signal on the incoming side, and test access. While line units typically support two wires (the local loop), trunk units can support two-, four-, six-, or eight-wire circuits—with as many variations of signaling as the imagination can create.

Digital Switch Requirements and Examples

Digital switches must fill several important requirements. The first is high capacity—that is, the switch should be capable of terminating large numbers of lines and trunks and of switching a large number of calls per hour. The second major requirement is modularity. A modular switch is a reasonably low-cost switch that serves a small number of lines. Its modularity also supports incremental growth of terminations and call-handling capacity by adding modules. This permits a single architecture to serve a wide range of office sizes. In addition, in this era of shortened equipment lifetimes, a modular switch allows for the graceful incorporation of new technology. For example, as new processor technology comes along, only the processor module must be replaced to upgrade the switch.

Another major requirement of digital switches is that they be capable of supporting new services as needed and available. This ability derives from the modularity of the switch as well as the ability to define new services in software. Examples of the services include Integrated Services Digital Network (ISDN), Signaling System 7 (SS7) signaling, support for the Advanced Intelligent Network (AIN), and support for broadband switched services such as cell relay and frame relay.

Finally, digital switches must evolve to support voice and other services over packet switching in addition to circuit switching. The clear trend in the industry is to migrate the public telephone network from circuit switching to packet-switched networks using either Asynchronous Transfer Mode (ATM) or the Internet Protocol (IP). Switch vendors are also busily planning for switch architectures to support packet telephony.

The two major switch vendors in North America are Lucent Technologies and Nortel Networks. Their respective switch platforms are the 5ESS and the DMS-100. Examples of both companies' digital switches are covered in the next few sections, along with two other types of digital switches: a privately-owned class of switches that support local switching requirements and the Digital Cross-Connect System (DCS) class of switches.

The Lucent 5ESS and 5ESS-2000

The Number 5 Electronic Switching System (5ESS) is Lucent Technologies' product for local and local/toll office use. The first 5ESS switch was deployed in 1982 in Seneca, Illinois, and is now an integral part of the telecommunications networks in more than 50 countries, supporting more than 108 million lines and 48 million trunks. The switch was designed with a flexible and evolvable architecture.

The latest version, 5ESS-2000, was introduced in the mid-1990s and can support more than 200,000 lines and 1 million calls per hour. The 5ESS-2000 platform supports future applications requiring enhanced processing capacity, performance, reliability, and operations, administration, maintenance, and provisioning (OAM&P). It uses a modular architecture with functions spread among three module types: one administrative module, one or more switching modules, and a communications module. While many small offices don't need communications modules, switches having more than one switching module must include such a module.

The 5ESS employs a distributed processing architecture with most call processing functions performed in the switching module, where lines and trunks are terminated. The administrative module processor is the main computer of the switch and handles administrative tasks. The communications module is part of the switched network that enables talk paths to be established between switching modules. It provides the TMS function when two TSIs are connected between different switching modules. The communications module processor (CMP) and switching module processor (SMP) each participate in call processing functions.

Switch user groups are not always large and can be scattered; these types of groups can be served by a small version of the switch, a remote switching module (RSM), an extended switching module (XSM-2000), or a remote line interface. The compact digital exchange uses a small version of the communications module (CM-2) in combination with switching modules or SM-2000s to support 2000 to 15,000 lines. Very Compact Digital Exchange (VCDX) is possible using a workstation to replace the administrative module—it supports up to 14,000 lines.

The 5ESS-2000 supports multimedia services, including teleconferencing, using ISDN. It also offers a direct interface for SONET/SDH transmission facilities. With the addition of the 5ESS AnyMedia Multiservice Module, it can also interface with IP and ATM networks.

In a typical 5ESS-2000 installation, the wire center houses the administrative module (AM), communications module-2 (CM-2), and one or more switching modules (SM). Remote switching modules (RSM) can be connected to the host by T-1 systems on copper or by fiber links. In the latter case, the module is called an optical remote module (ORM). SLC-2000, the next generation subscriber carrier system, can be connected to any SM whether in the wire center or remote.

5ESS Small Exchanges   The 5ESS-2000 Compact Digital Exchange (CDX) is a small switch configuration suitable for use in campus, office park, and rural locations. The switch can serve 100 subscribers or up to 15,000 local access lines and 15,000 remote access lines with the same service as a full-sized 5ESS-2000. It supports POTS, ISDN, Centrex, wireless, and SS7 services.

The switch can be reconfigured to a full 5ESS-2000 should growth require it. Its reduced size and cost allow advanced digital services to be brought to rural areas.

The VCDX is the smallest configuration in the family. It replaces the AM and CM with a UNIX workstation. The workstation supports a single SM or SM-2000, and the entire switch can fit in two, 6-foot cabinets. Such a configuration could be attractive to small office buildings, competitive local exchange carriers (CLECs), or cable TV providers.

5ESS Evolution to 7 R/E

A clear trend in the telephone industry today is the evolution of voice from circuit-switched to packet-switched networks. As communications have evolved from simple voice telephony to e-mail, facsimile, packet data networking, and the Internet, we have witnessed the rise of special purpose networks to handle data traffic using packet switching. The reason is that packet networks allow users to share network resources at the expense of incurring some delay. The result is higher network utilization and lower cost. While data can tolerate the delay to achieve the cost benefits, voice is very delay sensitive and has traditionally demanded the use of circuit-switched networks.

However, the dream of network designers has been to develop a single network architecture that could support all traffic types, including voice, and to reap the economic benefits of reduced infrastructure costs. The rise of telephony over the Internet and other IP-based packet networks has been at the forefront of a movement spurring development of technology to allow all traffic types to ride on a common network. The importance of this trend has not been lost on vendors of circuit-switching equipment for telephone carriers. The risk of losing their customer base to other vendors of packet technology has prompted more vendors to evolve their switch architectures to support packet networking.

Lucent has developed a follow-up to the 5ESS architecture: the 7 R/E. The 5ESS will continue to serve as a major platform for telephone switching for many years. Its modular elements are not likely to disappear anytime soon. However, instead of growing switches by adding 5ESS switch elements, the 7 R/E offers the opportunity to grow the switch with an entirely new set of switching elements that support packet networking. For a carrier, this means the ability to grow and incorporate new technology at the same time without replacing existing switching equipment.

The elements of the 7 R/E include gateways for converting between circuit- and packet-switching environments for both voice information and signaling. They also include a number of feature servers that execute the programs to support traditional and new services. At the heart of the architecture are switches that support both the ATM and IP environments.

The Nortel DMS-100

The Digital Multiplex Switch-100 (DMS-100) is the premier digital switch from Nortel Networks. Variations of the switch can be used for local, local/toll, and toll applications. The DMS-100 employs a modular architecture that supports up to 100,000 lines.

In the DMS-100 model, each office must have a SuperNode and two or more network modules. Line group controllers and line concentrator modules are provided in sufficient quantities to terminate the lines and trunks and to afford sufficient traffic capacity.

SuperNode   DMS-SuperNode performs call processing and administrative functions, such as billing and traffic data collection. SuperNode consists of three major components:

▼   The DMS core is the processor. The product computing module load (PCL) program runs here to control switch operation.

■   The DMS bus is a fiber bus that connects all parts of the SuperNode together and to the network modules.

▲   The link peripheral processors (LPP) are adjunct processors employed to support packet switching, SS7 operation, and other services.

Network Modules   The network modules (NMs) furnish the talk paths between lines and trunks. The architecture is that of a four-stage time-space-time-space (TSTS) switch. NMs are traffic engineered to furnish sufficient capacity.

Evidence of the advantage of modularity is shown by Nortel Networks' replacement of the standard NMs with an enhanced network (E-Net). No other components require replacement. E-Net has high-speed fiber links to the line group controllers (LGCs) and uses circuit boards to connect to the fiber buses. The circuit boards comprise a single stage time switch that performs time slot rearrangement between the input and output links. This assures a predictable 125 microsecond delay through the switch. E-Net also allows a reduction in footprint from 32 bays to 2 bays.

DMS-100 Internet Access   Much attention has been focused recently on the loads placed on telephone switches by users dialing up modem connections to ISPs (see the sidebar earlier in this chapter). The long holding times of such connections can cause switch overloads. Users are also dissatisfied with the slow speed of Internet access at 28.8 or 33.6Kbps. Of course, 56K modems can potentially help in the user's quest for faster downloads, but they do nothing about the switch overload problem.

The ADSL has emerged as a method for Internet access that offers much higher speeds than traditional modems. Nortel has integrated the technology into the line concentrator module (LCM) using a new data-enhanced bus interface card in each LCM drawer and a new 1.3Mbps modem line card for each affected subscriber line. Once the LCM is equipped, the user needs only the Nortel Networks 1Mbps modem and standard Ethernet connectivity. No splitter is necessary at the customer location.

DMS-100 Typical Applications   Nortel Networks is evolving the DMS-100 to a SONET and broadband-capable switch. The SuperNode includes the network modules, line group controllers, and line concentrator modules. Additional constituent parts are the DMS transport node and DMS access node.

▼ The transport node is an OC-48 level SONET-compatible node with integral digital cross-connect and add/drop capability.

▲ The access node is a remote fiber terminal that supports digital subscriber carrier systems and RSCs.

DMS-100 Evolution   Nortel Networks has also designed an evolutionary path for DMS switches to support packet networking using ATM. The Succession Network solution is a series of modules that support ATM networking and interface with the circuit-switched public telephone network.

The DMS-100 is evolving from a traditional architecture consisting of the SuperNode, E-Net, and peripheral modules; the switch can evolve to a distributed structure with ATM switches at the core of the network. Gateways and call servers will support interfacing with circuit switches, and SS7 and will perform the call setup function.

This evolutionary plan offers a way for carriers to protect their existing investment in TDM switches while implementing support for packet networking. It is modular, which allows for scalability, and by supporting both old and new environments it reduces total cost of ownership.

Private Branch Exchange

A private branch exchange (PBX) is a privately owned switch that is typically located on a business customer's premises and is connected to the telephone network by T-1 carrier lines. Modern PBXs are digital switches that bear a striking resemblance to their ancestors, the digital switches used in the public telephone network. PBXs, however, typically employ software that allows them to offer business customers a variety of useful features, such as four- or five-digit dialing, conference calling, and automatic completion of a call to a busy telephone once it becomes idle. Two of the primary vendors are, again, Lucent Technologies and Nortel Networks.

The Definity Enterprise Communications Server is Lucent Technologies' PBX offering. It is a digital switch that supports from 20 to 25,000 stations. It can function as a telephone feature server on a LAN and provide an access link to more bandwidth for voice and data traffic being routed to ATM networks. It can also provide call center features such as automatic call distribution and computer/telephone integration. Lucent also has other, smaller PBXs such as the Merlin.

Nortel Networks offers the Meridian SuperNode digital PBX, capable of serving 4000 to 36,000 stations. The maximum size does not include any ISDN lines—if 25 percent of the lines are ISDN, the maximum line size is 27,000. Like the Lucent offering, it hosts a full range of business features.

It should be noted that most of the modern PBXs offer a gateway to connect to voice over IP (VoIP) networks. In fact, some vendors offer VoIP PBXs with gateways to the PSTN! The details of this type of system are beyond the scope of this book. For anyone interested in delving into the world of VoIP, a good starting point is to examine Cisco's AVVID platform.

Digital Cross-Connect System

Between wire centers, extensive facility networks carry trunk circuits between switches as well as private line circuits used by business. With digital transmission systems such as T-1, it used to be typical to terminate a transmission line at a channel bank, demultiplex the signal, and bring a separate wire pair for each channel to a distributing frame. There it could be cross-connected to a switch appearance or, in the case of private lines, connected to another channel bank to go out of the wire center.

One problem with private lines is that they encounter considerable "churn." That is, a business might change its private line topology on a monthly, weekly, or even daily basis. Making these changes used to require a service order to change the cross-connects, which had to be manually changed on the distributing frame. This method of rearrangement was labor-intensive, time-consuming, costly, and error-prone.

These problems were addressed with the introduction of the DCS, which is similar in form to the central-office voice switch; however, it is simpler as it does not have to handle all of the voice features or the volume of requests. Rather, it is used to reroute digital circuits either due to a change in requirements or in the case of a network failure.

The DCS was developed to avoid churn in the physical cross-connect field on a distributing frame. It employs a TSI (introduced earlier in this chapter), a basic architectural component of digital switches. While switches use the TSI to switch every call, the DCS uses it to make the less frequent changes required in private line configurations.

Digital transmission facilities may be terminated on a DCS, which then uses a mapping arrangement to decide how to rearrange incoming time slots onto outgoing facilities. Control of the mapping arrangement is provided by a maintenance terminal. In some cases, customers have been given control so they can make rearrangements as required without the need for service orders or telephone company intervention.

The details of how this occurs are not important at this point. What is important for you to know is that the DCS affords a way of reconfiguring lines without demultiplexing the signal, converting it to analog, and rearranging physical cross-connects.

The reconfiguration performed by a DCS includes "groom and fill" and *hubbing* functions. *Grooming* is the selection and removal of channels from a multiplexed digital signal for rerouting to another facility. The DCS usually segregates circuit-switched traffic from channel-switched, or "nailed-up," lines. The former are typically connected to a central-office switch for dial-up routing. Special services such as leased lines, now off-loaded from the switch, are then combined (*concentrated*) to fill up a minimum number of long-haul trunks to the remote digital facility. Similarly, a customer might funnel several different services through a premises-based DCS to one or more remote sites over a single transmission line.

While the word *concentrated* was used in the preceding paragraph to refer to bandwidth efficiency (hence, lower costs) along a point-to-point path, hubbing also increases economic efficiency by reducing the number or lengths of transmission lines and by reducing equipment costs. By homing in on a centralized site, metropolitan area traffic could be concentrated onto a single DCS for long-distance transport to remote campuses or offices. Alternatively, traffic could be backhauled to access services available only from equipment at a particular location.

Transmission costs are reduced over a full-mesh configuration, but reliability could be compromised in a simply hubbed network. A more survivable architecture would likely implement a ring structure or another route-redundancy topology. Digital cross-connects allow locally hubbed nodes to be connected into economical, reliable, and reconfigurable networks.

SUMMARY

In this chapter, we looked at the basics of traffic engineering and the components of switches that are being engineered. We covered the history of PSTN switches from manual switchboards to fully digital switches capable of handling hundreds of thousands of subscribers. We finished the chapter with a brief look at private switches and DCS systems that are used today. In the following chapters we are going to look at the signaling systems that are in use in and between these PSTN switches.

CHAPTER 12

Common Channel Signaling System Seven

Interoffice call *signaling*—the conveyance of destination address information—has evolved from human operators and melodic tones to a message-oriented system known as *Signaling System 7* (SS7). Originally, these signals carried only the identification of the destination telephone. Today's signals can carry just about any type of information pertaining to a call, the users, or the instruments used in placing the call.

SS7 is an out-of-band signaling system that utilizes an overlay network of interconnecting telephone switches (see Figure 12-1). The layered structure of the signaling messages supports call setup and administrative functions. These attributes provide a robust signaling system with the flexibility to meet the demands of current telephone systems as well as most new services that are being or will be introduced. In this chapter, we will briefly discuss the evolution and structure of the SS7 network, how it enables the intelligent network, and what advantages this signaling provides the end user.

Since the introduction of the telephone, signaling and its partner, *supervision* (the conveyance of local terminal information), have undergone many revisions. Regardless of the telephony technology and the type of the telephony device, signaling and supervision are required in all on-demand scenarios.

In plain-old telephone systems (POTS), supervision was used to determine when a device was busy (off hook) or idle (on hook) and signaling was used to pass information such as dialed digits, busy tones, and ringing between the telephone devices. Together, signaling and supervision allowed telephony devices to establish and release connections.

Signaling and supervision are found in two instances during a telephone call. The first instance is between the telephones and the originating and destination telephone switches; this is the signaling and supervision familiar to those who use telephones. The second instance is the signaling and supervision that takes place between telephone switches. This behind-the-scenes, interoffice signaling and supervision is the subject of this chapter.

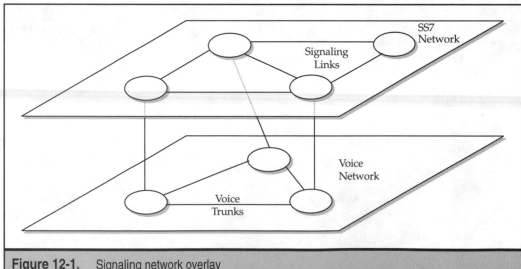

Figure 12-1.   Signaling network overlay

Signaling in today's telephone network can be found between all of the elements of the telephone system. Signaling is found between the subscriber and the network, within an interoffice domain, and between subscribers. The last of the subscriber-to-subscriber signaling requires end devices to communicate over the telephone network. Today, the most common subscriber-to-subscriber signaling method is the use of keypad tones between a user and a service provider. The main purpose of the subscriber-to-network signaling systems is to request network services for users and to deliver services from the network. Both subscriber-to-subscriber signaling and subscriber-to-network signaling systems are not defined further in this chapter. The last signaling use, interoffice signaling, is responsible for the delivery of services, the location of subscribers, and the operational support systems of the network.

The final signaling domain exists between subscribers of the telephone network. This final signaling relationship, called the *intelligent network*, encompasses all the elements mentioned in the previous paragraphs, plus a few others. The intelligent network uses the interoffice and subscriber-network signaling systems to provide an expanding set of advanced call features for fixed and mobile subscribers.

A HISTORY OF SIGNALING

Interoffice signaling has undergone radical transformations since the introduction of direct-distance dialing-capable switches. One of the early international standards of analog interoffice signaling is the Consultative Committee on International Telephony and Telegraphy (CCITT) Signaling System No. 1 signaling system. This system, introduced in 1934, uses a 20Hz ring tone and a 500Hz ring-back tone. The R1 regional system was a manual international supervisory system only; the international attendants provided all the signaling information necessary for call completion. This system was transformed through dial-pulse systems (CCITT Signaling System Nos. 2–4) and tone-signaling systems (CCITT Signaling System No. 5 and the regional signaling systems R1 and R2). The latest versions of these systems, still in use today, not only carry call signaling information but also a limited amount of billing and network management information.

NOTE: One of the problems associated with tone signaling systems is fraud. An end user has access to the signaling bandwidth (the voice channel). If a user sounds like a switch, the distant end has no way of knowing whether the call is legal or not.

In 1976, AT&T introduced a radically new telephone signaling system, based on CCITT Signaling System No. 6, called Common Channel Interoffice Signaling (CCIS). This new system used messages, as in the X.25 network (see Chapter 16), rather than tones, to establish and terminate telephone calls. This standard defined the use of a variable length, multi-layer message to establish a virtual circuit through a packet-switched network. The call request packet contained mandatory fields and a list of optional elements that depended on the type of service being requested by the users. This structure allowed the provisioning of a broad range of packet options, all accessible from a single interface standard. The use of a digital message rather than a set of tones to convey signaling and supervision information decreased the amount of time required for call setup and increased the information content of the signaling sequence.

Another change was the separation of the signaling information from the voice traffic. The signaling was performed on a signaling channel common to all voice channels. This common channel signaling (CCS) system gave the telephone industry the speed and flexibility of the X.25 network and a cost savings over a per-trunk registry type signaling. A single CCS processor could perform the signaling for all the voice channels on a given switch.

CCITT Signaling System No. 6 (CCS6) offered another capability, that of nonassociated signaling. Because the signaling was being performed processor to processor, rather than switch terminal to switch terminal, the signaling messages no longer had to follow the voice call. A call setup message could travel from the originating processor through an intermediate processor and finally terminate at the destination processor. Conversely, the voice traffic would pass through the local exchange and one or more tandem exchanges, finally terminating at the destination exchange. In this version, the only common points between the signaling and the voice call are the originating office and the terminating office. For the rest of the trip, the voice traffic and the signaling traffic are not associated with one another.

A final advantage of the CCS system was its ability to carry more than just call signaling and supervisory information. The CCS also carried billing and calling card information, which allowed carriers to automate calling card, third-party billed, and collect calls. Finally the CCS system eliminated toll fraud from the so-called "Blue Box Bandits." By taking the signaling out of the voice bandwidth and creating digital messages rather than tone sets, toll fraud became difficult, if not impossible. The structure and capabilities of the CCITT Signaling System No. 6 system provided the cornerstones for the next signaling system, CCITT Signaling System No. 7 (a.k.a. CCS7, or SS7).

SS7 DESIGN AND IMPLEMENTATION

When the CCITT study group was tasked to create a new CCS system, the basic flaws of the predecessor CCS network were a starting point. The CCS6 network had a number of limiting factors:

▼ The message structure was monolithic, which did not allow for easy addition of new features and capabilities.

■ The transport structure was based on analog technology, and the 2400 and later 4800bps signaling rates were too slow for growing call volumes.

■ The service set was limited to call processing and database lookup; this limited the capabilities of the network.

▲ The network was designed to handle voice traffic only. Non-voice services could not be offered on these networks.

These limitations set the path for the following design goals of the CCITT SS7 study group:

▼   Any new system must be built on a modular, layered protocol.

■   Any new system will be based on a digital technology, preferably, the 64Kbps channel.

■   Any new system will provide a full set of network services that can support a full range of network capabilities.

▲   Any new network must be capable of supporting both voice and non-voice traffic, as well as handling new traffic sources as they are introduced.

Technological advancements also added to the design of the new signaling system. High-speed computers allowed the distribution of the intelligence throughout the network and packet-switching technology allowed the creation of regional and national packet-switched networks with high degrees of reliability and redundancy.

SS7 Components

The SS7 network is a packet-switched network and as such is composed of users, switches, and links connecting switches. The users and the switches are considered signaling *points*, and the links are considered signaling links. Each of these SS7 elements is defined in the following discussion.

Signaling points are defined as the SS7 user and switching nodes. The functions of these nodes determine the actual *designators* (see Figure 12-2) of a call. SS7 switches are

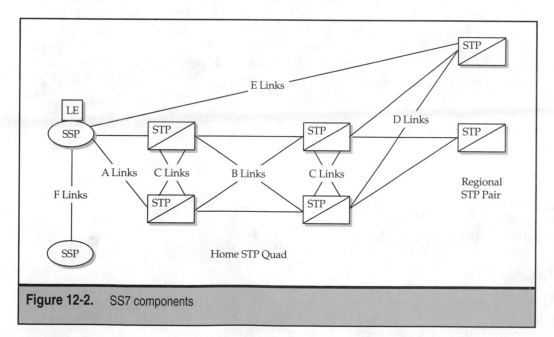

Figure 12-2.   SS7 components

called *signal transfer points* (STPs), SS7 call origination points are called *service switching points* (SSPs), and SS7 databases are called *service control points* (SCPs). Each signal point has a defined role in the SS7 network.

STPs are the heart of the SS7 network. These ultra-reliable packet switches are deployed in mated pairs. Each STP load shares traffic with its mate and can carry the full load of traffic in the event of a failure. STPs perform three functions:

▼ The routing of messages between other signal points.

■ **Global title translation** The STPs maintain a list of database subsystem identifiers and their point codes. When a SSP receives a called number not locally translatable (800/888 calls), the number is sent to the STP for a global translation lookup. The STP performs the translation and routes the message to the destination.

▲ **Gateway screening** When employed, the STP can act as a firewall for messages not originating from the home network by blocking or allowing messages to pass between adjacent networks. STPs are deployed in one of four roles within the SS7 network: a home STP is connected to switching points of a single network only; a regional STP connects local SS7 networks; an International STP connects national networks; and the Gateway STP provides translation between SS7 implementations.

SCPs provide access to the databases maintained for the SS7 network (see Figure 12-3). Like the STPs, SCPs are typically deployed in redundant pairs, each maintaining a mirror image of the other. When 800-number translations must be performed, the SCP accepts the requests, passes the queries to the databases, and formats the responses. SCPs are typically part of the SS7 network, but they can be privately owned entities. Some examples of the databases that an SCP may house are the line information database (LIDB) that contains calling card information, the 800/888 database, and the call feature database that contains subscriber profiles, such as who is paying for call waiting or caller ID.

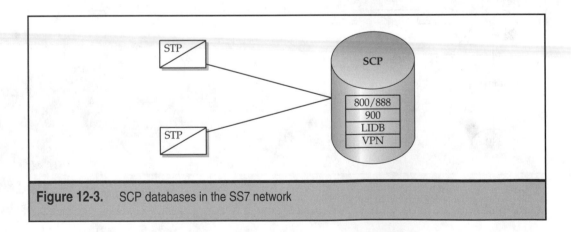

Figure 12-3. SCP databases in the SS7 network

SSPs are the entry points for SS7 traffic. SSPs are associated with telephone local exchanges and access tandems. In cellular networks, the SSPs are associated with mobile switching centers. The SSPs generate two types of SS7 messages: call-related traffic and database-query traffic. For most call-related traffic, the SSPs maintain routing tables of other SSPs. For database-query messages, the SSPs rely on the global title translation capabilities of the STP. For smaller SS7 networks, the SSPs can also act as an STP. In this case, the SSP becomes a signal relay point (SRP).

SS7 Protocol Architecture

The SS7 architecture can be divided into two functional groupings: the *network service part* and the *user parts*.

The network service part is responsible for the reliable end-to-end delivery of messages between signaling points. This part offers full connection-oriented service and connectionless service, as well as a number of service options that fall between these two extremes. The network service part is further broken down into three functional areas:

▼ **Signaling data link**   Responsible for the signaling transmission link. All physical, electrical, and functional characteristics of full-duplex digital transmission links fall within the responsibility of the signaling data link. The functions of the signaling data link are defined in the standards as the first layer of the message transfer protocol (MTP). The MTP itself encompasses layers 1, 2, and 3 of the OSI Model and is used to route higher layer messages responsible for the actual setup and release of a call.

■ **Signaling link control functions**   Provides the functions and protocols necessary for the reliable transfer of user part information over a signaling data link. The signaling link control functions include error recovery and flow control. The second layer of the MTP carries out the functions of the signaling link control area.

▲ **Common transfer functions**   Provides the means to pass user part information over a SS7 network under normal and erroneous conditions. The common transfer functions rely on the signaling link control functions to guarantee link-to-link reliability. The common transfer function provides end-to-end guarantees to the user parts when required. The functions of the common transfer area are carried out by the protocols of the third layer of the MTP and the signaling connection control part (SCCP), which acts as a layer 4 transport protocol.

The user parts are responsible for the handling of signaling point data over the network service part. As with the network service part, the user parts offer a number of services, including call information transfer, database queries, and network management information, to name a few.

The user part is divided into application-specific groups. Each group supports the protocols and function necessary for its specific purpose. The telephone user part and

ISDN user part groups are responsible for the setup and release of circuit-switched connections between serving central offices. The message set and sequences are specialized for this responsibility. Other user part groups include mobile users part, transactions processing, and operations and maintenance functions. The user part function is performed by application and user part protocols.

SS7 Scenarios

The definition of the SS7 protocols and components is somewhat obtuse if their operation is not also covered. In the following paragraphs, a number of SS7 signaling scenarios are presented.

Whether the call is a POTS call or an ISDN call, the message sequence is similar (see Figure 12-4):

1. A request for service is recognized and called digits are collected.

2. The originating local exchange instructs the SSP to generate an initial address message to the destination of the call. This message contains the pertinent information about the call.

3. When the destination switch receives the initial address message, it rings the destination terminal and sends address complete message or answer/charge message back to the originating switch.

4. When the called party answers, an answer message is sent.

5. When the call is terminated, the releasing party's SSP sends a release message that is acknowledged by a release complete message.

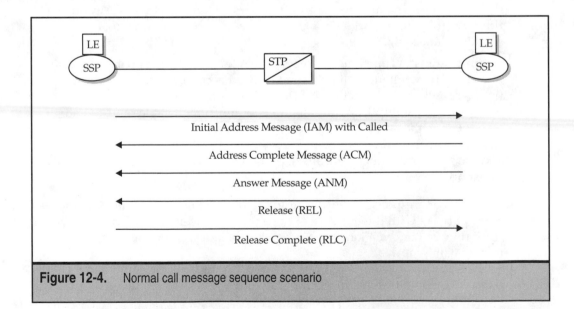

Figure 12-4.   Normal call message sequence scenario

The next scenario illustrates a calling card call:

1. The originating station dials a 0+ call and enters a calling card number at the prompt.

2. The SSP generates a query for the calling card authorization. The query contains the calling card number and a global title translation request.

3. The STP performs the translation and routes the request to the proper SCP and subsystem.

4. The SCP forwards the query to the database and receives the response.

5. The SCP forwards the response to the originating SSP. When the SSP receives the response, if the call is allowed, it is processed as a normal 1+ dialed call.

The difference between calling card, collect, phone card, and credit card calls are the databases queried. 800/888/900 calls are all special billing arrangements or alternate billing services (ABSs) facilitated by SS7. The call scenarios for this type of call are shown in Figure 12-5.

1. The originating party dials a 1+800+ type number.

2. The originating SSP initiates a number translation query to the SS7 network. The query contains the dialed number and a global title translation request.

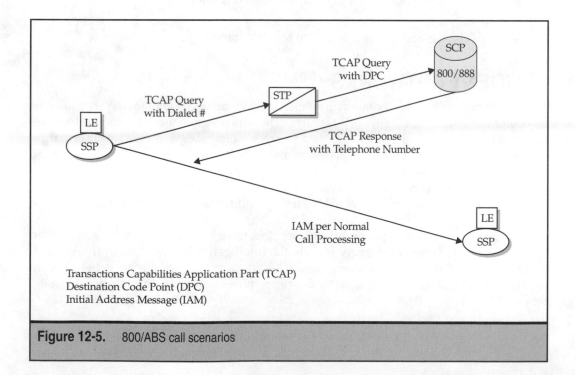

Figure 12-5.   800/ABS call scenarios

3. The receiving STP routes the request to the proper subsystem.

4. The receiving SCP passes the query to the appropriate number translation database (800/888/900) and receives a response.

5. The response is sent to the originating SSP.

6. The originating SSP initiates a call as a normal 1+ dialed scenario.

Advanced call features will become the most common type of SS7 exchange due to the introduction of distributed switching and the Advanced Intelligent Network (described in the next section). An example of an advanced feature is automatic callback—if the initial call attempt reaches a busy signal, the originating caller can have automatic callback place the call when the desired party's line is idle. Here's how automatic callback works:

1. The originating SSP sends a query to the destination SSP to verify that the feature is allowed.

2. The destination SSP sends a response acknowledging the acceptance of the feature and indicates that the called party is still busy.

3. The originating SSP sends a query requesting notification when the called party becomes idle or a timer is reached.

4. When the desired line becomes idle, the destination SSP sends a response to the originating SSP indicating that the station is idle.

5. The originating SSP initiates a call to the original station, and if the station answers, the SSP places a 1+ call to the destination.

INTELLIGENT NETWORKS

An *intelligent network* (IN) contains distributed processors that facilitate the creation and provisioning of advanced network features. These processors incorporate the logic required to execute advanced features and to access databases that contain information about a customer's specific network design and its features.

An IN is an excellent example of the growing merger between computers and communications. Computers provide advanced services to specific customers, and a communications network (such as SS7) provides access and transport capabilities between the users and the network-based processors.

The IN separates basic call processing services from enhanced services. Rather than requiring feature software in every switch, the intelligence is contained in the SS7 network and is passed to switches as required. This, of course, means that switches must be able to recognize that handling a particular call requires additional information from the service control point (SCP). Such *triggering events* must be handled in each switch.

The IN is an architectural concept for the creation and deployment of network services. These services can be characterized as distributed, portable, modular, standardized, and partially under user control.

▼ **Distributed**   The services are removed from direct control of the serving local exchanges and centered around a series of distributed databases and their associated processors.

■ **Portable**   Services created in this architecture are portable between vendor's equipment, operating systems, and lower layer protocols.

■ **Modular**   The service architecture is constructed on a building-block approach, where components are reusable between service implementations.

■ **Standardized**   The service components, the interfacing protocols, and the architectural elements are all standards based.

▲ **Partial User Control**   Services are based on the concept that the user can have direct control of the service and the creation of the service.

The evolution of the IN and its successor, the Advanced Intelligent Network, stem from the operation of the first telephone switchboards. In that era's service definition, control and implementation were based on a user's requests. When an operator serviced a call, she performed whatever service was necessary for the satisfactory completion of the call. The introduction of computer-controlled switches allowed for a better level of service. The switch itself responded to the user requests for service and through intelligent programming could meet the changing needs of the user.

Examples of IN Services

When the complexity and demands of today's local exchange network are analyzed, the concepts of the IN become a requirement rather than an option. The network requirements to support personal communications services, multimedia, and mobile computing must be provided by the local provider or customers will take their business elsewhere. IN makes possible the provisioning and deployment of new services to meet these requirements.

As services are required in a pre-IN network, the service provider would have to purchase software feature packages for each switch offering the service. For unique services, a service provider would have to contract with a third-party vendor to attach an adjunct processor to the switches. Both of these options require large amounts of money and lead times, often measured in years.

In an IN environment, as requirements are received from the customer, they can be entered into a service management system that is responsible for service creation at the SCP. The telephone switches will be programmed to act on customer-defined needs and will request call processing instructions from the SCPs via the SS7 network. This feature implementation mechanism can reduce lead time for customer services from years to days. It is foreseeable that as this service creation environment matures, customers will be responsible for the creation and management of their own services using the switching and transport provided by the exchange company. Using the services provided from a central network facility (SCP), service interactions, partitioning, and billing will be controllable.

The following list describes some of the more common offerings of the IN:

▼ **Virtual private networks** By using custom dial plans and closed user groups, a private network can span multiple service providers and geographic boundaries.

■ **Alternate billing service** Allows the use of phone cards, credit cards, third-party billing, and other forms of call charging.

▲ **800-based calling** The array of alternative dialing plans and 800/888/877/900/700/500/976 all are possible because of the IN call model.

Other IN applications include wireless network integration, local number portability, and enhanced messaging.

Virtual Private Networks

Consider the case of a multistate corporate customer that leases facilities from a telephone company to build a private network. This involves leasing part of a telephone switch and having the telephone company install special software. It also requires that the corporation lease trunks between switches, which are accessed on a switched basis by the corporate customer and are unavailable to normal callers.

The IN allows the construction of a similar service using public facilities at a greatly reduced cost to the corporate customer. Such a service is called a *virtual private network* (VPN) because it gives the customer the *illusion* of a private network, including a private dialing plan, but eliminates the need to lease facilities.

Building the VPN network involves two essential steps. First, the local switches must be configured to know that calls originating on certain lines require the switch to access a database in the IN to obtain call-handling instructions. Second, the database must be built to identify the services available and the corresponding information needed by switches to process calls. For example, if the user dials a seven-digit private network number, the database query on this number returns the full number that would ordinarily be dialed, and it uses this complete number for routing.

Information in the database can also be used to route calls differently based on time of day or class of service for the individual line. This service exists today. AT&T calls its service Software Defined Network. MCI calls its service Vnet. While the implementations differ, the idea is the same. Corporate customers that used extensive private networks in the 1980s have been attracted back to the public network by substantial cost savings using VPNs.

Alternate Billing Service

Alternate billing service (ABS) refers to the billing of calls to a number other than the originating telephone number—this includes credit card calls as well as collect and third-party billing. The service provider is triggered to an ABS condition by either the service profile of the telephone or a dialing sequence. The request for ABS can be triggered by a dialing sequence of 0+ (for example, 10-10-220 and other such memorable digit sequences). This dialing plan indicates to the telephone switch that some sort of alternate billing option is to be used.

The requests for ABS are forwarded, along with the call information, the IN service control point (SCP). Once accepted, the SCP validates the call information and starts a billing record or requests additional information, such as a credit card number. When the call is terminated, the billing information is completed and the paying customer's account is charged. This type of service has given rise to alternative calling plans, calling cards, and the ever present television ads for $1.00 calling and 10 cents a minute, no contracts.

800-Based Calling

The 800-based calling services are based on the concept that the number called is not the number to which the call is connected. The calling party number and the dialed number are sent to a database (SCP) that performs a lookup of the actually routed telephone number. This number is then forwarded to the originating serving office and the call is placed as normal. The request, lookup, and response are all supported by the IN.

As an example of how an intelligent networking affords new, enhanced services for business, let us consider the hypothetical National Pizza Company. Businesses know the value of giving customers simplified access to their products and services, and National is no exception. They know that many businesses offer single, nationwide 800 numbers with an easily remembered mnemonic to allow customers to place telephone orders or access customer service. The problem is that if National Pizza advertises a single number, they face the potential burden of having a service center answer the call, take the order, and then pass the order to the store closest to the customer. The solution to the problem is to use the intelligent network (IN). To use an IN, it is necessary to establish a database that associates a particular store with the telephone numbers close to that store. This is simple enough, because not all store telephone numbers need to be listed. Instead, only the area code and three digit office codes of the offices closest to each store need to be entered in the database. If the database is available to the signaling network, only the customer's telephone number needs to be forwarded to the database for lookup. This is easily done because the protocol used in SS7 automatically includes the calling and called numbers in the initial packet. The only thing left to do is to include recognition in the telephone switches so that they require call-handling instructions from the database to set up a call to a particular number. The database query returns information to the switch about how to route the call to the closest National Pizza store. Calls from subscribers not close to a National store could be routed to an announcement indicating that they are not in a served area.

The Future of the Intelligent Network

The physical entities and functional capabilities of the IN are still evolving. The next step involves the integration of existing IPs with the telephony network. The popularity of IP voice and the changes in quality of service capabilities of the Internet are paving the way to an integration of an Intranet/WAN switched network that handles both data transport and IN data traffic. This evolution is also creating a new family of intelligent customer premise equipment (CPE) devices that can act as an adjunct for IN services, a private branch exchange (PBX) for switched traffic, and a data router for Internet/intranet traffic.

As the use of wireless technologies expands, the IN will evolve to include this technology as well. The wireless intelligent network (WIN) will allow a subscriber to attach to the network and regard all affiliated cellular providers as the home area. This ushers in the next generation of personal communications, the "follow-me wherever" capability. The foresight to create an out-of-band, modular signaling system will allow the current network to continue to meet these evolving user needs and those not yet considered.

SUMMARY

This chapter covered all the relevant aspects of modern day signaling in the telecommunications network. It started with a brief description of the roles that signaling performs in the network and followed with the types of current signaling systems. One of the most modern and feature rich signaling systems, Signaling System 7, was then defined, its architecture, components, and protocols. The chapter concluded with a definition of a new use for SS7, the use of the signaling network as a service delivery platform, rather than a support platform. The components and services of this intelligent network were covered as well as applications that are being supported on the IN.

CHAPTER 13

Integrated Services Digital Network

Integrated Services Digital Network (ISDN) is a networking technology that enables the integration of voice and nonvoice services over a limited set of user-to-network interfaces. It supports both circuit-switched and packet-switched services for data and enhanced signaling services for telephony as well as data. This chapter covers the fundamental aspects of ISDN, including a short history, an explanation of terminology and ISDN components and signaling capabilities, and a look at the future of ISDN. Although ISDN is a mature technology, it still sees new deployments, and worldwide sales—though minimal in North America—continue to grow.

INTRODUCTION TO ISDN

ISDN is an extension of the digital transmission network and the out-of-band signaling network used by telephone companies worldwide to bring telephony services to users. It was designed as an evolution, rather than a revolution, of the telephone network. As an evolution, it uses existing telephone company resources when possible. The user-to-network connection is via existing local loop, T-1, or E-1 facilities. The basic building block of information transport (the 64Kbps channel) is also compatible with existing telephone company resources.

To maximize the capabilities found in the telephone network's Signaling System 7 (SS7) network, ISDN supports a similar (and compatible) message-oriented signaling scheme. This is a major advantage of ISDN.

ISDN is available in two versions: Basic Rate and Primary Rate. The Basic Rate version, called Basic Rate Interface (BRI), uses the local loop to provide physical connections between the user and the network. It supports data rates and services that were intended to fulfill the requirements of homes and small offices. The higher bandwidth Primary Rate version, called Primary Rate Interface (PRI), uses T-1 or E-1, depending on the country, and supports applications typically associated with businesses.

Why ISDN?

ISDN is a 1984 creation of the telecommunications establishment. ISDN completed the move from an all-analog phone network to an all-digital phone network. Before ISDN, phone companies used digital switches, and most interswitch transmission was digital (or being converted to digital). The only analog transmission system remaining was the connection to the customer. ISDN also offered the first glimps of the concept of *convergence*, in which all the communication requirements of a location are placed on a single facility, rather than spreading them between service providers and transmission systems.

This convergence includes the extension of the features of the SS7 network, used inside phone networks, to the customer. These features might be ISDN's greatest advantage. The signaling extension allows the customer and the network to exchange specific information about the nature of a call, without interfering with the call.

ISDN might not be a state-of-the-art technology now, but when it was developed it was the best available. Technology has continued to improve while the ISDN standard has been stable. Because ISDN is an industry standard, multiple products purchased from a variety of vendors are able to work together. Today's newer services, such as xDSL (a generic abbreviation for the many flavors of DSL, or digital subscriber line, technology) and cable modems, present strong competition to ISDN in the residential and small business markets.

ITU-T ISDN

ISDN is a network whose operation and interactions with users are in compliance with the International Telecommunication Union-Telecommunication Standardization Sector (ITU-T, formerly the CCITT) I- and Q-Series recommendations. Regional ISDN specifications from American National Standards Institute (ANSI) and European Telecommunications Standards Institute (ETSI) have followed the ITU-T recommendations. These regional specifications were developed after intial implementations of ISDN used different subsets and/or extensions leading to incompatabilities. Now that implementations follow the ANSI and/or the ETSI specifications, incompatability issues are a thing of the past. The following list of ISDN specifications is derived from the ITU-T I-Series Blue Book to show the various categories of recommendations. Q-Series recommendations that deal with signaling issues are also listed where appropriate.

- ▼ **I.211**   Defines the services attributes
- ■ **I.330**   Defines the ISDN addressing and numbering procedures
- ■ **I.430 and I.431**   Define two distinct physical ISDN interfaces
- ■ **Q.920 and Q.921 (a.k.a. I.440 and I.441, respectively)**   Define link access procedures at the user-to-network interface
- ■ **Q.931 (I.451)**   Defines the OSI layer 3 call-control procedures at the user-to-network interface
- ▲ **Q.932 and Q.933**   Provides implementation details on the use of Q.931 messages

Another series applicable to ISDN but not shown on the preceding list are the E-Series recommendations that concern the telephone network and ISDN:

- ▼ **E.163 and E.164**   Provides addressing and ISDN numbering schemes
- ▲ **E.401-E.880**   Provides quality of service and network management

What Went Wrong with ISDN?

While ISDN, in theory, is a standard, it has suffered for many years from too many *versions* of the standard. As a result, equipment manufactured by one company could not be used with a switch from another company, and features that worked on one system would not work on another. While much of this is not true in the industry today, the memory of past problems lingers. Consistent ISDN standards were slow in the making. Some say that if a *true* standard had been in place in the mid-to-late 1980s, ISDN might be far more utilized today. Clearly, early disagreement about the standard did not help deployment.

ISDN is an expensive option for Internet access when compared to the cost of cable modems and asymmetric digital subscriber line (ADSL) services, which offer high speeds at a low cost.

To analyze what went wrong with ISDN, we should focus on three aspects of the technology: cost, capacity, and complexity.

▼ **Cost**   ISDN has never been viewed as a low-cost service. Frequently, ISDN data services are sold based on usage. Because you pay more the more you use it, ISDN is not ideally suited to today's Internet-based browser society's long hold times.

■ **Capacity**   ISDN's capacity to handle data traffic is limited to a 64Kbps channel, or some combination of channels. While at one time a desirable data speed, particularly from a home system, ISDN's speed has not advanced as much as that of xDSL or cable modems. Also, traditional modems have advanced to provide near ISDN speeds without the cost or complexity of ISDN.

▲ **Complexity**   ISDN has remained a complex technology.

ISDN EXPLAINED

To understand how ISDN is deployed, provisioned, and equipped, one must understand ISDN terminology. The following list and the accompanying sections define the principals of ISDN.

▼ **Functional devices**   Types of devices with specific functions and responsibilities; their actual implementation may or may not follow the functional separation.

■ **Reference points**   A conceptual point at the conjunction of two non-overlapping functional devices. In nonstandards jargon, the reference points define the characteristics of a wire between devices and are commonly used for design and test definitions.

■ **Channels**   A designated part of the information transfer capability having specific characteristics and being provided at the User-to-Network Interface.

- **Interfaces** The common boundary between two associated systems; the interface between the terminal equipment and the network at which the access protocols apply.
- **Services** A feature provided to a user under agreement with the service provider. These services might apply to transporting data or might be value-added services.
- ▲ **Signaling** The mechanism for requesting services from the network and delivering services from the network.

ISDN Functional Devices

Figure 13-1 displays ISDN functional devices; short descriptions of these devices follow.

- ▼ **Local Exchange (LE)** An ISDN-equipped central office switching platform
- **Network Termination Type 1 (NT1)** Provides the physical interconnection to the local exchange, such as ISDN data service unit/channel service unit (DSU/CSU)
- **Network Termination Type 2 (NT2)** A service distributor (such as a PBX or multiplexer)
- **Terminal Equipment Type 1 (TE1)** ISDN-compatible equipment
- **Terminal Equipment Type 2 (TE2)** Non-ISDN equipment
- ▲ **Terminal Adapter (TA)** The interface device used to connect non-ISDN equipment (TE2) to an ISDN

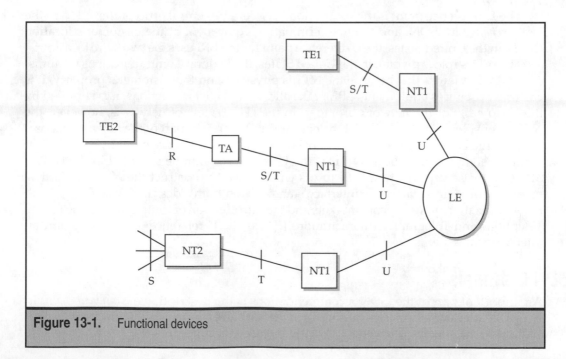

Figure 13-1. Functional devices

ISDN Reference Points

ISDN reference points (sometimes called *interfaces*) are the specifications for communication between functional devices. Figure 13-1 shows several examples of ISDN reference points. The international ISDN standards specify only the S and T reference points. The R reference point is not actually an ISDN standard; instead, it is determined by the requirement of the TE2, such as RJ-11, EIA-232, and so on. The U reference point is not specified by the ITU-T, but it has been specified by the ANSI for use in the United States and adopted the ETSI for use in the EuroISDN standard.

The R reference point allows non-ISDN equipment to be interfaced to an ISDN via a terminal adapter (TA). The TA converts existing non-ISDN protocols to ISDN protocols; it therefore must be tailored to the specific equipment. The R reference point is the traditional reference point now found in networks. This reference point allows the enormous variety of existing equipment to use an ISDN. EIA-232-E is an example of an R reference point.

The S and T reference points define the boundary between user equipment and network equipment and describe access points for ISDN services. The choice in naming a particular reference point "S" or "T" depends on ownership, presence, and form of provision of the communications equipment.

When an NT2 is present, the S reference point is the interface between the NT2 functional unit (such as PBX) and the ISDN-type end device, which can be either a TE1 or a TA. The T reference point exists between the NT2 and the NT1. When an NT2 is absent, the interface between the NT1 and the ISDN-type equipment (TE1 or a TA) can be an S, a T, or an S/T.

The U reference point is the connection between the central office, referred to as the *local exchange* in ISDN, and the line terminating equipment at the customer's location (NT1). In the United States, the U reference point for the BRI uses the two-wire local loop.

The NT1's role is pivotal. In the United States, the Federal Communications Commission's (FCC) view is that because the NT1 is physically on the customer's property, it is customer premises equipment (CPE). Like all CPE in U.S. jurisdictions, it is supplied by the customer, not the network provider. In the ITU-T view of networks, however, the NT1 is considered the endpoint of the network and therefore part of the network. Consequently, the U reference point is not considered an ISDN reference point by the ITU-T. It is extremely important, however, that the U reference point be specified in the United States so that a variety of manufacturers will be able to construct the NT1. To do this, manufacturers must know the interface standards on both sides (FCC vs. ITU-T). With the privatization of the telecom markets and the introduction of competition to the Postal, Telephone and Telegraph Administrations (PTTs), the U reference point has become an international issue.

ISDN Channels

An important part of the ISDN recommendations is the description of *standard channels*. An ISDN interface is an access line between the user and the network, with one of several standard line speeds. The capacity of the interface is time-division multiplexed into several *channels*, which are the individual bit streams derived from the interface's aggregate data rate. These channels allow several network services to be provided simultaneously over a single physical interface.

B-Channels

B-channels are digital 64Kbps circuits intended to provide access to network services. An ISDN's ability to provide a variety of services over a single access line is dependent on the B-channel's ability to carry the various voice and data services. The B-channels provide a point-to-point connection between two users or a user and a network service. B-channels do not carry signaling information and are therefore 64Kbps "clear" channels.

A rate of 64Kbps for digitized voice was state of the art when the initial work on ISDN standards began. Since then, alternative techniques allowing significantly lower speed voice encoding have been developed. ISDN CPE may support the further demultiplexing of B-channels into several lower speed derived circuits, but this is transparent to the ISDN and the B-channel.

B-channels are time-division multiplexed over the interface to the serving switch, where they may connect to a number of different subnetworks within the ISDN. These networks include a circuit-switched voice network or a packet-switched network.

B-channels provide a dedicated path from the user to the local exchange, and onward to another user. Any protocol above the Physical layer employed on this channel is an issue between the user and the end connection; the ISDN connection may be located between two users or between a user and a packet-switched network. In either case, the B-channels do not assure the compatibility or the reliability of the information transferred.

Rate Adaption   Any CPE that utilizes the ISDN B-channels must do so at 64Kbps. If the CPE does not operate at 64Kbps (and few types of CPE do), the CPE rate must be rate adapted to the 64Kbps B-channel rate. CPE may also derive multiple sub-64Kbps data channels from a B-channel; however, these multiple streams can be delivered only to a single user-to-network interface. Multiplexing and demultiplexing must be handled by the user's equipment.

Two rate adaption schemes are identified in the ISDN recommendations: V.110 and V.120. Both offer multiplexing, signaling, and control services. Terminal adapters might support one or both of these schemes. Both ends of a call connection must use the same rate adaption scheme. The V.120 scheme has won popularity in the United States. Most ISDN terminal adapters support V.120 rate adaption, whereas only about one-third of them (based on an informal study) support V.110.

D-Channels

The D-channel is a key element in ISDN operation. The user and the network use this channel to convey information to each other regarding service requests.

Carrying the signaling on the D-channel removes signaling from the other channels, freeing them for other services. In addition, the other channels (Bs and Hs) are "clear," because in-band signaling information is not required.

Because the D-channel is used for signaling, and signaling in the ISDN is packet oriented, the D-channel by design is a packet-mode-only channel. In other words, all traffic that passes between the customer's location and the local exchange on the D-channel must be in packet format. Any excess capacity on the D-channel can be used to pass user packet information in X.25 formats. This excess capacity provides the customer with another bearer channel. This excess capacity can be used for such applications as automated meter reading and point-of-sale devices. These applications can be combined with voice and circuit-switched data services or used on their own.

D-Channel Protocols   The ISDN standards describe the three levels of protocols carried by the D-channel for signaling between a user and the network. The Physical layer defines the electrical and mechanical characteristics of the ISDN interface. The Data Link layer is a bit-oriented protocol and performs the traditional Data Link layer function of ensuring that the messages from the Network layer are sent and received correctly over the physical link. The third layer, the Message layer, defines the specific messages to be used for requesting and accepting services and transferring user data. The signaling and message set are covered later in this chapter in the "ISDN Signaling" section.

H-Channels

A third type of channel, the family of H-channels, were defined because applications require data rates in excess of 64Kbps. Table 13-1 depicts the specially defined H-channels and their speeds.

NOTE: H_{10} and N x 64 are defined by ANSI but are not included in the ITU-T recommendation.

The term *wideband*, referring to a bandwidth greater than 64Kbps, is often applied to H-channels. H-channels might require rate adaption to the channel speed. For these

Channel	Bandwidth (Mbps)	Equivalent Bandwidth (B-channels)
H_0	0.384	6
H_{10}	1.472	23
H_{11}	1.536	24
H_{12}	1.920(ETSI)	30
N x 64Kbps	.128–1.536	N

Table 13-1.   H-Channel Types

channels, no standards have been adopted by the ITU-T, so each CPE vendor uses a local method for rate adaption. The end-to-end compatibility is the user's responsibility.

H-channels are used for dialable wideband applications, such as high-speed Internet access and video conferencing. For Internet access, the interfacing router uses a primary-rate interface to provide access to the Internet at speeds from 128Kbps to 1.536Mbps. For video conferencing, H-channels can be established on demand to various locations. The video and audio signals are adapted to the H-channels by digital codecs and transmitted. Channel speeds of 384Kbps offer good quality video conferencing.

N x 64 Channels   In 1991, Telcordia released a specification defining a dialable bandwidth capability over an ISDN interface. This capability is called *N x 64* channels. This channel type has been included in the U.S. ISDN specifications. The ISDN subscriber identifies the size of the channel at call initiation. Included in the call setup messages is an identification of the number *N* (2–24); this identifier indicates to the local exchange the number of time slots to be used during the requested call. The local exchange allocates this number of 64Kbps time slots between the originator and the destination.

The principal users of the N x 64 channels are devices referred to as *dialable wideband terminal adapters*. These devices can request an amount of bandwidth and demultiplex a higher rate data stream into the time slots of the N x 64 channel. A similar device at the destination is responsible for remultiplexing the incoming time slots into the original data stream. These devices, in combination with the N x 64 channels, offer a form of bandwidth on demand for data communications users.

NOTE: The difference between an N x 64 channel and an H-channel is the granularity of the channel size. H-channels are fixed in size; N x 64 channels range from 128Kbps to 1.536Mbps, in increments of 64Kbps.

ISDN Interfaces

One of the key aspects of the ISDN recommendations is the definition of a limited set of user-to-network interfaces. This small set of interface definitions allows interoperability between equipment vendors. One of the problems that ISDN faced is that not all vendors could agree on all the aspects of these interfaces. As such, multiple versions of both interfaces exist today.

The ITU-T ISDN standards define two user-to-network interfaces:

▼   **Basic Rate Interface (BRI)**   Carries three channels and competes with today's Centrex, business, and dedicated lines.

▲   **Primary Rate Interface (PRI)**   In North America, carries 24 channels and is equivalent to a T-1 carrier. The PRI provides network access to CPE requiring this number of channels (such as PBX) or the higher capacity H-channels.

These two interfaces were the first defined by the ITU-T and are now sometimes called the *narrowband ISDN interfaces*. They differ from each other at the Physical layer but are otherwise intended to use the same higher layer (signaling) protocols. Both are implemented worldwide.

Basic Rate Topology

The BRI offers two B-channels, each operating at 64Kbps, and one D-channel operating at 16Kbps. The U reference point defines the reference point between the LE and the NT1. The BRI U reference point uses a single twisted pair (two wires).

The S/T reference point (see Figure 13-1) is a four-wire passive bus. The ITU-T specification allows up to eight devices to be attached to this multipoint circuit. Some individual implementations specify that the eight devices can include a maximum of two voice terminals. The devices can be either a TE1 that communicates with ISDN protocols or a TA that converts a non-ISDN terminal (TE2) to the appropriate protocols.

Basic Rate Channel Structure   The BRI provides three time-division-multiplexed channels for full-duplex, user-to-network information exchange. The three channels are multiplexed at the S/T reference point: 1 bit from the D-channel is interleaved between each group of 8 bits of B-channel information (the multiplexing at the U reference point has a different interleaved pattern). The two B-channels operate at 64Kbps and provide either circuit, packet, or frame service. The D-channel operates at 16Kbps and is used for ISDN signaling and user packet data. Support exists for two forms of user data: the X.25 standard Packet-Layer Protocol and ISDN signaling messages.

The two B-channels can be used independently to access different network services. Each B-channel is "owned" by a user device for the length of the call (an exception to this is when packet service is used). However, a user can use CPE to multiplex multiple data streams onto a single B-channel. In addition, CPE can be used that allows the B-channels to be multiplexed together, thus providing a single 128Kbps channel.

Although the ITU-T recommendations specify that the BRI will support up to two B-channels and one D-channel, provisioning options have specified different configurations: for example, 1B+D specifies that only one of the B-channels is active on the interface; 0B+D provides low-speed data service only. Another option forbids the use of the D-channel for data; the nomenclature often used for the BRI without the use of D-channel data is *2B+S* (*S* indicates that the D-channel is only a signaling channel). In Internet digital subscriber line (IDSL) services, another flavor of ISDN is 2B+0, in which 128Kbps of bandwidth is provided without signaling capabilities.

Primary Rate Topology

The PRI is intended mainly for volume users, such as businesses that require more than a few lines. It can terminate into equipment such as a PBX or a multiplexer that can distribute the individual channels to the appropriate user. In North America, the PRI supports 23 B-channels and a single D-channel, all operating at 64Kbps. A second PRI at the same location and attached to the same equipment could use the D-channel of the first interface for signaling and employ all 24 channels for bearer services.

The U reference point in the PRI is a four-wire system that looks like a standard T-Carrier interface. Outside North America and Japan, this interface might provide a different capacity because of the way other digital carrier systems operate. In particular, the European PRI is configured over E-1 facilities with 30 B-channels and one D-channel, and it operates at 2.048Mbps.

Primary Rate Channel Structure The channel structure on the primary rate is variable. Depending on the tariff and the provisioning options, the PRI can contain B-, D-, and H-channels. In T-Carrier systems, the PRI supports 24 time slots that are assigned on an as-needed basis for the various channel types. A D-channel must be present at each interface. The exception to this is when multiple PRIs are grouped between a customer location and a common LE. In this case, a single D-channel can be used to signal for the group of PRIs. The European PRI follows the Conference of European Postal and Telecommunications Administrations (CEPT) standard for Level 1 (E-1) digital time-division multiplexing. The ETSI defines a 32-time-slot digital carrier system in which time slot 0 is used for framing and synchronization and time slot 16 is used for signaling. The other 30 time slots are used as traffic carrying channels.

Primary Rate Interface Applications PRIs are deployed for applications requiring either numerous B-channels or for those with higher bandwidth than just the two B-channels of a BRI. These applications, in addition to mainframe access, Internet access, and videoconferencing, are supported on CPE that can terminate these higher speed interfaces. Examples of such CPE follow:

▼ **Private branch exchanges (PBXs)** A PRI allows flexible calling groups and calling line identification (CLID), coupled with high-speed call setup and clearing.

■ **Front-end processors (FEP)** An FEP allows cost effective bandwidth on demand for mainframe interconnection.

■ **Inverse multiplexers** A PRI offers dialable wideband connections for multimedia conferencing and remote video.

■ **Routers and remote access servers** A PRI offers digital access, caller validation (CLID), and bandwidth on demand for remote LAN access.

▲ **Internet access** PRIs offer an Internet service provider (ISP) many benefits not found in other services. Flexible calling, analog and digital termination, CLID, and faster call setup are all offerred over the PRI. This application accounts for the largest growth of PRI in North America.

A PRI provides digital access (higher speed), packet-oriented signaling (enhanced services), and termination of digital and analog call requests. PRIs can replace traditional T-1 or E-1 service for network access.

ISDN Signaling

The user-to-network signaling defined for ISDN differs greatly from today's plain-old telephone systems (POTS) operations. ISDN uses signaling in a packet format between the user and the network. The ISDN standards for signaling, which have been structured on the Open Systems Interconnection (OSI) Model, describe three levels of protocols, as discussed in Chapter 2. Here they are in a nutshell:

▼ **Network layer protocol**   A set of user-to-network signaling packets defined by the ITU-T that are used to originate, manage, and terminate ISDN service requests.

■ **Data Link layer protocol**   Provides error-free transmission services for the Network layer. It is crucial that the packets are sent and received correctly between the user and the network.

▲ **Physical layer protocol**   Defines the electrical and mechanical characteristics of the interface, and the ISDN signaling channels.

The ITU-T named the signaling protocols used in ISDN *Digital Subscriber Signaling System No. 1* (DSS #1). DSS #1 is defined in ITU-T Recommendations Q.920 and Q.930 for the Data Link and Network OSI layers, respectively.

The signaling packets used in DSS #1 define various aspects of the ISDN call:

▼ Type of service requested (circuit, packet, or frame)

■ Channels used for transmission

■ Service characteristics (the how and why of the requested service)

▲ Call status

Various signaling schemes are used in a complete ISDN. The signaling used between the ISDN users and the serving local exchange (defined as the ISDN loop) is the associated mode, out-of-band DSS #1. These signaling packets are conveyed on a separate channel that is time-division multiplexed on the ISDN loop.

ISDN and SS7

Signaling between central offices in a complete ISDN uses SS7 (see Chapter 12). The SS7 network supports non-ISDN services as well as ISDN services. The signaling information travels on a network that is totally separate from the user data transmission networks. The signaling network can convey a limited amount of user information (such as passwords). SS7 is typically thought of as an interoffice signaling system that allows calls to be quickly established and broken down without using the bandwidth of the call channel. Another important feature of the SS7 network is that it passes information related to ISDN supplemental services between ISDN local exchanges. Without this SS7 link, the

ISDN supplemental services are usable only on a single local exchange. With SS7 support, the size and complexity of the ISDN becomes transparent to the users.

The impacted ISDN supplemental services are user-to-user signaling, closed user groups, calling line identification, direct inward dialing, and call forwarding. Each of these services require the passing of information between the originating and terminating local exchanges and possibly the originating and terminating terminals. Listed here are the supplemental services and information to be passed over the SS7 network:

▼ **User-to-user information (UUI)** Additional information (up to 128 octets) entered by one terminal to be delivered to the peer terminal. The information can be passed in the call processing messages or during the actual call. UUI data will be passed over the SS7 network as long as its inclusion does not cause the SS7 messages to exceed maximum length criteria.

■ **Closed user group (CUG)** Implemented by all users having the same interlock code. When a call is initiated between terminals, the interlock code must be passed to the destination exchange where it is compared to the code for the destination terminal.

■ **Calling line identification (CLID)** CLID presentation and restriction, as well as the CLID, must be sent over the SS7 network to allow for proper handling of the call at the destination.

■ **Direct inward dialing (DID)** Requires that the destination local exchange be informed of the number of digits and manner of sending the digits to the destination.

▲ **Call forwarding** The information to be passed during call forwarding is the original called party, the redirected telephone number, the redirecting reason, and the number of redirects (possibly including an original redirecting reason).

Due to the deployment of regional SS7 networks, most ISDNs are supported by SS7. This does not imply that full ISDN services are offered end-to-end in these networks. What is missing from these national and international ISDNs is a seamless SS7. For a truly seamless environment, each regional SS7 network must connect with adjacent SS7 networks that provide the same level of support for ISDN as found internal to the regional SS7 network.

In the United States, SS7 interconnection and interoperability was required when the FCC mandated 800 number portability. To support this capability, TCAP (transaction capabilities application part) interoperability was required. For many networks, TCAP was the only interoperability provided, so an 800 number could be verified, but an ISDN supplemental service to that 800 number would not be completed. An example is higher layer compatibility information. This data, entered by the originating terminal, is transported through the SS7 network and delivered to the destination in the outgoing Call Setup message. Without ISDN User Part (ISUP) interconnection, this information would be lost.

Sample ISDN Call

The signaling packet process that might be used to establish a voice call via an ISDN BRI follow. All the packets are carried on the D-channels at the respective interfaces.

1. The calling user initiates a call by sending a setup packet containing attributes specifying B-channel #1 for a circuit-switched voice call using μ-255 Pulse Code Modulation (PCM) encoding.

2. The originating LE asks for the called party's number (ISDN address). This occurs only if the LE can satisfy the initial request for a B-channel.

3. The calling user sends the number to the LE.

4. The LE sends an "in-progress" message back to the calling user, and it sends relevant portions of the request (attributes and number) to the destination LE. A signaling network, such as SS7, can handle the last step.

5. After the destination LE receives the request, the LE broadcasts it to *all* ISDN devices associated with the designated interface. When the call is offered, the attributes are delivered with the call setup packet. In this example, B-channel 2 is used for A-law PCM encoding (European).

6. Station 12 (presumably an ISDN telephone), determines that it can handle the request and responds with its identifier and an acknowledgment. Station 12 also alerts its user, perhaps by audible ringing. (An X.25 station would ignore the request because it doesn't "understand" A-law or circuits.)

7. The destination LE sends an "alerting" message back to the originating LE, which passes it to the calling station. When the calling station receives the alerting packet, it generates an audible ringing signal.

8. The called party picks up Station 12's headset, causing the station to send a connect message to the destination LE.

9. The destination LE confirms connection to both Station 12 and the originating local LE, which in turn informs the calling user's station.

10. The two parties carry on their conversation on the interconnected B-channels.

CURRENT STATUS OF ISDN

ISDN now competes against ADSL, cable modems and cable telephony, fixed wireless, and mobile wireless for networking customers' communications dollars. ISDN is frequently neither the least expensive nor the highest speed service available. In the presence of such competition, many customers will opt for the cheaper/faster service.

Industry focus has also changed from ISDN to the newer services. Most companies hardly advertise their ISDN offerings, while many try hard to convince customers of the benefits of new service offerings such as cable or DSL. Even so, ISDN PRI service still plays an important part in many customers' communications plans. PRI offers advantages currently unavailable in any other offering and thus has a niche market. In addition, BRI service still provides higher speed access than a modem. Customers who do not have access to either fixed wireless, cable modems, or ADSL service will still find ISDN BRI desirable.

BRI also plays a role in voice communications. Many customers use Centrex service, and ISDN BRI provides a reasonable growth path. BRI allows for attractive features such as caller name and number display and for multiple call appearances at a phone (useful for call pickup). This area must be watched, as IP telephony is showing strong growth in the business market and could be competitive with ISDN BRI for the delivery of basic voice service in an office setting. The advantage of the IP-based solution might be using the LAN to deliver voice traffic, thus eliminating another interface (the twisted pair for today's standard analog or ISDN BRI).

SUMMARY

ISDN was a breakthrough concept for the integration of multiple services offered via a single interface. While ISDN has not gained the penetration levels once expected, it is still an important technology. BRI is still in common usage for Internet access and video conferencing, yet there is market erosion where ADSL and/or cable modem service is offered. PRI is still extremely important. In the voice world, the signaling capabilities are widely used for PBX connections. In the data world, PRI remains the interface of choice for terminating dial-up connections for ISPs.

CHAPTER 14

Wireless Telephony

T his chapter provides an overview of wireless telephony. It begins with a brief discussion of the discovery and use of radio waves in communication, as well as their inherent problems. The chapter also describes the principles of cellular telephony, the basic structure of the cellular network, and the concepts of handoffs, signaling, and roaming. Multiplexing techniques and the Global System for Mobile Communications are also covered. Finally, the chapter introduces third-generation wireless systems.

RADIO WAVES

We use air for breathing and for communicating. In fact, the spoken word carried as a pressure wave by the air is the most basic form of wireless communication. The process begins at the speaker's vocal cords and ends at the listener's eardrum. Unfortunately, the pressure wave attenuation makes this form of communication impractical for long-distance communication.

In 1876, Alexander Graham Bell solved the long-distance communication problem when he invented the telephone. In the telephone, the voice pressure wave is converted to an electrical signal that is carried over a twisted pair of copper wires. When the signal attenuates to a certain level, it is amplified and then continues to its destination.

In 1888, Heinrich Rudolf Hertz transmitted and received "invisible waves of force." His experiment transmitted an interrupted carrier wave (approximately 100MHz) from a spark transmitter and detected it by observing the spark produced across a small gap in a single-turn coil that was dimensioned to tune to the transmitted frequency. These waves were transmitted through the air, a process now often referred to as a "spark-gap radio." In honor of Hertz's discovery, the unit of frequency was changed from cycles per second to Hertz in the 1960s.

In 1896, Guglielmo Marconi managed to send electromagnetic waves over several kilometers. The information was telegraphy signals (Morse code); Marconi called the procedure *radio*. In 1901, Marconi transmitted the first transatlantic message.

In 1905, Reginald Fessenden used radio for speech and music. People tuning into the various Morse code broadcasts were surprised to hear Fessenden's voice describing the outcomes of the New York yacht races. In 1921, the first commercial radio broadcasts occurred in the United States.

People soon realized that if the use of radio frequencies were not regulated, chaos would result. The Radio Act of 1912 required anyone setting up a radio transmitter to register with the Department of Commerce, but it did not establish any regulation toward the use of the spectrum. The Radio Act of 1927 and the Communications Act of 1934 established the Federal Communications Commission (FCC) and provided broad regulatory power for the U.S. government to control the use of the spectrum in the United States. The National Telecommunications and Information Administration (NTIA) administers the federal government's use of the spectrum while the FCC administers all other uses.

Today most of the easily usable spectrum has been allocated for the United States. In the 1980s the FCC reallocated the spectrum for UHF television channels 70–83 for mobile services. The original spectrum for cellular telephony was licensed by hearing procedures. Today licenses are granted via a competitive bidding process.

The spectrum from 1850 to 1990MHz was originally assigned to privately owned, fixed, point-to-point microwave systems. The FCC determined that these users could be transferred to other frequency bands and designated the band from 1850 to 1990 for use by emerging technologies. Other emerging technology bands are 2110–2150MHz and 2160–2200MHz. Personal communications services (PCS) licenses are distributed via an auction. If there is an incumbent microwave user in the area, the PCS licensee must negotiate arrangements to relocate the incumbent to another spectrum band or to a different technology.

No matter which frequencies are used, *modulation* is the foundation of radio communication. When a radio announcer reads into a microphone, a diaphragm inside the microphone vibrates in response to the changing air pressure caused by the announcer's voice. The vibrating diaphragm causes a weak electrical signal; the amplitude of the signal corresponds to the amplitude of the sound wave. This weak electrical signal is amplified and sent to a modulator. The modulator combines a carrier wave—a high frequency sine wave—and the voice signal by altering either the amplitude, frequency, or phase of the carrier signal.

When the carrier signal has been modified by the voice signal in the modulator, it is sent to a transmitter, where its power level is set and applied to an antenna. Electrons vibrate up and down the antenna in relation to the signal. Electrons moving in an antenna create an electromagnetic field around that antenna, and that electromagnetic field radiates through space at the speed of light.

The electromagnetic field hits a receiving antenna and generates an electric current there. The current is proportional to the fluctuation in the electromagnetic field. The receiver takes this very weak current, amplifies it, removes the carrier signal portion, and extracts the voice signal via demodulation. The demodulated signal is amplified according to the volume knob on your radio and sent down the speaker wires, which causes the speaker cones to vibrate in proportion to the signal. The vibrating speaker cones cause the air pressure to change, which generates a sound wave moving at the speed of sound to your ear.

While radio waves are extremely useful, several things can happen after their transmission:

▼ **Freespace loss**   The power of the signal diminishes as the distance from the transmitter increases, in accordance with the inverse square law. For example, a signal of 100 watts at one mile will be 25 watts at 2 miles (i.e., one fourth the power).

■ **Blockage/attenuation**   The signal can be blocked or absorbed by some feature of the environment. The amount of signal loss is often a function of the signal frequency.

■ **Multipath propagation**   Radio waves can be reflected by the environment and thus create multiple paths from transmitter to receiver. With multiple paths, one has to deal with problems such as delay spread. Delay spread occurs because each of the paths can be of a different length, and therefore, the signals will arrive at the receiver at different times. Once again the amount of delay depends on the radio signal frequencies.

- ■ **Rayleigh fading** Radio waves are attenuated as they pass through various components of their environment. Each component (e.g., leaves, stones, concrete, wood) has its own unique fade characteristics.
- ▲ **Doppler shift** The frequency of radio waves can be shifted due to the relative motion between the transmitter and receiver. Police radar uses this property to detect speeders.

Mobile Radio

Radio-based communications may be fixed or mobile. Each has its own set of issues and problems that must be addressed. Our focus is on the unique set of problems associated with mobile radio, but we will address some fixed-radio systems as required.

In general, fixed wireless is characterized by directional antennas. The most popular type of directional antenna is a parabolic dish, which focuses the radio energy in a single direction. Mobile systems typically use omni-directional antennas. However, recent developments in smart antenna technology are starting to blur this distinction.

CELLULAR SERVICE

It can be said that cellular was born in the 1940s, planned in the 1960s, and marketed in the 1980s. It was developed to overcome the congestion and poor service quality of mobile radio. However, it was not really a change in the basic technology because FM transmission was still the backbone technology. The cellular architecture was just a change in the system philosophy.

The essential principles of cellular radio are listed here:

- ▼ Low power transmitters and small coverage zones
- ■ Frequency reuse to increase spectrum usage efficiency
- ■ Cell splitting to increase capacity
- ▲ Handoff and central control to provide cell-to-cell mobility

Cellular architectures increased the capacity of mobile telephony systems but never solved the problems associated with the capacity of a single analog radio channel. To increase the capacity of a channel, cellular radio is rapidly converting from analog to digital.

Cellular Network Structure

When constructing their networks a decade or so ago, the cellular service providers approached the problem of covering a large territory in a different manner from that used by older mobile telephone systems. Rather than cover the area using a single transmitter, which severely limits the number of channels that can be supported, the area was divided into small broadcast areas called cells. Each cell was only a few miles across and each had its own low-power transmitter. The frequencies used in neighboring cells are different to ensure non-interference.

By reducing the coverage area, it was theoretically possible to reuse the frequencies in nonadjacent cells and thereby create more channels for simultaneous users. There are 4, 7, or 12 cell reuse patterns. A typical 7-cell pattern, with no frequencies reused in adjacent cells, is shown in Figure 14-1.

An example of a 12-channel system will illustrate the significance of frequency reuse. Imagine that all frequencies could be reused in every cell. Thus, we would have 12 channels in every cell rather than just 12 channels for the total coverage area. If there were 100 cells, each about 10 miles across, 1200 usable circuits would be found in the city instead of only 12. Problems with interference prohibit adjacent cells from using the same frequencies. Frequency ranges can be reused in nonadjacent cells to increase effective channel capacity.

Despite this approach, however, current cellular systems are saturated, or nearly so, in many areas. The creation of additional capacity has been the major impetus behind the shift from analog technology to digital technology in the cellular arena. Newer digital systems can increase capacity within a given cell between three and eight times.

Figure 14-2 represents a generic cellular phone system. At the center of the system is the set of cells that make up the coverage area for service. Each cell has a base station antenna that communicates with the mobile user using traditional techniques. The base station is connected by a wireline trunk to the mobile switching center (MSC), sometimes referred to as the mobile telephone switching office (MTSO).

The switching center is the heart and the brains of the cellular telephone system. In this office, users are authenticated, calls are completed, handoffs are set up, and the billing information is captured. This is also where the mobile system is connected to the rest of the world. There is a trunk from the MSC to a tandem or end office of the wireline provider and thus a connection via the Public Switched Telephony Network (PSTN) to the rest of the world. A connection to the Signaling System 7 (SS7) network deals with issues associated with roaming between mobile carriers.

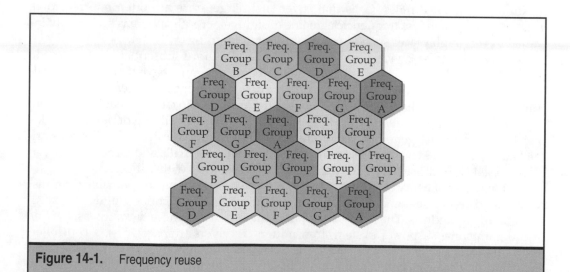

Figure 14-1.　Frequency reuse

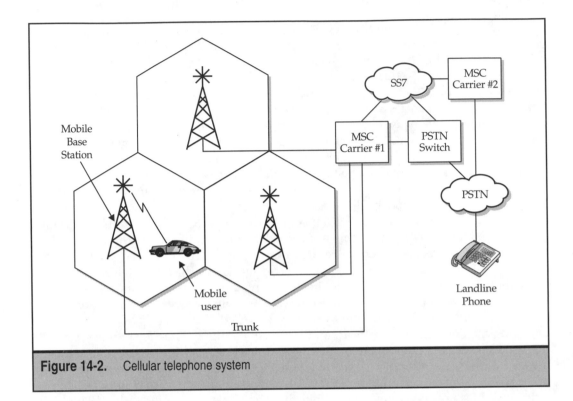

Figure 14-2. Cellular telephone system

Cell size is determined by the power the transmitter applies to an antenna. The maximum cell size radius for 800MHz cellular is about 30 miles; for 1800MHz systems, the maximum is about 6 miles. Cells that cover such large areas are often referred to as *macrocells*. Since wireless frequencies are line of sight, macrocell antennas are usually installed in very high places such as mountains or tall buildings.

Macrocells can be divided into *microcells* to handle more subscribers or to provide better coverage. When microcells handle subscribers, they must have their own controllers called *base stations*. When microcells provide better coverage to an area, they are often controlled by another base station and use the frequencies allocated to that base station.

Very small cells are called *picocells*. These tiny cells cover a building or part of a building. The antennas are generally just a few inches long, about the size of a cell phone antenna. Picocells are ideal for high traffic areas like shopping malls and airports; they can also be used in an office to replace traditional wired services with wireless.

At the center of every cell is the base station. The base station is responsible for transmitting and receiving signals over the air, assigning channels, connecting to the MSC, and initiating handoffs. The necessary equipment to perform these tasks consists of one or more antennas, a power system, transmitters/receivers (transceivers), a controller, and a communications link to the MSC. The base station equipment is usually housed in a small building or waterproof cabinet at the base of the antenna tower or pole. Some of the equipment is described in greater detail in the following list.

▼ **Transmitter**   This subsystem largely consists of a modulator and an amplifier; it takes a signal received from the MSC, modulates it onto the assigned channel frequency, and applies that modulated signal to the antenna.

■ **Receiver**   This subsystem consists of a radio receiver and a demodulator; it receives a signal from the mobile unit on an assigned channel and demodulates that signal for transmission on the base station's communication link. When the receiver for a channel is unused, it monitors signal strength received in preparation for possible handoffs.

■ **Controller**   This subsystem is the brains of the base station; it monitors the status of all the base station's equipment, setting up radio channel assignments, monitoring control channels, sending control messages out on control channels, and communicating with the MSC. The controller also converts all calls to the wireline telephony standard of 64Kbps.

▲ **Communications link**   This link between the base station and the MSC is often called the *backhaul*. The backhaul facility itself can be wireless using a point-to-point microwave link, or it can use a landline connection such as a T-1 line or a fiber-optic link.

Most base stations are connected to the local power company. If a power failure occurs, they are often equipped with a battery backup and/or a generator.

In information technology, the term *backhaul* refers to the communication link to get data onto a network backbone from an offsite location. In wireless communication systems this link is usually between the base station and the MSC, although it can also refer to the link between the MSC and the PSTN.

The backhaul facility can be either wireline or wireless. The most common wireline connection is a T-1 connection, although fiber-optic links are being used more and more. The most common wireless connection is a point-to-point microwave link, although infrared laser links are now being used by at least one PCS carrier. Infrared links do not have to be licensed by the FCC.

The MSC is the center of a wireless network. It controls all connected base stations, provides authentication, and switches users to the PSTN as needed. Additionally, the MSC maintains databases for billing, roaming customers, and home customers. For new digital systems, the MSC also provides features such as voice mail, call forwarding, caller ID, and messaging.

The MSC's switching functions are similar to those of the PSTN, and many manufacturers use the same switch for both. However, the control functions added to the MSC make it a very complex system. The MSC exchanges control information with all of the base stations, and in some cases, it controls individual calls at each base station.

The MSC contains software to authenticate callers and track usage for billing purposes. Caller authentication works in conjunction with the *home location register* (HLR), a database containing user information for each local customer. A similar database called the *visitor location register* (VLR) is maintained for roaming callers in the MSC's service area.

Telecommunications: A Beginner's Guide

MSCs communicate with each other using messages defined in Interim Standard 41 (IS-41). The interconnection medium is usually SS7 with X.25 as an alternative.

The HLR is a customer database maintained by the MSC. This database contains each customer's phone identifiers: electronic serial number and mobile identification number. The latter number is usually the customer's phone number. Additionally, the HLR contains data on the features each customer selects such as call waiting, call forwarding, and caller ID. In some cases, the HLR will contain data on where the customer was last located.

The HLR database is consulted on each call to perform caller authentication and routing. When a customer is roaming, the MSC (in the roaming area) contacts the roamer's home MSC and transfers customer data. The MSC in the roaming area stores the customer's data in the VLR.

The purpose of the VLR is to provide the same level of service to users while they are roaming. The VLR maintains a user's entries while that user is active in the roaming area so the local MSC and the user's home MSC do not have to exchange database entries each time the user makes or receives a call.

Handoffs

If all cells in a serving area were large, the problem of moving from one cell to another, or handing off, could be avoided. Unfortunately, only in rural areas are cell sizes so large that handoffs are usually unnecessary. Moreover, gaps or holes within a coverage area might require handoffs.

Although handoffs might appear obvious, they are difficult to implement. The challenge is to hand off at the correct time to a cell that has the capacity to accept an additional call. A handoff to a cell with no capacity results in a dropped call. The issue is low voice quality versus no call. Designers create extensive models of the handoff probability in a cell based on cell size, the mobile unit's speed, and call holding times. Handoff probabilities combined with dropped call probabilities lead to the system's grade of service (for example, one percent of all calls will be dropped at handoff).

The cell site processor constantly monitors the voice channel between the mobile unit and the cell site. Once the cell site processor detects that the signal is at a threshold, it requests a handoff from the MSC. Moving from one cellular system to another is only slightly more complex than an intrasystem handoff.

Signaling

Mobility is the ability to move from place to place. Thus a mobile telephone allows a user to be mobile while conversing with a distant party. A cellular system provides mobile telephony, but how much mobility is required?

Intracell mobility is available as long as the user doesn't lose the radio signal. Intrasystem service is provided by adding a handoff process to the service. But intersystem mobility is more than just a handoff and requires a set of procedures known as *roaming*.

The capabilities of roaming are fairly straightforward. First, you have to figure out how to get paid by the roamer in your system, and second, you have to locate the roamer. The first is an accounting and contracts issue; the second is an interconnection issue.

To roam from one mobile system to another requires a set of interconnection procedures that all the cellular systems agree upon. These interconnected systems must pass messages that facilitate intersystem handoff, verify roamers, deliver calls, and provide uniform services. The interconnection system used in North America and elsewhere is ANSI-41, and the interconnection medium is the SS7 network. SS7 is a fully layered architecture, conforming closely to the structure of the OSI Model, although it does not employ OSI standard protocols. SS7 is covered in Chapter 12.

The ideal interconnection system would be independent of the air interface. As a practical matter, this is virtually impossible to achieve in North America because a variety of systems—frequency division multiple access (FDMA), time division multiple access (TDMA), and code division multiple access (CDMA)—intertwine through the continent. ANSI-41 and Global System for Mobile Communications (GSM) implement roaming procedures that are independent of air interface, but the actual handoff is a function of air interface procedures.

The Telecommunications Industry Association (TIA) was formed to develop new land mobile telecommunications standards. One of the most significant of its products is Interim Standard 41 (IS-41), which was developed to support Cellular Radiotelecommunications Intersystem Operations. The importance of the standard in supporting cellular mobility is hard to overemphasize. Most recently, Revision C of IS-41 has been adopted as ANSI Standard 41 (ANSI-41).

The standard was developed to support three main components of cellular mobility. The first component is *intersystem handoff*. This refers to the handoff of a cellular subscriber from one MSC to another while a call is in progress. In such a scenario, signaling must occur over the radio link between two base stations (BS) and the mobile system (MS), as well as over the wireline link between the two MSCs that control those BSs. ANSI 41 deals with signaling between the two MSCs.

The second component is *automatic roaming*. This refers to the functions necessary for a subscriber to obtain services in a seamless fashion while roaming. The services include the ability both to originate and receive calls and the ability to use supplementary call features such as call waiting, call forwarding, and short messaging service. The ability for customers to obtain such services outside their local serving area in the same way as within their local serving area is what is meant by the term *seamless*. That is, the customer can make and receive calls and use supplementary services without having to take any special action or dial special codes. ANSI-41 defines the signaling necessary to support seamless roaming.

The third component, *operations, administration, and maintenance* (OA&M), refers to the functions necessary to oversee, record, and manage the network. Most of these functions deal with testing, blocking, and unblocking trunks in the network. ANSI-41 also defines the signaling necessary to support OA&M.

Let us look at an example of intersystem operation. In this case, a mobile station is moving out of range of one base station (BS) and needs to be handed off to another BS that is controlled by a different MSC. If the subscriber were moving to another BS homing on the same MSC, there would be no need for signaling between MSCs, and ANSI-41 would not apply. Many vendors use proprietary signaling in such cases.

Before this handoff can be made, the controlling MSC must determine which of several candidate MSCs is the appropriate one to receive the call. To make this determination,

MSC A (also known as the *anchor* MSC) will send messages to the candidate MSCs asking for reports of signal quality from the MS. This is done using SS7's Transaction Capabilities Application Part (TCAP) messaging. Once MSC A receives the signal quality reports, it determines which of the candidate MSCs is the best and then must identify an idle inter-MSC trunk to that MSC. It will then begin the signaling necessary for the handoff.

To achieve the handoff forward requires signaling between the MSCs. Once MSC A has decided that B is the MSC to which the call should be handed, it initiates the handoff by sending a facilities directive to MSC B. The contents of the directive include the identity of the reserved trunk between the MSCs. MSC B then has to identify an idle frequency pair served by the new BS. MSC B acknowledges it has initiated the handoff by returning a facilities result. This result includes information on what frequency pair the MS should use in the new cell.

At this point MSC A knows that the handoff can occur and must signal the MS. It commands the MS (through the BS) to switch to the new frequency pair in the new cell. Once it has done so and is detected in the new cell, MSC B is notified. MSC B can now complete the voice path by connecting on the inter-MSC trunk and completes the transaction by sending a mobile on channel message to MSC A. This completes the handoff.

Another handoff procedure would be used if the MS went back to a cell served by MSC A. In this case, rather than extending the link by adding another trunk between MSCs A and B, the existing trunk connection between them would be released. The MSCs must be aware that a trunk can be dropped rather than added to complete the connection.

When an MS roams to a new cellular system, it must first register and then be authenticated before being allowed service. Authentication is performed using the cellular authentication and voice encryption (CAVE) algorithm. Briefly, the algorithm uses an authentication key (A-key) that is permanently stored in the MS and is known only to the MS and the authentication center (AC). Each MS has a unique A-key. A calculation is performed by the AC using the A-key, the electronic serial number (ESN), and a random number. The same calculation performed by the MS using the same inputs should yield the same result, which is termed the *shared secret data* (SSD).

The AC (collocated with the HLR) generates a random number, executes the CAVE algorithm, and then requests that the serving MSC challenge the MS. The AC furnishes the MSC with both the random number and the calculated result. The MSC then sends only the random number to the MS (over a radio channel) and asks it to authenticate. The MS executes the CAVE algorithm and sends its calculated result to the MSC. The MSC compares the calculated result with the SSD it received from the AC to verify that they match. It then reports the result to the AC.

Roaming situations occur when an MS moves from one serving MSC to another. In this case, the HLR must be notified of the new location to which calls destined for the MS should be sent.

When an MSC detects a new MS in its serving area, it notifies the HLR with a registration notification. If the HLR has information that the MS is registered with another MSC, it sends a registration cancellation to the old MSC. The old MSC responds with a registration cancellation result message. The HLR can now complete the original transaction by sending a registration notification result message to the new MSC.

MULTIPLEXING TECHNIQUES

Spectrum can be impossible to get and is very expensive when it is available. The wireless industry uses three distinct techniques to allow multiple users to use the same spectrum:

▼ *Frequency division multiple access (FDMA)* is the simplest to conceptualize. In the U.S., the available spectrum is divided into channels with a given bandwidth of 30KHz. Each user is given one channel (two if full-duplex is used), and the receiver and transmitter are tuned to that frequency. Radio and television are allocated spectrum via this technique.

■ *Time division multiple access (TDMA)* is the process by which each user is given a time slice for his/her conversation. In digital cellular systems, information must be converted from analog (the human voice) to digital data (1s and 0s). A coder/decoder (codec) performs the analog-to-digital-to-analog conversion. The better the codec performs, the more time slices are available to be shared. If the human voice could be compressed at a rate of 5:1, for example, five time slices would be available; without compression/decompression, only one would be available. TDMA is usually used in conjunction with FDMA. Current implementations in the U.S. place three time slots on each 30KHz channel.

▲ *Code division multiple access (CDMA)* is a fundamental shift from FDMA and TDMA. Rather than divide users in time or frequency, each user gets all of the spectrum all of the time. Current CDMA implementations use a channel bandwidth of 1.25MHz compared to the 30KHz channels used in FDMA. A user's signal with a bit rate of 9.6Kbps is spread to a rate of 1.23Mbps when digital codes are applied. Multiple conversations can occur over the same channel, and the codes can then extract individual conversations at the base station. CDMA has several benefits including greater capacity, more security, and better call quality.

FDMA

As the name implies, FDMA is simply a form of frequency division multiplexing. The available bandwidth is split into a series of channels that are allocated either for control, signaling, and the like, or for voice. Channels are dedicated to a single user, except the control channel, which the users share. Conventional analog systems use this technique, where each channel is 30KHz in width.

FDMA solves the problem of channel assignment, but it does not address the problem of duplex operations. *Frequency division duplex* (FDD) solves the two-way problem by using two simplex channels that are separated from each other to reduce the interference at the end points of the circuits.

FDMA by itself works if the transmit and receive channels are separated. If only a single chunk of spectrum is available, FDMA with *time division duplex* (TDD) is used. FDMA/TDD replaces the typical separation in frequency with a separation in time.

The TDD time slots require that all the cell sites be synchronized—a property that pure FDMA systems do not require. Moreover, TDD usually does not provide an increase in spectrum efficiency from a traffic/capacity view because the guard time between TDD slots would be longer than the guard time between slots in a traditional TDMA system or the turnaround interval in an FDMA system.

TDMA

TDMA refers to the provisioning of time division multiplexing within a channel. To provide for TDMA, there has to be sufficient bandwidth to allow multiple users to share the channel. TDMA is usually combined with FDMA. Several channel sizes are defined in various cellular specifications.

The North American TDMA (NA-TDMA) standard specifies a 30KHz channel to maintain compatibility with the analog systems. The European TDMA standard (Global System for Mobile Communications, or GSM) specifies a 200KHz channel. The number of time slots per channel is also different between the two systems.

Duplexing in TDMA systems is done using frequency division duplex (FDD). In North America, the NA-TDMA channels are the same as those used for the Advanced Mobile Phone Service (AMPS) channels. In Europe, a different spectrum was initially used for the GSM channels because the analog channels were smaller than the GSM channels. The analog channels are being phased out and replaced with GSM channels.

CDMA

CDMA is based on the idea that if you know where to listen and what to listen for, you can pick an individual conversation out of a room filled with many low-level conversations (background noise). In CDMA, many users can share a single frequency range without "stomping" on each other's conversations. By giving each user a unique tool for identifying the correct conversation in the bunch, the available spectrum is much more intelligently utilized.

The current system for CDMA is the Interim Standard 95 (IS-95) that is used in North America. It uses the same spectrum as AMPS and IS-54. The channel size is 1.25MHz, allowing the spreading of 128 users (128 users at 9.6Kbps are spread to a channel of 1.2288M chips/sec). IS-95 requires two channels for duplex operation.

A Direct Comparison of FDMA, TDMA, and CDMA

Figure 14-3 compares the three multiple access techniques used in wireless communications.

In wireless communication systems, the radio link from the base station to the mobile receiver (the downlink) is separated from the uplink (mobile to base). Such a separation might be performed by the use of separate frequencies for the two links known as frequency division duplexing (FDD), or especially in instances where available frequencies are limited, through the use of time division duplexing (TDD). In the TDD case, a single band of frequencies are used with the uplink and downlink allocated certain slices of time into which they may transmit their information. In FDD systems, where the transmitter

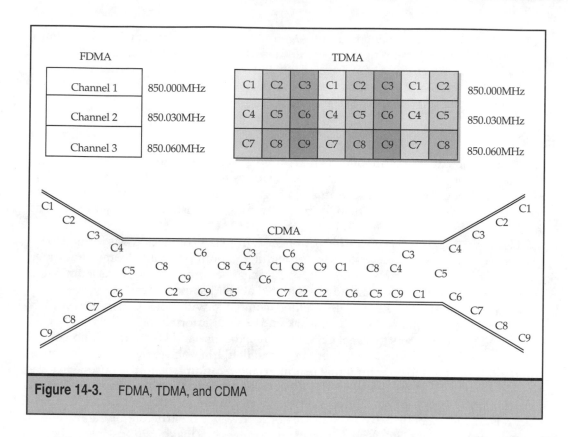

Figure 14-3.   FDMA, TDMA, and CDMA

and receiver are simultaneously active, they must be widely separated in frequency (45MHz in first-generation analog cellular) or highly selective filters must be used to protect a mobile unit's receiver from its own transmissions. When limited bandwidth is available, TDD is frequently chosen because the mobile's receiver is not operating while it is transmitting, which eliminates the need for tight filtering. By contrast, FDD eliminates the need for stringent timing requirements.

FDMA and TDMA operate in manners very similar to FDD and TDD. The primary difference is that for an FDMA system, each mobile will be allocated its own set of separated uplink/downlink frequencies. In the TDMA scheme, the mobile will be allocated its own time slice, which it performs the uplink/downlink transmissions within.

CDMA uses orthogonal (non-correlated) signaling sets to allow different users to share the available channel simultaneously. Individual users are selected by mathematical processing of the noisy waveform. By using the entire spectrum, CDMA cellular telephone systems typically achieve a spectral efficiency of up to 20 times the analog FM system while serving the same area with the same antenna system when the antenna system has three sectors per cell. In comparison to narrowband FDMA systems, the interference effects caused by other information transmissions at "next-door" frequencies are minimized in CDMA through the use of correlation processing of the transmitted signals.

Although adjacent channel and co-channel interference slightly affect CDMA transmissions, overlapping timing sequences would destroy the information in a heartbeat. In other words, since CDMA uses all frequencies, the frequency reuse efficiency is approximately 2/3 compared to 1/7 for narrowband FDMA.

GSM

The pan-European digital mobile cellular radio system, formerly known as Groupe Speciale Mobile (GSM) and now called Global System for Mobile Communications (GSM), represents the closest approximation to global roaming. Since GSM licenses have been allocated to 126 service providers in 75 countries, it is justifiable to refer to GSM as the global system for mobile communications.

It is interesting to reflect on the history of GSM. In 1982 a committee was formed to investigate the feasibility of a system for providing a common mobile communications platform for mobile communications across Europe. The committee's efforts culminated in the 1987 release of 13 recommendations aimed at GSM standardization.

The choice of TDMA as the access method, combined with other parameters such as numerous base stations and GSM network and service information delivery, allowed equipment manufacturers to target GSM areas of expertise and begin installation of the GSM network's backbone. By 1991, laboratory and field trials had shown successful operation, thus opening the door for initial commercial operation. The initial European locations to receive GSM service were the logical ones—airports and large cities. Tower construction along major pan-European roads began with an easing of tower siting restrictions in 1991 and 1992. By 1993, the major roadway infrastructure of the European continent had placed GSM servicing transmission towers, making it possible to hold a seamless (but expensive) conversation as one traveled across the continent. By 1995 coverage was extended to the rural areas.

North America made its late entry into the GSM world with a derivative called PCS 1900, or GSM 1900. In North America GSM operates much the same except with a different spectrum allocation.

The GSM uses time division multiple access (TDMA) with frequency division duplex (FDD). The channels are 200KHz wide with a data rate of 270.8Kbps. Using a 13Kbps speech encoding algorithm yields 8 slots per frame or channel. Half-rate coders yield 16 slots per frame. GSM uses frequency hopping, meaning that the mobile station can move between a transmit, receive, and monitor time slot within one TDMA frame, usually on different frequencies.

To increase the battery life, a GSM mobile phone has a discontinuous receive (DRX) operational mode—the transmitter is turned off during silent periods of a conversation. This mode allows the mobile station to synchronize its receiving times to a known paging cycle of the network. Typical power requirements can be reduced by up to 90 percent with DRX operation.

GSM has provisions for roaming and also makes use of short message service (SMS). SMS allows the GSM phone to send and receive messages of up to 160 characters (i.e., a message can be stored and forwarded when a phone is able to accept it). This service allows the GSM phone to act as a pager.

GSM is implemented at 900MHz, 1800MHz, and 1900MHz. The 1800MHz system is called Digital Cellular System 1800 (DCS 1800), and the 1900MHz North American system is called Personal Communications Services 1900 (also known as PCS 1900 or GSM 1900).

The GSM network is illustrated in Figure 14-4.

The GSM network consists of three basic parts: the mobile station or end-user device (such as mobile phone or personal digital assistant, or PDA); the base station subsystem, consisting of base station controllers (BSC) and towers; and the network subsystem, consisting mainly of the MSC, which performs call switching.

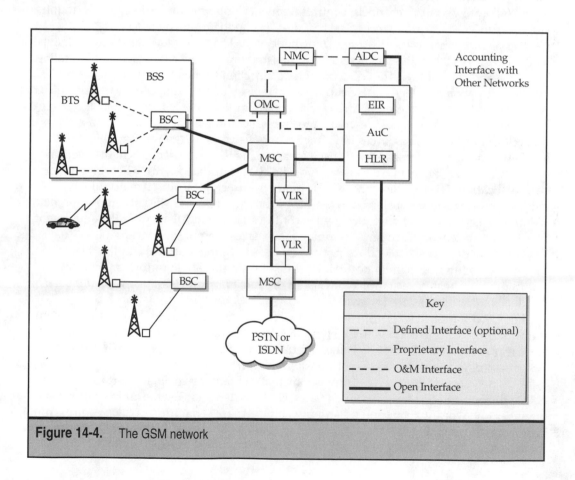

Figure 14-4.   The GSM network

The *mobile station* consists of the mobile equipment and a smart card known as the subscriber identity module (SIM). The SIM's purpose is to provide users with access to subscribed services even when they don't have a specific mobile station. By inserting the SIM card into another GSM mobile unit, the user can access subscribed services from any received calls through any GSM terminal. The SIM card makes it difficult to steal numbers or make fraudulent phone calls. It is protected with a password or personal identification number and contains an independent international mobile subscriber identity (IMSI) number, which is used to identify the user to the system.

The *base station* consists of two parts: the base transceiver station (BTS) and the BSC. The BTS defines cells and handles radio link protocols with the mobile station. The BSC handles radio channel setup, frequency hopping, and handoffs for one or more BTS. The BSC serves as the connection between the mobile station and the MSC. When the mobile user initiates a call, the mobile station searches for a local base station. The BSC handles radio resources in its region (allocates/controls channels, controls frequency hopping, undertakes local handoffs, and measures radio performance). The BSC also sets up the route to the MSC.

The *network subsystem* acts as a switching node of the PSTN or Integrated Services Digital Network (ISDN). It handles functions such as registration, authentication, location updating, handoffs, and call routing. The MSC provides connection to the PSTN or ISDN using SS7. The home location register (HLR), visitor location register (VLR), equipment identity register (EIR), and authentication center (AuC) are also part of the network subsystem.

When the mobile user initiates a call, the mobile station searches for a local base station. The BSC handles radio resources in its region and also sets up the route to the MSC. The MSC authenticates the mobile user's IMSI by accessing the user's records in the user's network HLR. When the mobile unit powers on, its HLR is updated with the mobile station's current location. The MSC uses the IMSI to query the user's HLR, and if everything is in order, this information is transferred to the VLR. From the HLR, the MSC sets the equipment identity register of the mobile equipment to control stolen, fraudulent, or faulty equipment. The TDMA system and frequency hopping make eavesdropping very difficult.

In call setup mode, GSM does not assign radio channels until the call is answered, thereby reducing the channel occupancy time of a single call. Handoff is performed by finding the six best base stations and using the most appropriate, or available, base at handoff. Dropped calls are unusual.

The GSM standard was proposed by the Conference of European Postal and Telecommunications Administrations (CEPT) in 1982. In 1989, responsibility for GSM specifications was given to the European Telecommunications Standards Institute (ETSI), and the first phase of the specification was published in 1990.

ETSI developed GSM with the philosophy that it should be an open interface and available to the public. GSM specifies a complete cellular system, not just the air interface. It is an intelligent network with an open distributed architecture separating service control from switching, thus making full use of SS7 as the signaling communications infrastructure. All interfaces are clearly defined and specified in ETSI/GSM recommendations.

3G WIRELESS

Early cellular systems (such as Nordic Mobile Telephone and Advanced Mobile Phone Service, or AMPS) used analog technologies. Frequency division multiple access (FDMA) allocated a pair of simplex channels to a user for a two-way voice conversation. Frequency modulation (FM) transmitted the signals between the user's handset and the base station. Services included voice and some low-speed circuit mode data.

In the 1990s, mobile telephony moved into the digital world to increase the capacity of wireless systems. Voice signals are digitized and mapped into a timeslot in the channel allocated for the call. This process allowed multiple calls to share the channel occupied by, perhaps, a single call in an analog system. TDMA and CDMA are the digital systems used in cellular telephony today. Services include voice and a combination of circuit-switched data and some packet-switched data.

In the twenty-first century, we are moving into third-generation (3G)—a.k.a. International Mobile Telecommunications-2000 (IMT-2000) by the ITU—digital systems that promise to deliver broadband digital services. A wideband version of the CDMA technology will likely be the air interface. Once again, we will find the venerable voice services, but now we have a definite push into the world of data services. Even more impressive is the 3G promise of multimedia services.

On the surface, 3G wireless offers the promise of mobile network technologies capable of delivering up to 2Mbps between the mobile handset and the network terminal. It is probably more impressive to look at the vision behind the promise and analyze how this vision will drive 3G systems forward.

First, the 2Mbps data rate will allow the delivery of mobile multimedia services—the actual data rate will depend on the user's mobility requirement. Second, these services will be delivered over a variety of mobile network types. So we might use something as simple as a cordless telephone network or as complex as a global broadband satellite network. Third, the system will be ubiquitous with consistent service offerings around the globe. Last, there will be harmonized operation between the different network standards. It would be ideal to have a single international standard, but that does not appear possible—at least not today. The compromise is to have harmony among the competing standards as a way to ensure interconnectivity and interoperability.

While this vision might be compelling, there are still challenges surrounding the implementation and delivery of mobile data services. Some important issues include the fact that the price per bit is higher in a mobile system than in a fixed line system, transmission quality depends on the level of interference that affects the radio signal, and contemporary handsets have limited data mobility. Perhaps the biggest danger is that the mobile data market has not proven its worth nor has it exhibited any significant growth. It might be several years until the success, or failure, of 3G systems can be determined.

The motivation to develop a third generation of wireless technology can be summarized in three key initiatives.

First, and perhaps foremost, is the need for higher bit rates for multimedia applications. First- and second-generation mobile communication technology was oriented toward voice conversation. Data communications was a second priority, if accommodated at all. However, during the development of second- and third-generation wireless technology, the size of computing devices decreased dramatically. It is now possible to carry a fully functional computer on your wrist. Consequently, we now seek transmission speeds for mobile computing devices that we expect from their wired counterparts.

Unfortunately there is currently no consensus about how to implement these higher transmission speeds, and we have a number of incompatible "standards." Therefore, a second key initiative is to harmonize, as much as possible, these different methods. Everyone agrees with this goal, but few can currently agree about how to accomplish it. Various aspects that must be harmonized include the modulation scheme, the air interface, and the spectrum.

Finally, in most cases 3G wireless technology will not be implemented in the absence of legacy (first- and second-generation) technology. Some wireless users want to keep their first- or second-generation service and do not want to change. Meanwhile, many wireless operators want to convert their existing spectrum to 3G capability. A key factor in implementing 3G wireless will be backward compatibility with second-generation systems to smooth the transition.

The ITU is developing a suite of standards known as IMT-2000 for 3G wireless systems. Part of the IMT-2000 specification includes minimum transmission speeds. The minimum transmission speed is inversely proportional to the moving speed of the wireless device.

There are three minimum transmission speed categories. For the category of high-speed movement of the mobile device, the minimum speed is 144Kbps. This category includes automobiles traveling at highway speeds and bullet trains. The second category is that of slow-moving mobile devices, such as those used by pedestrians or bicyclists. The minimum transmission speed for these devices is 384Kbps, which is deemed the minimum for acceptable quality video. Finally, stationary wireless users are to be able to connect at transmission speeds of at least 2Mbps.

An important facet of 3G wireless and the IMT-2000 implementation is the ability for current systems to either evolve into a 3G system or to migrate from its current structure to the 3G structure. As we discuss the 2G to 3G moves, note that about 40 percent—give or take 10 percent—of the current mobile users still have 1G systems. This means that the move to 3G will occur in waves. As the current 2G users move to 3G, the current 1G users will have to move to 2G and then to 3G. At least that is the hope today.

The evolution/migration strategy stems from the idea that the current wireless providers will not dump their 2G systems and run to 3G. The investments in 2G infrastructure are huge, and in many cases, these costs have not been fully recovered. Many speculate that the slow movement of U.S. providers to 3G is due to their current unrecovered investments.

The reality is that 2G systems are available, and the number of subscribers continues to grow. For many users, the technologies of 2G systems might provide all the features and benefits they will need for the foreseeable future. As a way to hedge their bets, recover current investments, and convince users that there is a better world of wireless ahead, the evolution strategy contains a step into a 2.5G system.

At 2.5G, users can get more services, higher data rates, and a glimpse of the multimedia future at a more reasonable cost. Infrastructure can be incrementally upgraded and much of the current user equipment will find a place in the 2.5G world.

At 3G, both the infastructure and the users' terminals will significantly change. Much of the infrastucture changes will be in the radio systems and the change to ATM for the backbone. From the users' perspective, the most dramatic change will be in the display technology and data-entry capability. Multimedia and high-speed data need a different user interface than that found in voice telephony.

The last, and perhaps most significant, change in the users' equipment will be increases in both processor speeds and memory capacity. 3G applications will be more sophisticated and by some estimates will require three times the processing power of a 2G terminal. Moreover, many of the user/system attributes will be downloaded (the amount of random access memory will have to grow). Finally, the digital signal processors for wideband CDMA (W-CDMA) protocols are much more sophisticated than those today.

SUMMARY

Wireless communication is increasingly becoming an expected component of everyday life throughout the U.S. and the world. As more wireless solutions become available, it becomes more important to understand the strengths and capabilities of each technology. Cellular telephony is certainly not as mature as traditional wired communications. As such, wireless is still undergoing an evolution as multiplexing and signaling standards, as well as services, are developed and accepted globally. The capabilities promised by 3G systems will revolutionize the way we work and play.

PART IV

Data Networks

CHAPTER 15

The Local Area Network

The local area network (LAN) was developed as a way to interconnect computers. Although LANs were first developed in the context of the minicomputer world, the LAN came into its own when personal computers (PCs) became the norm in most networking environments.

In the mainframe and minicomputer worlds, resource sharing is relatively straightforward: Users typically sit at a terminal—a keyboard and monitor connected to the mainframe or minicomputer—to complete their work. All users are sharing the processing power, memory, and resources of the same computer (or a small set of minicomputers).

PCs, on the other hand, are stand-alone devices. Each has a processor, its own memory, and an array of input/output (I/O) devices. Before LANs, a user at one PC could not access a printer connected to another PC directly. PC users had to copy whatever they wished to print to a disk, walk over to a computer attached to the printer, and use that system to print the document. If the hard disk on one PC was reaching saturation, a user could not access unused disk space on another computer without employing floppy disks to move files. And a user at one computer could not run an application located on another computer.

LAN networks were seen as the solution to the resource problem. Computers could be fitted with a new I/O device (the LAN adapter) and directly connected to one another. Initially, LANs were deployed to support shared printer access and to facilitate the movement of files between systems. It did not take long, however, to harness this platform to support other applications, such as e-mail.

LAN COMPONENTS

A properly functioning LAN requires the integration of many different components that determine how the devices are connected and what medium will be used, what format the data will be transmitted in, and how to ensure that multiple stations can transmit at the same time. This section details each of the major elements of the LAN and some of the options commonly available to network designers.

Media Type Considerations

The Physical layer is concerned with moving bits (1s and 0s) among *devices* in a network The *medium* is the infrastructure over which these bits are moved. When selecting the medium to be used in deploying a LAN, the issues that are most frequently considered are cost, transmission rate, and distance limitations. Distance limitations and transmission rates are factors related to the ability of the medium to preserve signal integrity as the signal propagates down the length of the cable. Numerous factors can contribute to signal degradation along a medium, including attenuation, noise, and signal spread.

Attenuation is the loss of signal power between the point of transmission and the point of reception. Like a garden hose, a medium tends to leak—hoses leak water, but media leak electrons or photons. The loss of electrons or photons along a particular medium results in a corresponding loss in signal power. We can determine (based on rating schemes that have been developed by testing various media) the best way to use a particular transmission medium. Beyond a defined length, an insufficient number of electrons or photons can be recovered and the original signal cannot be received. To further complicate matters, copper media tend to resist the flow of electrons, aggravating the signal loss problem. Finally, some electrons or photons are reflected back toward the source by imperfections in the medium.

Noise is the introduction of random, unwanted signals in a communication system. Noise can originate from a wide variety of sources. Electric motors and fluorescent lights are noted for introducing noise, and it can also originate from another nearby media; this type of noise is called *crosstalk*.

Signal spread occurs when a single signal comprises several frequencies. As the signal propagates, lower frequencies outpace higher frequencies and the signal "blurs." Over sufficient distances, adjacent signals bleed into one another, making it difficult or impossible to distinguish individual signals.

Although distance limitations and transmission rates (which are affected by noise, attenuation, and signal spread) are factors in considering an appropriate medium, cost is more often the primary consideration. It is interesting to note that the cost differential between the various media is fairly minimal; most media cost just a few pennies per foot. However, different media require different interfaces to a computer system, and it is often the cost of these interfaces that proves to be of substantial difference. For example, the optical/electronics required for connection to optical fiber media is significantly more expensive than the simple electronic circuitry required for copper-based media. The installation cost of different media can also vary widely.

NOTE: Network administrators are not free to "tweak" these parameters for a particular network environment. The administrator knows the required transmission rates, the distance requirements, and the potential for noise in the environment. Based on these restrictions, an appropriate medium can be selected.

Wired vs. Wireless

Today's networking media fall into two broad classes: *wired* and *wireless*.

Wired media are those in which the signal is guided along a physical substance—typically copper, glass, or plastic (almost any conductor can be used as a wired medium). The characteristics of the signal are defined by examining the nature of the original signal and the nature of the medium itself. The characteristics of the signal can be directly impacted by the nature of the medium.

By way of example, consider a physical phenomenon of some transmission environments known as the *skin effect*. The skin effect describes the tendency for high-frequency electronic signals to migrate out to the edges of a conductor. If a conductor is small, the circumference of the conductor, which is where high-frequency signals tend to migrate, is small. Because the resistance encountered by a signal is inversely proportionate to the size of the conductor along which it travels, high-frequency signals encounter a great deal of resistance, while lower frequency signals encounter less resistance.

Shannon and Nyquist proved that the maximum data transmission rate of a medium is coupled to the bandwidth of the medium and the signal-to-noise ratio. Therefore, if high-frequency signals cannot be used because resistance too seriously attenuates them, the maximum transmission rate is also reduced. To counter the skin effect, the diameter of the medium must be increased (the cable must be made thicker). This has the effect of increasing the available range of frequencies and/or the distance the medium can be extended. The lesson: The thicker the pipe, the faster and further it can go.

Wireless describes all transmission environments in which the signal is not guided through a specific media, not even air! Wireless is the quintessential medialess media—a true philosopher's dream. The characteristics of a given signal are governed by the nature of the transmitted signal and some environmental considerations. Unlike wired media, where a physical separation of signals and the characteristics of the media impact the ability of signals to interfere with each other, there is no physical separation in a wireless environment.

Wireless transmissions have a broad range of usable frequencies, from conventional radio frequencies (the AM and FM public radio stations) to the lower light frequencies (such as infrared).

Another wireless issue is *spectrum* allocation. Because the signal is essentially unbounded, it is necessary for users in a common geographical area to avoid using common frequencies, or they'll overlap (called *bleeding*). The range of a signal is determined by its frequency and the power with which it was initially produced. In the United States, for most of the frequency spectrum (barring infrared and a few specially allocated frequencies), the Federal Communications Commission (FCC) requires frequency licensing, thereby providing a degree of control over how the spectrum is used. Other nations have their own corresponding regulatory agencies. In the U.S., the most common wireless LAN technologies employ frequencies in one of the industrial, scientific, and medical (ISM) bands in the 2.4GHz range.

Many of the wireless LAN products on the market are designed to replace a portion of the wire in a conventional wired LAN. Others are truly wireless LANs, with only a pair of antennas between two LAN adapters.

Topologies

The type of media used within a LAN is only one of the Physical layer issues to consider. Another consideration is the fundamental organization of the medium. In a LAN, this organization can be described by the *physical topology* and the *logical topology*. The physical topology of a network literally describes where the wires go (through which walls, between which points, and so on) and their general configuration. The logical topology of a LAN describes the organization of the attached devices and the flow of signals between attached devices.

One of the defining characteristics of a LAN is its high speed. A high-speed transmission environment should have low end-to-end delay characteristics. The lowest possible end-to-end delay is achieved if two devices have a direct wire (or wireless) path to one another. If a wire directly connects devices, the end-to-end delay is simply a function of the propagation delay of the medium. This delay cannot be avoided (unless someone has the clout to repeal the laws of physics!). Propagation delay can be minimized, however, by limiting the length of the medium, and this is one of the reasons LANs must remain small.

LANs are, however, intended to connect more than two devices. They are intended to be multi-access environments (multiple devices are connected to the same LAN). The need to connect multiple devices while retaining the low delay characteristics of the LAN led to a limited number of physical and logical topologies. Each of these defines a different approach to connecting the devices while retaining end-to-end delay.

Common methods of organization for both logical and physical topologies are bus, ring, and star. Each of these is described in this section.

Physical Topologies

The *physical bus* topology describes an environment in which the cable plant is an expanse of cable, terminated at both ends, to which LAN adapters are passively attached. The cable plant usually extends through the ceilings and/or floors of a building. Every LAN adapter is attached to the cable via a connector, also known as a *medium-dependent interface* (MDI). The specific nature of the MDI differs depending on the specific medium being used. A LAN adapter transmits a signal onto the bus, where the signal propagates in both directions. The terminator at each end of the bus is a resistor that is responsible for preventing a signal from reflecting back onto the bus, where it would corrupt subsequent signals.

The use of the word *passive* describes an environment in which no attached device (such as a LAN adapter) has responsibility (at the Physical layer) for regenerating or repeating another adapter's transmission. They simply passively monitor the bus and are capable of transmitting signals onto, and reading signals from, the bus. As a consequence, the failure of any LAN adapter on a bus has no impact on the physical viability of the bus.

Because of the passive station attachment, physical bus topologies have historically been described as highly reliable environments. The argument suggested that the bus is a distributed environment and hence has no a single point of failure. The validity of this view is questionable, however. The loss of a single terminator or the presence of a single cable short can destroy the viability of the entire LAN. Not only is the cable plant a single point of failure, but it is a single point of failure that can be several hundred meters long and run through walls, ceilings, and floors! This makes it extremely difficult to troubleshoot when failures inevitably occur.

Historically, the most popular medium for a physical bus topology was coaxial cable—largely due to the early popularity of coaxial cable for deploying Ethernets. Although an amount of coaxial-based Ethernet is still deployed, physical bus topologies are disappearing as a strategy for LAN cabling to the desktop.

The definition of a *physical ring* topology is simple: devices are attached via a series of point-to-point links that form a closed loop. In most physical ring topologies, the links are typically simplex, resulting in transmissions that always move in one direction around the ring. Each device takes the signal it receives on its input link and repeats the signal to its output link. The devices attached to the ring are the LAN adapters, which provide this repeating function. Because of this feature, LAN adapters attached to a ring are said to be *actively attached*.

Note that a physical ring also constitutes a broadcast environment, even though it is a collection of point-to-point links. Every LAN adapter sees what every other LAN adapter transmits because each transmission is propagated around the ring, from adapter to adapter, until it returns to the originating LAN adapter. This original transmitter does *not* repeat the signal again, effectively "taking it off the ring."

It should be evident that the ring topology depicted in Figure 15-1 has serious problems. Notice that each of the links and each of the points of attachment is a point of failure. If any one of these links fails, the ring is *open* and transmissions can no longer flow completely around the ring. For this reason, physical rings are seldom deployed in production networks. Where they *are* used, most LANs implemented using physical rings use two rings, with a mechanism for using one of the rings as an alternative path. This is true of Fiber Distributed Data Interface (FDDI).

In a *physical star* topology, all the cabling extends from the attached devices (the LAN adapters) to a central point. This point of contact is a piece of equipment generically referred to as a *hub*. Physical stars are the most popular form of LAN cabling in use today. Although physical stars most commonly implement twisted pair or optical fiber cables, coaxial cable can also be used in this configuration. The hub at the center of a star topology is often known as a *switch*.

The bus, ring, and star physical topologies are shown in Figure 15-1.

Logical Topologies

In a LAN that implements a *logical bus* topology, the signal generated by a transmitter is delivered by the LAN to all attached devices. This is the true broadcast environment—no attached end device is required to regenerate a signal that other attached devices can receive.

Note that a *logical* bus topology is intrinsically a *physical* bus topology as well. In fact, in the early days of Ethernet and LocalTalk, it would have seemed somewhat odd to separate these concepts at all. However, in more modern LANs, logical bus topologies are more commonly associated with physical star topologies (compare the topologies in Figures 15-1 and 15-2 to see the similarity). That is to say, a logical bus is implemented over a physical star (a hub-based network). The hub is essentially a multiport repeater that is responsible for regenerating the signal to all attached systems. Although the cabling plant differs, the concept of the bus topology remains the same: No attached device (other than the repeaters themselves) regenerates another device's signal within the LAN (passive attachment).

The *logical ring* topology describes a LAN environment in which the signals originate from one attached device and flow to the other attached devices one at a time. The attached devices are located in a predefined order, and the last attached device completes the process by returning the signal to the original transmitter. Thus, the transmitter both initiates and terminates a transmission.

Clearly, a *logical* ring topology is most closely associated with a *physical* ring topology (again compare the two figures). Indeed, the two concepts seem the same. However, physical ring topologies suffer from the problem of having multiple points of failure (several places where an equipment problem can cause communications to fail). In some cases, a second ring is deployed to provide a backup function.

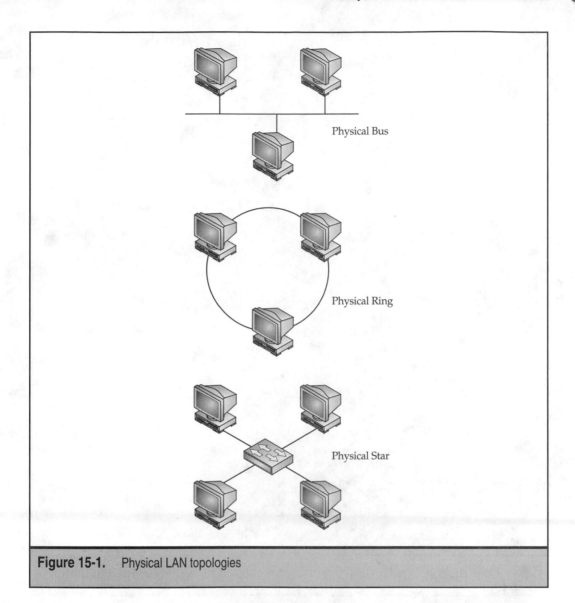

Figure 15-1.   Physical LAN topologies

A more popular way of deploying a logical ring topology is to use a *physical star* topology. In such an environment, the hub (more commonly called a *multistation access unit*, or *MAU*) passes the signal it receives on one port to the next port in the logical ring. The device connected to that port repeats the signal back to the hub, which passes it on to the next port on the logical ring, and so on. In fact, this is all the hub needs to do—take a received signal and pass it to the next port on the logical ring.

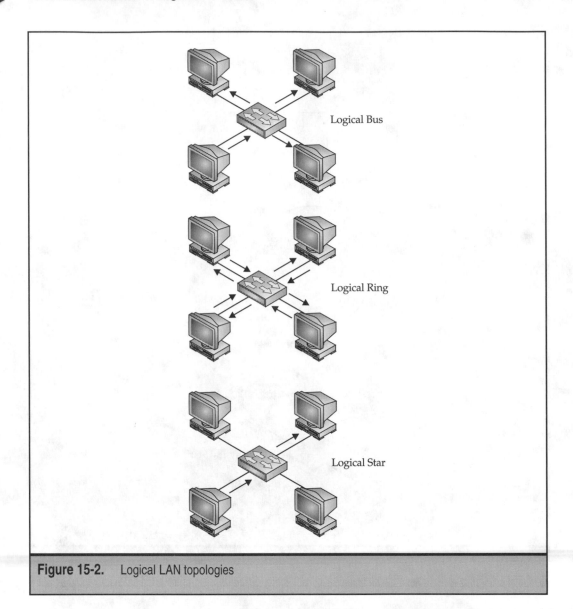

Figure 15-2. Logical LAN topologies

MAUs come in two flavors. The first is known as a *passive* MAU. This is nothing more than a box with ports to plug into. Each port's receiving media is wired to the transmitting media of the neighboring port. The passive MAU does not require power. As a result, the signal will eventually attenuate as it travels around the ring. To prevent this, the *nodes* (PCs and servers) on the ring have to regenerate the signal.

The other, more common, MAU is the *active* MAU, a device with power that regenerates the signal as it passes through each port.

In a *logical star* topology, the signal generated by a transmitter is switched by the LAN directly to the intended recipient. Thus, a logical star topology requires a switch at the heart of the LAN and is always implemented over a physical star topology. Every station attaches to the switch and the switch passes transmissions between any two devices that need to communicate. It should be noted that such a device is far more than simply a Physical layer device. It must have the intelligence to read destination addresses or respond to requests for circuit establishment. Such a network straddles the fence between being a LAN and being a collection of interconnected LANs.

Most early LANs implemented either a bus topology or a ring topology. Processors have become more powerful, and memory is far less expensive. Even the manner in which switching is performed has changed significantly, with advances like cell-based switching, cross-point matrix switches, shared-memory switching, and bus switching. In other words, we have learned to switch quickly and at a lower cost. The purchasers of technology recognized these advantages and began designing networks based on switches instead of multiport repeater hubs. This increased demand further reduced the price per port of switches due to increased vendor competition and economies of scale. As a result, the logical star topology with a switch at the center is the most popular LAN technology on the market.

LAN Adapters

Regardless of the topology implemented or the specific medium used, the cable plant cannot simply attach to the back of a computer. Some kind of interface is required between the computer system and the medium. This is the role of the LAN *adapter*.

A LAN adapter has two interfaces:

▼ One interface connects to the host computer. Associated with this interface is a parallel connector that physically attaches to the host and the circuitry to support exchanges between the host and the LAN adapter.

▲ The other interface connects the adapter to the LAN. Associated with this interface is the hardware to perform parallel-to-serial conversion, a serial connector, a line driver to boost the signal strength, and the circuitry necessary to support transmissions to and from the LAN.

Also found on the LAN adapter is one or more integrated circuits (ICs) responsible for controlling the operation of the adapter. One of the responsibilities of the LAN adapter is to gain access to the medium. This Media Access Control (MAC) protocol is implemented within the adapter. This means that each LAN adapter is suited to one type of LAN only, so an Ethernet LAN adapter can be used only on Ethernet networks and Token Ring LAN adapters can be used only on Token Ring networks. This minor element has been overlooked in some network migrations from one technology to another!

Finally, the LAN adapter must implement some amount of memory to be used as *buffers*. When a transmission is received from the LAN, it must be buffered until the LAN adapter can verify its correctness. When a transmission is received from the host to be sent onto the LAN, it must be buffered until the LAN adapter can gain access to the LAN.

The operation of a LAN adapter is straightforward.

1. If the host computer needs to transmit, it builds a frame that includes the address of the LAN adapter in the transmitting system, the address of the LAN adapter in the intended recipient, and the information to be transmitted. There might also be other control information.

2. The host computer passes this frame to the LAN adapter, where it is buffered. The LAN adapter then secures access to the LAN.

3. When access is secured, the LAN adapter begins transmission, calculating a frame check sequence (FCS) as it does so. The FCS is appended after the final bit of the data.

4. If a LAN adapter detects a transmission in progress on the LAN, it reads the transmission until it has read the entire destination address. If the LAN adapter recognizes its own address as the destination, it copies the entire frame from the LAN, calculating the FCS as it does so.

5. When the entire frame is copied, it checks its calculated FCS against the one appended to the frame by the transmitter. If they match, the frame is passed to the host computer; otherwise it is discarded.

ACCESS CONTROL (MAC SCHEMES)

Most of today's LAN environments are similar to the party lines of the telephone network, or to a group of people sitting around a conference table. Several devices, each equipped with a LAN adapter, share a common communication environment, and every LAN adapter can see what every other LAN adapter transmits. This is called a *broadcast environment*.

As with the infamous party line, or with the conference table scenario, some mechanism is required to ensure that only one party uses the facility at a time. With the party line, the mechanism is fairly simple. Someone wishing to use the phone simply picks it up and listens for a moment. If someone else is already talking, the interrupter puts the phone back down and tries again later. If the user gets a dial tone, the line is free and he or she can make a call. If everyone observes the rules, sequential access and a certain degree of privacy are ensured.

Several strategies can be used at the conference table. A rule of "don't interrupt while someone is talking" would result in an operation almost identical to that of the party line. On the other hand, one person could be designated as the chairperson, who would grant permission to communicate in an orderly fashion. Still another strategy would be to have people take turns around the table. Any of these approaches would ensure a certain degree of orderly access.

Rules specifying how communication takes place are called *protocols*. Each of the examples cited is a slightly different protocol. The protocol ensuring sequential access in a LAN is called a Media Access Control protocol, or MAC scheme. The MAC scheme is a function of the LAN adapter.

MAC schemes fall into two broad classes: *contention* and *polling*. In brief, contention schemes are a form of a controlled free-for-all, where each LAN adapter makes its own decision about transmitting. A polling scheme is an orderly access protocol, in which the LAN adapters explicitly interact to determine which LAN adapter gets to transmit next.

Although there are many possible contention-based schemes, the two most commonly implemented are *carrier sense multiple access with collision detection* (CSMA/CD) and *carrier sense multiple access with collision avoidance* (CSMA/CA). The former is implemented by all Ethernet LANs, and the latter by LocalTalk and many wireless LAN technologies.

Polling schemes can be further classified as *distributed* or *centralized*. In a distributed MAC scheme, the attached devices function as peers for the purpose of arbitrating access to the LAN. The most common form of distributed polling scheme is token passing, which can be performed over a ring topology (Token Ring) or over a bus topology (token bus).

In a centralized polling scheme, some device is identified as the primary (the *poller*), and the other devices play a secondary role (the *pollees*). Access to the LAN is arbitrated by the primary device. In a LAN, this type of MAC scheme is associated with star-wired LANs, and the hub plays the primary role. While functional LAN technologies were built around the centralized polling scheme, the market has long since left these technologies behind.

Today, CSMA/CD technologies dominate the LAN marketplace with Token Ring a distant second.

CSMA/CD

In CSMA, operation when the medium is idle is trivial. The LAN adapter simply transmits. But what happens when the medium is busy?

There are two basic flavors of CSMA, distinguished by their performance when a busy medium is detected. These are classified as *nonpersistent* CSMA and *persistent* CSMA. An analogy will underscore the difference between these two flavors: Every office environment requires a certain degree of interaction between the people who work there. It is fairly common to walk to a colleague's office to ask a question or pass on a message. What happens if the colleague is busy—on the phone or in conversation with another colleague? Some of you might casually walk by or check back later. Others will stand by the door, leaning on the door frame, until the person completes his conversation.

The first type of employee is practicing *nonpersistent* CSMA. In nonpersistent CSMA, a LAN adapter wanting to transmit, but detecting a busy medium, waits a random interval of time and checks back again later. The second type of employee is practicing *persistent* CSMA. When a LAN adapter wanting to transmit detects a busy medium, it continues to monitor the medium until it returns to the idle state (when the currently transmitting station finishes its transmission). The majority of LAN adapters on the market use some form of persistent CSMA.

What happens in a persistent CSMA environment when the medium returns to an idle state? In *1-persistent* CSMA, a LAN adapter waiting for the medium to return to the idle state begins to transmit as soon as this transition takes place, allowing for only a small interframe separation time. 1-persistent CSMA is simple and has reasonable performance characteristics. It also ensures no access delay when the medium is not being used.

Since LAN adapters in a CSMA environment can sense the medium, they can be designed to detect a collision as well as a busy or idle medium. While transmitting, the LAN adapters listen for a collision and, if a collision is detected, the LAN adapters detecting the collision send a short burst (32 bits), called a *jamming signal*, to ensure that all transmitting LAN adapters detect the collision. All transmitting LAN adapters then cease transmission and execute a "back-off" procedure. The back-off procedure requires each LAN adapter whose transmission is part of a collision to wait some pseudo-random period before attempting the transmission again. A maximum number of reattempts is defined, after which the LAN adapter signals the attached computer that it is unable to access the network.

IEEE 802 STANDARDS FOR LANS

By the early 1980s it became clear that LANs were becoming widespread enough to require standardization. The Institute of Electrical and Electronics Engineers formed the 802 committee in February of 1980 with the goal of standardizing LAN media and frame formats. Each LAN technology is given its own working group numbered in order of their formation. For example, Ethernet is formally described by the 802.3 committee, while Token Ring standards are the result of work by the 802.5 committee.

IEEE 802.3: Ethernet

In 1972, a team at the Xerox Palo Alto Research Center (PARC) was given the task of designing the "office of the future," in which information would flow effortlessly from machine to machine and, hence, from user to user. Two members of that team, Robert M. Metcalfe and D.L. Boggs, first outlined the idea for broadcasting messages over a local bus network, attaching headers to them for addressing and other control functions. Thus was *Ethernet* born.

The idea was perhaps a bit ahead of its time, but by the time the PC revolution began to take over the desktops of the world in the 1990s, Ethernet (since standardized by the Institute of Electrical and Electronics Engineers as *IEEE 802.3*) was ready to provide simple, inexpensive, and fast networking to the millions of users who demanded it. Only IBM's promotion of its Token Ring standard kept Ethernet from being the LAN standard.

The history of IEEE 802.3, commonly called Ethernet, is interesting. Ethernet began as a research project at PARC, where it was originally called the Alto ALOHA Network, a name that reflects the fact that Ethernet's CSMA/CD MAC scheme is contention-based, as was the scheme implemented by the University of Hawaii's ALOHA System packet radio network. It was renamed Ethernet in 1973 by Bob Metcalfe, one of the principals in its development. In 1980, IEEE formed the 802 Project, initially to develop LAN standards implementing CSMA/CD, Token Ring, and token bus. By this time, the PARC research project had become a product marketed and strongly supported by Xerox, Intel, and DEC. Version 2, called Ethernet II or Ethernet V2, was released in 1980.

Within the last few years, Ethernet has become the most popular LAN technology. Despite its humble origins, Ethernet has evolved into a high-speed switched architecture. No longer is Ethernet being used only on the LAN, but fiber-optic technology and full-duplex point-to-point links has allowed Ethernet to emerge as a strong contender for WAN access links, especially in metropolitan areas. Instead of the network being ATM from end-to-end as was envisioned years ago, there is a future where the network is all Ethernet.

Ethernet II vs. IEEE 802.3 Specifications

While the Xerox team at PARC was finalizing its Ethernet II product, the IEEE 802.3 committee was working on the standard, based on the Ethernet specification. By the time it was completed, a number of changes had been made. Thus, the two specifications, Ethernet II and IEEE 802.3, have many common aspects as well as a few differences.

The main differences can be attributed to the fact that the IEEE, being a standards organization, was bound to the Open Systems Interconnect (OSI) philosophy of a layered model. Specifically, in the OSI Model, each layer is transparent to the other. The Ethernet II product, being developed by a corporation, was not bound by such ideology. Both specifications include the use of 1-persistent CSMA/CD and both operate over a bus topology. Both can be deployed over twisted pair, coaxial cable, or optical fiber. In fact, although there are minor differences in the physical aspects of the two specifications, a single LAN adapter can be (and often is) designed to implement either of the two.

The most important difference between the two specifications is in the format of the frame. While essentially identical in structure, one critical field in the frame differs in meaning. Where IEEE 802.3 places a 2-octet *Length* field in the frame, Ethernet II places a 2-octet *Type* field. In IEEE 802.3, the Length field records how much data is being carried in the frame and allows the receiver to distinguish between Data and Pad when a frame carries less than 46 octets of data. In Ethernet II, the Type field records the specific protocol carried in the Data field. It is used by higher layer software and is not examined by the LAN adapter.

Although the differences seem trivial, they do pose implications for interoperability. If a LAN adapter implementing the IEEE 802.3 frame structure receives an Ethernet II frame, the Type field will be interpreted as a length and will be incorrect, causing a frame discard. If a LAN adapter implementing the Ethernet II frame structure receives an IEEE 802.3 frame, the Length field will never be examined. It will simply be passed up to the higher layers together with the Data field. If a Pad field is present, it will be passed to the higher layer as though it were data. This LAN adapter cannot distinguish between Data and Pad.

Ethernet Media Standards

IEEE 802.3 can be deployed as a physical bus (usually coaxial) or a physical star (twisted pair and optical fiber). The standard supports transmission rates at 1, 10, 100, and 1000Mbps, depending on the specific media option being employed. There are no longer any 1Mbps IEEE 802.3 products; the standard for this Physical layer option has fallen by the wayside, along with the standard for 10Mbps operation over a broadband (analog) infrastructure. Today, IEEE 802.3 LANs operate primarily at 10 and 100Mbps. The 100Mbps standard emerged in 1994 and is dubbed *Fast Ethernet*. The 1000Mbps standard (1 gigabit per second) is dubbed *Gigabit Ethernet*.

IEEE 802.3 has a number of associated standards describing various media options. Each standard has a descriptive name comprising three parts. The first part specifies the transmission rate of the medium in megabits per second. The second specifies the type of signaling used: baseband or broadband. The final part of the standard name originally indicated the maximum length of each segment (without repeaters) in hundreds of meters.

However, since 1990 a number of new standards have emerged, in which various maximum lengths are possible depending on the specific medium. For these standards, the final part of the name designates the general type of media.

Historically, two of the most widely deployed media were 10Base2 and 10Base5, both of which specify the use of coaxial cable. Although this media might still be deployed, the current hottest selling technology is twisted pair.

The 10BaseT standard describes how twisted pair can be used to implement an IEEE 802.3 LAN. 10BaseT has been the most widely implemented IEEE 802.3 Physical layer standard since its completion in 1990.

The next IEEE 802.3 standard (1994) was 100BaseT. Despite the *T* in the name, 100BaseT describes support for both twisted pair and optical fiber. A newer 1000BaseT specification (gigabit per second Ethernet) uses Cat 5 and Cat 5E, with all four pairs in full duplex. Other media support for 1000Mbps include multimode and singlemode fibers.

With 100BaseT firmly entrenched, the industry has turned its attention to pushing the speed boundaries again. Advancements in LAN applications, the workstations that run them, and the nature and amount of communication between servers and workstations have called for greater speeds in the networks that serve them. As 100Mbps networking becomes a common desktop technology, it defines a need for an even faster backbone service.

The Gigabit Ethernet Standard

Enter the Gigabit Ethernet standard. Essentially an extension of 10BaseT and 100BaseT technologies and defined originally in IEEE 802.3z, Gigabit Ethernet is capable of both half- and full-duplex transmission modes (see the sidebar "Half-Duplex and Full-Duplex Operation," later in this chapter); it is only found, however, as a full-duplex implementation. Gigabit Ethernet retains the traditional IEEE 802.3 frame type, which allows its simple integration into existing 10BaseT and 100BaseT networks. Initially, the physical plant used the same physical signaling topology as fiber channel, although 802.3ab, a later version of the standard, added Cat 5 UTP as the physical medium.

The IEEE has not set a maximum distance at which Gigabit Ethernet is expected to operate for the various physical media. Instead, it has set minimum distances that implementations must meet to conform to the standard. For 1000BaseSX, it is 2–275 meters for 62.5 micron multimode fiber (MMF) and 2–550 meters for 50 micron MMF. For 1000BaseLX it is 2–550 meters for MMF and 2–5000 meters for SMF. Vendors now offer versions of Gigabit Ethernet that operate over significantly longer distances over single mode fiber (SMF) than these. For example, Cisco's 1000BaseLX/LH operates over 10,000 meters of standard SMF, while 1000BaseZX operates over 100 kilometers via dispersion-shifted fiber (DSF). The implication of these distances is that the advent of Gigabit Ethernet is moving Ethernet technology from the office LAN into the MAN. This inexpensive, easily understood technology with quick deployment times is quickly becoming the solution of choice for broadband providers in metropolitan areas. The IEEE working group has also begun work on 802.3ad, the standard defining 10-gigabit Ethernet and already talk has begun of 100-gigabit Ethernet.

MAC Addressing Frame Structure

Communication across a LAN involves the transmission of a frame from one LAN adapter to one or more destination LAN adapters. For a LAN adapter to determine when a transmitted frame is destined for itself, some form of addressing is required. All the IEEE 802 standards use a common MAC addressing scheme.

IEEE 802 standards support two separate address formats: a 2-octet address and a 6-octet address. The 2-octet form is seldom (if ever) implemented and will not be reviewed here. The 6-octet form is universal.

A MAC address can represent a single LAN adapter or a group of LAN adapters. The latter provides support for multipoint addressing. To distinguish between the two classes of address, IEEE has given the first transmitted bit a special meaning. If this bit is set, it means the following address is a multicast address. One form of the multicast address used frequently by LANs is the broadcast address, which all LAN adapters that receive it will process.

To prevent the manufacturers of standards-compliant LAN adapters from creating LAN adapters with identical addresses, IEEE administers the address space. When a manufacturer wants to market LAN adapters, it applies to IEEE and is assigned a pool of addresses. The assignment is made by specifying a particular prefix that must be used in all LAN adapters marketed by that manufacturer. The prefix comprises the first three octets of the MAC address and is called the Organizationally Unique Identifier (OUI). The manufacturer is free to use any possible value for the remaining three octets—a total of some 16 million possible addresses. Manufacturers are free to apply for additional prefixes as the need dictates. The following illustration shows a sample MAC address.

Organizationally
Unique Identifier

┌─────────┐
00-40-05-DF-1C-D7
└─────────┘

Locally Assigned by
Card Manufacturer

Token-Ring Distributed Polling: IEEE 802.5

Token passing is a MAC scheme that ensures orderly, sequential access to the LAN for all LAN adapters. To ensure this, a single *token* is introduced into the LAN and passed from LAN adapter to LAN adapter. On receiving the token, if the LAN adapter has anything to transmit, it can do so. If the LAN adapter has nothing to transmit, it simply passes the token to the next LAN adapter. The token is not physical; it is a special pattern of bits that is recognized by all LAN adapters as the token.

While this concept is simple, the simplicity might be somewhat deceptive. The protocol implies a good deal, and much can go wrong. The protocol must guard against, prevent, or resolve many problem situations. For example, the protocol implies an ordering to the LAN adapters. Further, it implies that the LAN adapters are arranged in a closed loop to allow the token to continue to circle indefinitely between LAN adapters. If the LAN adapters are part of a ring topology, the ordering is provided by the hardware.

The protocol must also defend against one LAN adapter monopolizing the bandwidth. Rather than impose a maximum frame size, most token passing schemes define *timers*. Typically, the timer controls how long a given LAN adapter can retain the token before releasing it back to the LAN.

Token passing is essentially a collision-free MAC scheme. Use of the token restricts transmission to one LAN adapter at a time. Unfortunately, it also introduces some minor, unnecessary delay. If a LAN adapter needs to transmit, it has to wait for the token, even if no other LAN adapter is currently transmitting. This is known as *low-load delay* or *zero-traffic delay*. The maximum low-load delay is equal to the *ring latency*, the time it takes a signal to circumnavigate the ring. In a maximum size ring, this delay is approximately .000078 second. At 4Mbps, this translates to approximately 40 octets. At 16Mbps, the maximum is approximately 160 octets.

Recall that in a ring topology, a transmission is repeated from LAN adapter to LAN adapter around the ring. Unless a mechanism is found for ending the cycle, the transmitted frame will circulate endlessly. To ensure the broadcast nature of the ring and minimize delay, the transmitting LAN adapter is responsible for cleaning up after itself.

Notice the potential for problems. If the transmitting LAN adapter malfunctions before it removes its frame from the ring, the frame could circle endlessly. In a like manner, the token can be lost. To deal with these (and other) potential problems, various LAN adapters in the ring assume monitoring roles. The process for electing LAN adapters to provide these functions, and detecting if they are lost, is also fully distributed.

IEEE has standardized the Token Ring MAC scheme in IEEE 802.5, which defines a distributed polling scheme, specifying the use of a token passing protocol on a ring topology. The standard supports transmission rates of 1, 4, and 16Mbps, although 1Mbps token rings do not now exist in the marketplace. Token Rings can be deployed over shielded twisted pair (STP), various grades of unshielded twisted pair (UTP), or multimode optical fiber.

Half-Duplex and Full-Duplex Operation

Conventional Ethernet and Token Ring LANs are essentially half-duplex communication environments. In a half-duplex environment, transmission in one direction must be complete before transmission in the other direction can begin. This condition holds between any possible pair of systems in a conventional Ethernet and Token Ring LAN; this reality is imposed by the operation of the MAC.

Ethernet requires a transmitter to wait for an idle network before it begins transmitting. Token Ring requires a device to wait for the arrival of the token; because there is only one token, there can only be one transmitter at any given time. Although transmission can occur between any two devices in an Ethernet, only one device can transmit at a time; therefore, from the perspective of any two attached devices, Ethernet communication is *half duplex*. This aspect of Ethernet is a holdover from the days when Ethernet was deployed over coaxial cable. The broadcast nature of the cable plant, and the use of a single conductor pair, forced two simultaneous transmissions to collide in the cable plant, destroying both transmissions. Thus, access was forced to be strictly one at a time.

The development of 10BaseT fundamentally changed the nature of the cable plant. Each device is attached to the hub with two twisted pair—one for transmission and one for reception. In this environment, a collision is slightly redefined. Because the signals from two devices cannot collide in the cable plant (for example, a LAN adapter's transmitted signal occupies a physically different pair than its received signals), a collision is defined as "receiving a signal on the receive pair while transmitting on the transmit pair."

Given the existence of separate transmit and receive pairs in 10BaseT, the concept of collisions is meaningless and wastes time. Full-Duplex Ethernet is an Ethernet variant that permits transmissions and receptions to occur simultaneously. Devices can simultaneously transmit and receive, effectively doubling the total throughput of the network to 20Mbps; however, calling Full-Duplex Ethernet a 20Mbps environment can be grossly misleading because the transmission rate is still 10Mbps.

Although Full-Duplex Ethernet *does* improve LAN performance, it requires a change of equipment at both ends of the Ethernet segment. Each of these systems must implement a Full-Duplex Ethernet LAN adapter. Most NICs purchased today support this feature out of the box. This means, however, the typical "conventional" hub (which does not implement a MAC scheme) cannot be used in a Full-Duplex Ethernet environment. If a hub is the point of attachment, the hub needs to implement Full-Duplex Ethernet ports. This means the hub needs to be more than just a simple hub; it must also provide a bridging or routing function.

Migrating a LAN to a full-duplex environment has many advantages. Specifically, it could be used to enhance the performance of hub-to-hub links, and links to servers, routers, and other critical systems.

Wireless LANs: IEEE 802.11

IEEE has assigned the task of developing MAC and Physical layer standards for wireless LANs to the 802.11 working group. The working group formed in late 1990, but it produced little in its initial three years of life. The working committee, comprising almost 200 members, could not agree on foundational issues. The major disagreement centered around the issue of the type of MAC to be standardized.

Wireless MAC schemes can be coordinated in either a distributed or centralized fashion. The former is called *distributed coordination* and the latter is called *point coordination*. Distributed coordination requires the individual devices to share responsibility for arbitrating access to the LAN. This is the model used in most wired LANs. In point coordination systems, a centralized device controls access to the medium.

In November 1993, the issue was finally resolved when the 802.11 committee voted on a MAC standard based on the proposal jointly submitted by NCR, Symbol Technologies Inc., and Xircom, Inc. The conflict between distributed coordination and point coordination was resolved by including *both* models in the standard. The IEEE approved the 802.11 standard in June 1997.

The fundamental MAC scheme is distributed coordination function, which is essentially a contention-based MAC scheme implementing a version of CSMA/CA. In this scheme, collision avoidance is based around a Request To Send (RTS) and Clear To Send (CTS) exchange in which the stations are involved. After determining whether the medium is free and calculating an appropriate wait time, an RTS is sent to indicate the intention to transmit and the length of the transmission. If a CTS is returned, the transmission of the frame begins. All other stations are aware that the transmission is taking place and know the expected length of the transmission, thereby avoiding collisions. A point coordination function (PCF) scheme is also supported, which operates more on the centralized polling model (a central point controls access to the medium).

Developments in the standard have moved more rapidly since its early days. In September 1999 a revision, 802.11b, to the original standard was ratified. It is also called *802.11 High Rate,* as it has introduced much higher data rates while maintaining the original protocols.

The LAN is deployed over a wireless logical bus topology. Options available for the Physical layer include direct sequence spread spectrum (DSSS), frequency hopping spread spectrum (FHSS), and infrared. The spread spectrum operates in the 2.4GHz band. The FHSS option supports 1Mbps with an option of 2Mbps in "clean" environments. With the advent of 802.11b, the DSSS option supports the 11Mbps with fallback speeds of 5.5Mbps, 2Mbps, and 1Mbps.

The 802.11b standard provides for data rates up to 11Mbps with extensions to 22Mbps. The 802.11a standard uses the 5GHz band and provides for data rates up to 72Mbps.

While most people are familiar with an Ethernet network—they need look no further than their own home or business network—wireless LANs may not be so well known. For the most part, the design considerations of a wired Ethernet are the same as a logical star Ethernet. The difference is that instead of twisted pair running to each desktop, client machines are fitted with a wireless LAN adapter. These wireless LAN adapters transmit to a nearby access node that connects the client machines to the wired backbone. Taking range and interference from physical objects into account, access nodes can be dispersed in an environment in such a way that the user roams from room to room, floor to floor, or even building to building. Due to the large amounts of bandwidth required for most servers, the primary use of the 802.11 technologies will be for connecting client computers to a high-speed wired backbone and in small office/home office (SOHO) applications.

Logical Link Control: IEEE 802.2

The logical link control (LLC) is a layer of protocol that sits above the MAC layer and completes the capabilities of the Data Link layer, and providing functionality beyond the OSI definition. It has three basic functions:

▼ To provide connection oriented, connectionless, and acknowledged connectionless services over the intrinsically connectionless MAC scheme

■ To hide the specific nature of the LAN from the upper layers

▲ To provide logical service access points (SAPs) so that multiple logical connections can be supported across the same physical interface (LAN adapter)

For a number of years, the LLC layer was not widely implemented, for the simple reason that its three basic functions were not viewed as important. Virtually all of the networking protocol suites implement a Network layer providing connectionless service to the layers above it. Therefore, it made no sense to sandwich a connection-oriented layer between the connectionless MAC and Network layers. The interface between the MAC and the software implementing the Network layer was largely defined by the company creating the networking software (such as Novell or Banyan), and adhered to by the companies manufacturing the LAN adapters and writing the controlling software (the LAN drivers). This arrangement effectively "hid the MAC layer" without the need for LLC. Finally, few systems were deployed that could simultaneously support multiple protocols, so the multiplexing capabilities of LLC were largely unneeded.

Today, many systems support two or more protocol suites. Perhaps the system runs NetBEUI to communicate with the file servers and TCP/IP to gain access to the Internet. This trend in the market increased the importance of LLC. IP itself is generally transmitted in Ethernet II frames since it eliminates the need for overhead that the LLC adds. Most other LAN protocols, however, do use the LLC. The argument could be made that TCP/IP network does not need LLC at all, but even then some support protocols are going to be making use of the LLC. Although the LLC layer is responsible for providing SAPs, the number of available SAPs is small. Of the 8 bits that make up the SAP, two are reserved, leaving only 6 bits for a total of 64 unique SAPs. IEEE and ISO protocols reserve most of these. This creates a problem for some protocol environments, especially those that were originally designed to operate over Ethernet.

The Ethernet frame format includes a 2-octet Type field. This field records the particular protocol residing in the Data field of the Ethernet frame. This function is not unlike the function of the LLC destination SAP (DSAP) and source SAP (SSAP). However, with 2 octets, more than 65 thousand unique identifiers can be created. Some suites (such as AppleTalk) that made use of this rich set of identifiers found the limited number of SAPs in LLC a problem.

The use of SNAP is somewhat analogous to the four-digit extension added to the zip code several years ago. As it became clear that the five-digit number was not enough to represent all of the locations in the U.S., an extension was added to increase the number of zip codes available. The Subnetwork Access Protocol (SNAP) protocol discriminator effectively increases the number of SAPs available in an IEEE environment.

Many people find themselves confused by the array of options for Data Link layer protocols in a LAN. The options, and the consequences of choosing each, are relatively simple. We should begin at the bottom.

It is possible to configure many LAN software packages to use the native MAC frame without LLC or SNAP. When configured in this way, upper layer messages are carried directly in the Data field of the MAC frame. However, because the MAC frame has no protocol discriminator and no SAPs, a system configured this way is typically limited to supporting one protocol suite (such as NetWare, TCP/IP, or VINES) per LAN adapter. If multiple protocols are to be supported, multiple adapters are needed. Furthermore, only connectionless Data Link layer service is possible. Protocol environments that require a connection-oriented Data Link layer (SNA and NetBEUI) cannot be supported in this fashion.

It is also possible to configure the software to use the MAC frame together with LLC. In this case, the LLC frame is carried within the MAC frame, and the upper layer message resides within the LLC Data field. This configuration makes it possible to support multiple protocols, but the set of protocols that can be supported is somewhat limited. However, the Data Link layer can now provide connectionless, connection-oriented, and acknowledged connectionless services.

Finally, the software can be configured to use SNAP, which implies the presence of LLC (you cannot use SNAP *without* LLC) and the MAC frame. Protocol layers are nested one inside the other, with the upper layer message residing in the Data field of the SNAP frame. Using this configuration, an extended set of protocol suites can be supported.

BRIDGING/SWITCHING ARCHITECTURES

One of the classical performance problems of a CSMA/CD-based LAN is that increasing the amount of end stations increases the potential for collisions and delay. This problem can be solved through the use of a *bridge* on a network. A bridge is an intelligent device that is able to learn the location of end stations and logically segment traffic.

A bridge operates like the cubicle dividers in an office building. A group of people can have a conversation in one cubicle without disturbing those in another cubical. If there is a need to communicate with someone outside the cubicle, one can just shout over the top. A bridge, like the cubicle, allows multiple conversations on each of its ports. Data communication between PCs on one port follow the rules of CSMA/CD only on that port. PCs on another port have their own CSMA/CD independent of the others. By separating a LAN into two or more Ethernet segments and interconnecting them with a bridge, the performance on the Ethernet as a whole increases. This reduces the average number of devices attached to each Ethernet and therefore the average number of collisions and delay on each physical LAN segment.

The LAN switch takes this bridge-based approach to its extreme, in effect, giving each attached device its own private Ethernet connection. Often advertised as "dedicated Ethernet to the desktop," it allows the network administrator to boost the performance of the network without having to change any of the cabling or end station equipment.

The architecture of the LAN switch deserves attention. One of the greatest claims of LAN switches is their speed. Many of these products achieve latencies that are near wire-speed. In other words, the time between when the first bit arrives at the LAN switch and when the first bit is forwarded onto the destination link is only slightly more than the time it would take for that same bit to travel down a relatively short length of cabling. This is because LAN switches often support a cut-through feature, where conventional bridges are *store-and-forward* devices. In a store-and-forward device, the entire frame must be received and buffered before it can be processed and forwarded. In an Ethernet world, even if there is no other traffic in the network, it means that a 1500-octet frame will experience at least 1.2 milliseconds (the time it takes to transmit or receive a 1500-octet frame at 10Mbps) of delay passing through a bridge. A bit could travel 363 kilometers in that time!

LAN switches implementing a cut-through feature need only delay the frame long enough to read the destination address. Because the address is always one of the first things in the frame, this information is the first to arrive. In a cut-through Ethernet LAN switch, a frame would be delayed a mere .0048 milliseconds (4.8 microseconds). A bit can travel approximately 1460 meters in that time. A Fast Ethernet LAN switch has the potential to reduce that number to 146 meters!

However, cut-through comes at a cost. Most conventional bridges, once receiving the entire frame, can also process the frame check sequence (FCS) at the end of the frame and detect bit errors. If a frame contains errors, it is a waste of bandwidth to pass on the frame, so the bridge discards it. A LAN switch with a cut-through feature has already started forwarding the frame when the FCS arrives and cannot choose to discard the frame. In the early days of LAN switching, this led to a great debate over whether LAN switches should be cut-through or store-and-forward. The former camp pointed to the incredibly low latencies with minimal wasted bandwidth in the event of errors. The latter camp noted the devastation a malfunctioning LAN adapter would wreak on a network.

As with most things related to networks, the result was a new class of products that could do both. Normally, these devices operate as cut-through switches. However, each frame is error checked as it is forwarded. The frequency with which errored frames are detected is recorded. If it passes beyond a configurable threshold, the LAN switch reverts to store-and-forward mode until the error rate drops below the threshold. Some of these devices are able to make this change on a per-port basis. A modified cut-through switch waits until the frame being received has been determined not to be a collision frame. This wait time is time it takes to transmit the minimum frame size.

It should be noted that any LAN switch, even when operating in cut-through mode, is forced to operate in store-and-forward mode if the port to which the frame is to be forwarded is currently in use.

LAN switches tend to be particularly susceptible to congestion due to the nature of the switch and the manner in which many LAN environments operate. Whether operating in cut-through or store-and-forward mode, LAN switches are high-speed devices with extremely low latencies (compared to conventional bridges). As such, congestion conditions arise far more quickly. This is especially true if multiple end stations are accessing the same target device, a common situation in many LANs.

Consider a 24 port LAN switch with 22 workstations and two servers. If six users upload files to one of the servers simultaneously, the switch receives six traffic streams destined for the server's port. Assuming all ports are operating at the same transmission rate and the frames are approximately the same size, only one of the six frames is forwarded immediately; the other five are buffered.

If the network protocols support windowing for flow control (for example, the Transmission Control Protocol), the senders continue to transmit frames, even though the first frame has not been acknowledged. The second set of frames are all buffered, because the switch is still forwarding the first round of frames. If this continues for too long, the buffers fill and the switch is forced to begin discarding frames.

There are four approaches to solving the problem. The first is to implement a switch with differing port transmission rates. For example, an Ethernet switch could be equipped with two, 100BaseT ports and 22, 10BaseT ports. Servers attached to the 100BaseT ports are less likely to congest. This architecture, while extremely common in the marketplace, is effective only when the points of congestion can be reliably predicted. In a peer-to-peer environment, any device could potentially become a point of congestion. Furthermore, it opens the door to congestion in the opposite direction (when the system with the high-speed attachment begins to stream frames to a system with a slower link).

A second approach is to increase buffer capacity in the LAN switch. Congestion tends to be a momentary thing. If the switch has buffers that are sufficiently large, it can survive periods of congestion without dropping frames. There are two problems with this approach. First, it is difficult to predict how much memory is required to survive any possible congestion event. Under normal conditions, this extra memory is virtually unused. This approach is analogous to buying a high-bandwidth service to deal with wide area traffic that is only occasionally heavy. Most of the time, the bandwidth is a wasted expense.

For Ethernet switches, a third alternative has emerged—*backpressure*. Backpressure is nothing more than a simulated collision. If the switch is congested, it generates a jamming signal when a new transmission is received. This fools the transmitting LAN adapter into believing that a collision has occurred. This LAN adapter stops, waits, and tries again a few milliseconds later. This gives the switch time to forward some of the frames out of the congested buffer.

Transparent Bridging

Today's LANs are commonly broadcast environments. This is certainly true of Ethernet and Token Ring networks. In these environments, every LAN adapter attached to a single LAN receives what every other LAN adapter transmits.

One consequence of a broadcast environment is the requirement for sequential transmission; LAN adapters must take turns using the media. The need for sequential access introduces the problem of access delay. As more and more LAN adapters need to transmit, it becomes more likely that they will be required to wait until other transmissions are completed. The greater the total offered load to the LAN, the greater the chance that a frame will have to wait in a queue on the LAN adapter until the network becomes available. As the LAN grows (the number of attached devices and/or total offered load increases), this access delay can become unacceptable. Users begin to notice performance problems.

Studies of human communications within a multidepartment company have shown that most of the communicating is done within a single department. This tends to be true of memos, personal conversations, and phone conversations. It is also true of computer communications. If a company has a single LAN, an analysis of the communication patterns would most likely show that most of the communication is occurring within workgroups or departments. This is especially true if each department has a server and has local control over its files and applications.

This tendency of systems to communicate in groups indicates a way to improve the performance of the LAN—by breaking up the LAN by department (or workgroup). We need to introduce a device that directs traffic between the different LANs. Its job is to monitor all LAN traffic and allow only interdepartmental messages to move between the various LANs in the network. This is the role of a bridge.

This type of bridge is highly dynamic. Once configured and installed, it will begin to learn about the attached LANs and collect information concerning the location of various LAN adapters in the network. The operation of a bridge is relatively simple. For each frame that a bridge receives on any given interface, it must only decide whether to forward this frame to another interface and thus another LAN segment or to discard the frame. This decision is made based on the current status of each port and information stored in a special table, called the *filtering database*.

The filtering database is somewhat misnamed. This "database" is actually a table listing all known MAC addresses. Associated with each address is an age (indicating when the entry was made in the table) and a port identifier. The port identifier specifies the port to which a given frame should be forwarded so as to reach the specified destination MAC address. Thus, the "filtering database" might more appropriately be named a "forwarding table."

If the MAC relay entity determines that a frame is to be filtered, the frame is simply cleared from the buffers and the bridge takes no further action. If the MAC relay entity determines that a frame is to be forwarded, the frame is passed to the appropriate MAC entity (the LAN adapter). The MAC entity then secures access to the attached LAN and transmits the frame. If there are more bridges in the network, the process can be repeated until the frames arrive at the LAN to which the destination LAN adapter is attached.

The Learning Process

A key part of the filtering/forwarding decision is based on the filtering database, which is basically a table containing MAC addresses and related port identifiers. The table is keyed by the MAC address. The MAC relay entity knows the port on which a given frame arrived and the MAC address in the Destination Address field of the frame. Armed with this information, it queries the filtering database to find out what it should do. The filtering database contains two types of entries: static and dynamic.

Static entries are manually entered by a network administrator, or are governed by a network management system (NMS). From the perspective of the transparent bridge, these are unchanging entries (changing these entries requires intervention by some outside force). Static entries have the advantage of being highly configurable.

Static entries are inflexible in their need for manual (or external) configuration. The bridge automatically maintains dynamic addresses.

A dynamic entry is a simple data triplet: a MAC address, a port ID, and a time stamp. The dynamic entry is accessed using the MAC address. The related port ID indicates the port on which the frame is to be forwarded. The time stamp is a mechanism the transparent bridge uses periodically to clean old information out of the table.

A transparent bridge has a mechanism by which it gathers information from the network to maintain the dynamic entries in the filtering database. It is because of this capability that this type of bridge is sometimes called a *learning bridge*.

A learning process maintains the dynamic entries in the filtering database. When a frame arrives at a transparent bridge, the bridge first learns everything it can from the frame. It then filters or forwards it according to the current port status and the contents of the filtering database.

The bridge first extracts the source address from the arriving frame. If there is a static entry for this MAC address, the bridge does not create or update a dynamic entry; it simply filters or forwards the frame based on the destination address. Static entries are manually configured and are not changed by the learning process.

If there is no static entry for the source address, the bridge is required to learn. The first step is to determine whether the source MAC address is already present as a dynamic entry in the filtering database. If the address is not present, it is added to the database, together with the ID of the port on which the frame arrived and a time stamp. The bridge has now learned the "location" of the transmitting LAN adapter in the network. If something coming *from* that LAN adapter arrived on a particular port, it is reasonable to assume that anything destined *for* that same LAN adapter should be forwarded on the same port (the "it's out there somewhere" approach, sometimes called *backward learning*). This operation is illustrated in Figure 15-3.

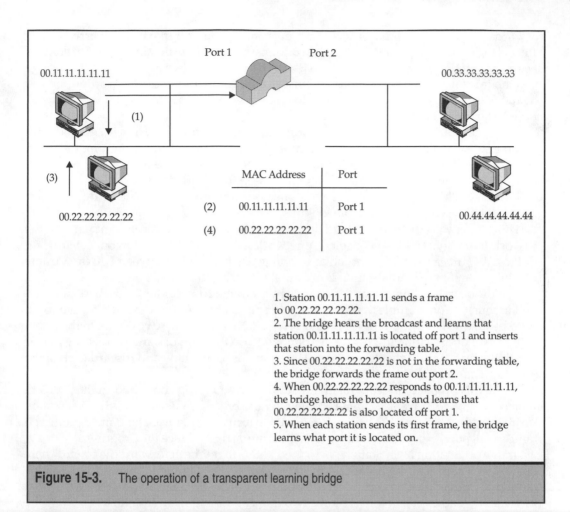

Figure 15-3.   The operation of a transparent learning bridge

If the source address is already present in the filtering database, the resident port ID is compared to the ID of the port on which the frame just arrived. If these IDs are different, it means this LAN adapter has been moved since its last transmission. Perhaps the entire system was relocated; or perhaps the LAN adapter was removed from one machine and placed into another machine elsewhere in the network. Whatever the reason for the move, the bridge updates the port ID in the database and resets the time stamp. If the source address is present in the database and the port ID is consistent with the ID of the port on which the frame arrived, no further learning is required. The bridge simply updates the time stamp.

With the learning process complete, the bridge can now access the filtering database to discover what should be done with the frame. Now it is acting on the destination address. If the bridge has no static entry and has never seen a frame from the destination station, it literally does not know what to do. In the absence of information, the bridge forwards the frame on all ports except the port on which the frame initially arrived. The effect is to "flood" the frame into the network, ensuring that the destination LAN adapter sees a copy. If there is an entry for this MAC address in the filtering database, the bridge uses the information in the database to forward the frame to the correct port or ports. If the database indicates that the frame is to be forwarded to the LAN from which it was copied, the frame is filtered.

The time stamp is important. If a bridge does not see a frame from a LAN adapter within a set period of time (typically 5 minutes), it clears the information from the filtering database. If the database is full and a new address is seen, the time stamp is used to clear the oldest entry (the one that has not been seen for the greatest period of time) from the database. This process minimizes the memory requirements of the bridge and improves delays associated with table lookups.

This strategy is effective and sufficient as long as there are no redundant paths in the network topology. The visual depicts a network with two LANs, interconnected using bridges. As you can see, this particular topology has two paths between LANs. A frame traveling from Station A to Station B might take either path.

The problem is a simple one. What prevents a frame from taking both these paths simultaneously? For example, a frame originating at Station A and destined for Station B could be forwarded by both Bridge 1 and Bridge 2. Now there are two copies of the frame on the destination segment. This wastes bandwidth in the network and processing power at both the bridges and the destination. The latter has to examine and discard each of the duplicate frames.

While this problem appears relatively serious, it is nothing compared to the problem created by broadcast frames (frames containing the broadcast address in the Destination Address field). A transparent bridge is required to forward broadcast (and most multicast) frames to all ports except the one on which the frame arrived. Trace the progress of a broadcast frame originating from Station A. Bridge 1 and Bridge 2 both forward the frame. There are now two copies of the broadcast frame on Station B's segment. Bridge 2 will see the frame that Bridge 1 sent and copy it to the other segment. Bridge 1 will see the frame that Bridge 2 sent and copy it. Now we have two frames running in an infinite loop.

This situation is referred to as a *broadcast storm*, and it can completely disable a network in a short period of time. If the network has a single loop, broadcast frames will circle the loop indefinitely. Each new broadcast will add traffic that never disappears to the network, and the network will collapse under the load within a few minutes. If the network has multiple loops, every time a frame arrives at a bridge with more than two ports or a LAN with more than two attached bridges, multiple copies of the frame are forwarded. This network collapses in a few seconds.

One way to eliminate broadcast storms is to eliminate the redundancy. However, there are many reasons why redundancy might be desirable. If a mission-critical application resides on one of the LANs in an internetwork, and only a single path exists between a given user and the application in question, the failure of any link or bridge along that path isolates the user from the application. Because LANs usually support more than one user, failures in such a network typically isolate entire groups. In fact, any single network failure divides the internetwork into two, unconnected entities. Users attached to one portion are not able to access applications on the other, and vice versa.

Clearly redundant paths make a network more tolerant of failure. However, redundant paths also lead to frame duplication at best, and broadcast storms at worst. Choosing between these two alternatives is a dilemma.

The solution taken by the transparent bridge is to temporarily disable redundancy. Transparent bridges are capable, if correctly configured, of locating and eliminating redundant ports on the various bridges in the network. The ports are not physically eliminated or disabled; rather, they are logically *blocked*. Bridges are prohibited from forwarding frames to these ports and cannot forward frames collected from these ports to any other port. They are also prohibited from learning on these ports. However, they do monitor the ports to verify that the attached LAN has not been isolated from the rest of the network. If such a condition arises, the bridges can reactivate one or more ports to restore the integrity of the network.

The protocol used by a transparent bridge to locate and eliminate redundant paths is called the Spanning Tree Protocol (STP). The name of the protocol comes from its effect on the network—STP takes a fully or partially meshed network and logically reduces it to a tree-like configuration. To maintain connectivity, the tree must span the entire network, hence the expression "spanning tree."

Source Route Bridging

Source route bridging (SRB) is a bridging strategy strongly advocated by IBM and commonly associated with Token Rings. The name describes how the protocol functions: The station originating the frame (the source) provides the route through the network's various bridges and rings. Bridges do not need to maintain a filtering database or to learn the location of stations (LAN adapters) within the network. In fact, the bridges are passive in the decision process. They simply examine information contained within the frame to determine whether they should filter or forward the frame.

This strategy has several advantages over transparent bridging. Since each frame contains a specific route, the problem of endlessly looping frames and broadcast storms is eliminated. This allows a type of load balancing as well. Frames sent from one LAN segment to another may end up taking different bridges. As a disadvantage, SRB begs the question of how client A finds the route to client B without ever having contacted that machine previously? The answer to this is explorer traffic. End stations in SRB generate explorer traffic that threads the network. In a network with a great deal of clients, this explorer traffic can become significant.

Like Token Ring, the market for SRB has greatly diminished in recent years. Even its largest proponent—IBM—is recommending a change of technologies to Ethernet LANs and transparent bridges.

SUMMARY

This chapter discussed the important technologies involved in the design and operation of the LAN. While not totally inclusive, it gives the reader an understanding of the variety of design decisions, media types, MAC schemes, and standards involved. Indeed, the nuts and bolts of LAN can be detailed, but fortunately for many budding network administrators, simply plugging devices together can configure most LAN networks. The details are hidden from the end user. The popularity of LAN technology has not only made LANs easier to use, but it's common enough that even those behind the technology curve can now create effective LAN networks in their own homes.

CHAPTER 16

The First PSPDN: X.25

By the mid-1970s, it had become apparent that statistically multiplexed packet-switching technology was well suited to applications with *bursty* transmission characteristics. In a bursty application, the length of the gaps between data transmissions exceeds the length of the transmissions themselves. Examples of bursty applications are credit card verification, banking transactions, and airline and hotel reservation applications. These applications tend to exhibit low-line utilization characteristics and are particularly ill suited to circuit-based networking approaches. Packet switching, with its ability to share network resources efficiently through statistical multiplexing, is the perfect technology for support of these bursty applications.

In 1974, a forerunner to today's Internet Protocol (IP) was operating successfully on the ARPANet. That same year, IBM began shipping Systems Network Architecture (SNA) technology with its System/370 line of mainframe computers. SNA was based on connection-oriented packet-switching technology. Also at this time, several national Postal, Telephone and Telegraph (PTT) authorities, most notably those in Canada and France, were offering packet-switched services on their public networks. Unfortunately, these national packet-switched public data networks (PSPDN) operated using proprietary protocols. At the request of the nations that were operating PSPDNs at the time, the Consultative Committee on International Telephony and Telegraphy (CCITT, now the International Telecommunication Union-Telecommunication Standardization Sector, or ITU-T) ratified an interim standard for access to a public packet-switched public network in 1974. That standard, Recommendation X.25, was officially ratified at the CCITT plenary session in Geneva in 1976.

Recommendation X.25 specifies a user-network interface to a PSPDN. More specifically, X.25 specifies a point-to-point, dedicated (private line), full-duplex interface between packet-mode data terminal equipment (DTE) and data circuit terminating equipment (DCE) in a PSPDN. The X.25 interface comprises three protocol layers that correspond roughly to the lower three layers of the OSI Reference Model. The layers of the X.25 protocol stack will be examined in detail in this chapter.

Between 1976 and 1984, numerous other standards, each in the form of an "X.dot" recommendation, were ratified by the CCITT for use in PSPDN environments. These include X.75, X.3/X.28/X.29, X.32, X.31, and X.121. This chapter will briefly discuss each of these standards.

You might wonder at this point why we are focusing on a set of standards whose use has been on the decline, especially in highly developed parts of the world (such as the United States and Canada), for some years. It is true that X.25 and related PSPDN technologies are gradually being replaced, both by newer technologies (such as frame relay and Asynchronous Transfer Mode, or ATM) and by contemporary technologies that are enjoying a renaissance (such as TCP/IP). But X.25 is important for reasons that go beyond the purely technical. From a standardization point-of-view, X.25 and its related recommendations represent one of the most comprehensive efforts ever made by a standards body. In fact, it is the completeness of the PSPDN standards that provided the impetus for carriers the world over to build PSPDNs and interconnect them into a truly global network for transaction-based data. In the same vein, the X.dot recommendations enjoyed near ubiquitous support by computer and other equipment vendors at one point in time. Finally, as an early effort, X.25 provides an excellent model for the development of a user-network interface (UNI). As we will emphasize in this chapter, many of the technologies that have become popular over the past few years have their roots in X.25. Having established the importance of X.25 on several levels, let's take a close look at it from a technology viewpoint.

ITU-T (CCITT) RECOMMENDATION X.25

Figure 16-1 illustrates the X.25 interface from two points of view. At the top of the figure, you can see X.25 as a point-to-point, dedicated duplex interface between a packet-mode DTE and a DCE in a PSPDN. In this context, a *packet-mode DTE* is a terminal device with the hardware and software capabilities necessary to implement the three layers of the X.25 protocol stack. That protocol stack is shown at the bottom of Figure 16-1.

At the Physical layer (layer 1), X.25 implements one of two protocols: X.21 or X.21*bis*. At the Data Link layer (layer 2), X.25 implements the Link Access Procedure-Balanced (LAPB) protocol. At the Network layer (layer 3), X.25 implements a protocol called the X.25 Packet-Layer Protocol (PLP). Note that as an interface specification, X.25 says nothing concerning DCE-to-DCE interactions within the PSPDN. The goings-on in the "cloud" shown in the figure are irrelevant from an X.25 point of view. As long as the procedural and operational aspects of X.25 are met at the interface level, a particular network can be characterized as X.25 compliant, regardless of its internal protocol implementation.

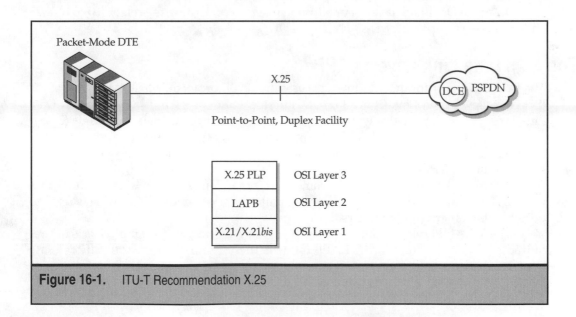

Figure 16-1.   ITU-T Recommendation X.25

The X.25 Physical Layer: X.21/X.21bis

The physical circuit over which the X.25 interface operates must be a point-to-point dedicated facility. In addition, it must support duplex transmission (see Chapter 4). The standard makes no mention of the specific physical facility used to implement the interface. Data rate, however, may be limited by the use of a specific Physical layer protocol. Common physical facilities that have been used in X.25 settings include analog private lines (such as 3002 circuits) and digital private lines (such as DDS circuits and T-1/E-1 circuits). Speeds from 300bps to 1.544Mbps can be found in X.25 environments.

The X.25 Physical layer protocol may be either X.21 compliant, or compliant to a variant of that Recommendation, X.21 bis. ITU-T Recommendation X.21 is a Physical layer protocol that is commonly used outside North America for access to Circuit Switched Public Data Networks (CSPDN). (X.21 is covered in Chapter 4; the reader is directed to that chapter for additional detail.)

X.21*bis* is used commonly in North America (United States and Canada) in place of X.21. For all practical purposes, X.21*bis* is equivalent to EIA-232-E. You will recall from Chapter 4 that EIA-232-E is limited to a data rate of 20Kbps (although it is often driven to higher rates with the use of special cabling). When rates in excess of 20Kbps are required in North American environments, ITU-T Recommendation V.35 is typically employed. Finally, X.25 is encountered in ISDN environments, both on the B-channel and on the D-channel. When X.25 runs over the ISDN B-channel on the Primary Rate Interface (PRI), V.35 is the typical Physical layer protocol. When X.25 runs over the B-channel on the Basic Rate Interface (BRI), the BRI Physical layer (2B1Q) is employed. When X.25 runs over the D-channel on the BRI, 2B1Q is used at the Physical layer along with a change from LAPB to LAPD at the Data Link layer. (For additional detail concerning the relationship of X.25 and ISDN, see Chapter 13.)

The X.25 Data Link Layer: LAPB

The Data Link layer for the X.25 interface uses the LAPB protocol, which is a subset of the ISO High Level Data Link Control (HDLC) protocol and, as such, is a bit-oriented Data Link layer protocol. Specifically, LAPB operates in Asynchronous Balanced Mode (ABM) in two-way alternate fashion. LAPB is designed to run over point-to-point duplex physical facilities. LAPB provides connection-oriented, sequenced, acknowledged bit-error-free service to the X.25 PLP.

The format of a LAPB frame is the same as that of an HDLC frame (see Chapter 5, Figure 5-2). The frame is delineated by the flag pattern at both ends. Bit stuffing is used to preserve the integrity of the flag pattern.

Because LAPB operates only on point-to-point facilities, addressing *per se* is not required. The address field in the LAPB frame is therefore used merely to differentiate Data Link layer commands from Data Link responses. When the DTE sends a command, the address field is encoded as *00000001*; when the DTE issues a response, the address field is encoded as *00000011*. The reverse is true for commands and responses issued by the DCE.

The control field indicates the LAPB frame type. Figure 16-2 illustrates the frame types that are allowed in LAPB, along with the encoding of the control field for each frame type. LAPB uses the information transfer frame type, several supervisory frame types, and several unnumbered frame types.

Information transfer frames are used to carry X.25 packets. Their length will be dictated by the length of the embedded packet. A rule in X.25 is that only one layer 3 packet can be contained in a LAPB frame. Information transfer frames also carry send sequence numbers, or $N(S)$, and receive sequence numbers, or $N(R)$, in their control fields.

Supervisory frames include receiver ready (RR), receiver not ready (RNR), and reject (REJ) frames. All supervisory frames contain receive sequence numbers for acknowledgement purposes. In addition, the RR and RNR frames are used to implement flow control over the LAPB link.

Unnumbered frames are used primarily for link establishment and release.

Format	Command	Response	1	2	3	4	5	6	7	8
						Encoding				
Information transfer	I (Information)		0	$N(S)$			P	$N(R)$		
Supervisory	RR (Receive Ready)	RR (Receive Ready)	1	0	0	0	P/F	$N(R)$		
	RNR (Receive Not Ready)	RNR (Receive Not Ready)	1	0	1	0	P/F	$N(R)$		
	REJ (Reject)	REJ (Reject)	1	0	0	1	P/F	$N(R)$		
Unnumbered	SABM (Set Asynchronous Balanced Mode)		1	1	1	1	P	1	0	0
	DISC (Disconnect)		1	1	0	0	P	0	1	0
		DM (Disconnect Mode)	1	1	1	1	F	0	0	0
		UA (Unnumbered Acknowledgment)	1	1	0	0	F	1	1	0
		FRMR (Frame Reject)	1	1	1	0	F	0	0	1

Figure 16-2. LAPB frame types and control field encoding

The X.25 Network Layer: Packet-Layer Protocol

At the Network layer, the X.25 PLP provides connection-oriented service to its upper layers. As such, procedures for establishing virtual calls, transferring data within virtual calls, and terminating virtual calls are specified at this layer. Virtual calls at the PLP are identified by a logical channel number (LCN) in the header of each packet belonging to a particular call. The X.25 PLP is statistically multiplexed. That is, multiple simultaneous virtual calls can be established over a single LAPB data link. Each virtual call is differentiated from all others by a unique LCN.

X.25 PLP Packet Types, Formats, and Encodings

All X.25 packets share three common octets in their headers. These common fields are shown in Figure 16-3. The general format identifier (GFI) comprises the most significant 4 bits of octet 1. The GFI indicates the modulus of a virtual call. The X.25 PLP can operate using a modulus of 8, 128, or 32,768. For PLP data-transfer packets, the GFI also includes two special bits—the D-bit and the Q-bit. D- and Q-bits will be discussed in more detail in the sections "End-to-End Acknowledgement: The D-Bit " and "Sending 'Qualified' Data: The Q-Bit" later in this chapter.

The least significant 4 bits of octet 1 and all of octet 2 make up the LCN of a particular virtual call. Because 12 bits are available to number virtual calls, up to 4094 unique virtual calls can be used per X.25 interface (all 0s and all 1s are reserved values and cannot be used to identify virtual calls).

Octet 3 is the packet type identifier (PTI), which is encoded according to the specific packet type. Twenty-one packet types have been defined in X.25. Figure 16-4 shows each of the 21 packet types along with the encoding of the PTI field for each. X.25 packets can be divided into categories, as shown in the table.

Call setup and clearing packets, as their names imply, are used to establish and release a virtual call. Among other fields, these packets typically contain the full destination address of the called party. Sometimes these packet types also include the address of the calling party.

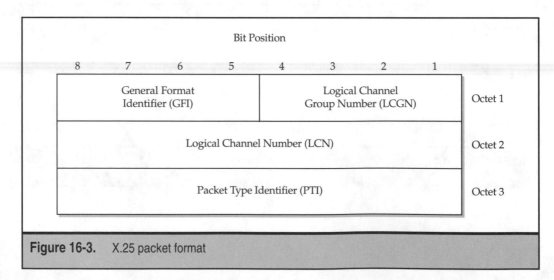

Figure 16-3.   X.25 packet format

Packet type		PTI Encoding							
From DCE to DTE	From DTE to DCE	8	7	6	5	4	3	2	1
Call setup and clearing									
Incoming call	Call request	0	0	0	0	1	0	1	1
Call connected	Call accepted	0	0	0	0	1	1	1	1
Clear indication	Clear request	0	0	0	1	0	0	1	1
DCE clear confirmation	DTE clear confirmation	0	0	0	1	0	1	1	1
Data and interrupt									
DCE data	DTE data	X	X	X	X	X	X	X	0
DCE interrupt	DTE interrupt	0	0	1	0	0	0	1	1
DCE interrupt confirmation	DTE interrupt confirmation	0	0	1	0	0	1	1	1
Flow control and reset									
DCE RR (modulo 8)	DTE RR (modulo 8)	X	X	X	0	0	0	0	1
DCE RR (modulo 128)[a]	DTE RR (modulo 128)[a]	0	0	0	0	0	0	0	1
DCE RR (modulo 32 768)[a]	DTE RR (modulo 32 768)[a]	0	0	0	0	0	0	0	1
DCE RNR (modulo 8)	DTE RNR (modulo 8)	X	X	X	0	0	1	0	1
DCE RNR (modulo 128)[a]	DTE RNR (modulo 128)[a]	0	0	0	0	0	1	0	1
DCE RNR (modulo 32 768)[a]	DTE RNR (modulo 32 768)[a]	0	0	0	0	0	1	0	1
	DTE REJ (modulo 8)[a]	X	X	X	0	1	0	0	1
	DTE REJ (modulo 128)[a]	0	0	0	0	1	0	0	1
DCE REJ (modulo 32 768)[a]	DTE REJ (modulo 32 768)[a]	0	0	0	0	1	0	0	1
Reset indication	Reset request	0	0	0	1	1	0	1	1
DCE reset confirmation	DTE reset confirmation	0	0	0	1	1	1	1	1
Restart									
Restart indication	Restart request	1	1	1	1	1	0	1	1
DCE restart confirmation	DTE restart confirmation	1	1	1	1	1	1	1	1
Diagnostic									
Diagnostic[a]		1	1	1	1	0	0	0	1

a) Not necessarily available on every network.

NOTE – A bit which is indicated as "X" may be set to either 0 or 1 as indicated in the text.

Figure 16-4.   X.25 PLP packet types and PTI field encoding

Although there are exceptions, most PSPDNs use the addressing format provided in ITU-T Recommendation X.121. We will examine the X.121 addressing format later in this chapter in the section "A Global PSPDN Addressing Scheme: Recommendation X.121."

Data and interrupt packets are used to transfer user data within an X.25 virtual call. At the PLP, maximum packet sizes may vary from a low of 16 octets of user data to a high of 4096 octets of user data. The default maximum packet size in X.25 is 128 octets of user data; all PSPDNs must support that maximum packet size.

Looking again at Figure 16-4, we see that the encoding of the PTI for data packets requires only that the least significant bit be set to 0. The encoding of the remainder of the PTI bits for a data packet is shown in Figure 16-5. Because X.25 provides acknowledged service, it is no surprise that data packets contain send sequence numbers—P(S)—and receive sequence numbers—P(R). Note that the send and receive sequence number formats shown in the figure are for a virtual call using modulo 8 sequencing. In addition, data packets contain a bit called the M-bit. We will discuss the role of this bit in the section "PLP Packetizing and Reassembly: The M-Bit."

Interrupt packets provide a mechanism whereby urgent data can be sent in X.25. Although not specified in the standard, most vendors' equipment maintains two queues for each output port—one for normal data and one for interrupt data. The typical rule is to be sure to empty the interrupt queue prior to servicing the normal queue. In this fashion, interrupt packets are expedited across the X.25 interface. Although interrupt packets are acknowledged (using the interrupt confirmation packet), they are not sequenced, because only one interrupt packet can be outstanding at any point in time. The length of the user data field for an interrupt packet can never exceed 128 octets.

Flow control is a critical aspect of X.25 service because of the guaranteed nature of virtual circuit operations. To ensure that no packets are lost, it is important to limit the number of outstanding unacknowledged packets across the interface. This limit is called the PLP "window size." Although any window size between the value of 1 and the modulus −1 is acceptable in theory, the X.25 default window size is 2. To simplify buffer management tasks, most PSPDN operators use the default value for the PLP window size.

While it is not always an absolute requirement that transmitters at the Data Link layer stop transmission upon the receipt of an RNR frame, transmitters at the PLP layer *must* stop transmitting when they receive an RNR packet. Only when a subsequent RR packet is received can a flow-controlled transmitter resume sending operations. As one would expect, flow-control packets are also used for acknowledgement purposes and contain a receive sequence number, P(R). In Figure 16-4, the 3 bits used for P(R) in modulo 8 environments are delineated by Xs.

Reset packets are used to signal failures at the PLP. Because no network can truly guarantee delivery of data units, the reset packets provide a mechanism by which the network can inform the user in the rare event that data is lost. Upon notification, it is the responsibility of an upper-layer protocol (such as the Transport layer) to affect recovery.

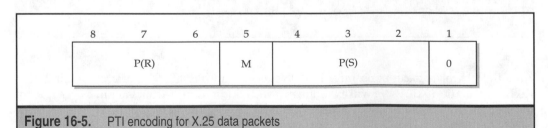

8	7	6	5	4	3	2	1
P(R)			M	P(S)			0

Figure 16-5.   PTI encoding for X.25 data packets

Restart packets are used to restart the PLP—these can be thought of as the "CTRL/ALT/DELETE" of the PLP. Whenever an unrecoverable error occurs across the X.25 interface, the interface is literally restarted by the use of these packets.

Diagnostic packets are used by the network to inform the user about network status information. They contain fields that indicate specific conditions that may exist in the network. Diagnostic packets always flow from the DCE to the DTE, and they always use logical channel number 0.

X.25 Call Establishment and Release

The ladder diagram in Figure 16-6 illustrates the process of establishing a virtual call across an X.25 interface and then clearing that call when it is no longer necessary.

1. The calling party DTE issues a call-request packet across the interface to its local DCE.

2. The DTE includes the address of the called party in the packet, and it also selects an unused LCN to apply to the call.

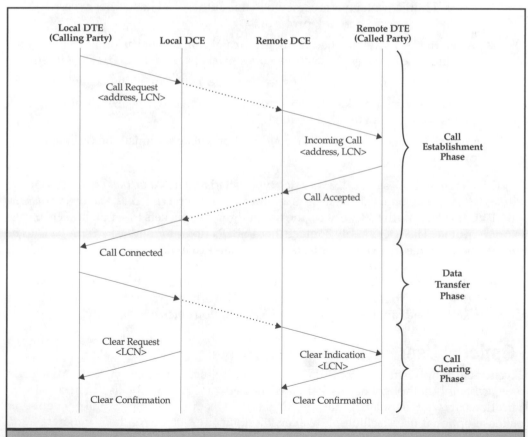

Figure 16-6.   X.25 call establishment and clearing

3. The call-request packet emerges from the network at the remote end as an incoming call packet, which flows from the remote DCE to the remote DTE.

4. The remote DCE chooses a LCN to apply to the call at the remote end. If the called party so desires, it accepts the incoming call by issuing a call-accepted packet to the DCE. The call-accepted packet uses the same LCN as the incoming call packet (X.25 virtual calls are duplex).

5. The calling party is informed of the acceptance of the call when its DCE issues a call-connected packet on the same LCN that the original call-request packet went out on.

6. The call enters the data transfer phase, wherein data (and possibly interrupt) packets and acknowledgements flow across both the local and the remote DTE/DCE interfaces. Flow control packets are used when necessary to prevent data loss. (The reader will note that, between the DCEs, nothing about the internal operation of the network is known—thus the use of dotted lines to indicate transport within the network. Interestingly, many different vendor's internal implementations are known to this author, and he has never seen X.25 used as an internal network protocol!)

7. When the DTEs are finished using the virtual call, one or the other initiates the clearing process by issuing a clear-request packet across the DTE/DCE interface.

8. Locally, the DCE clears the call by issuing a clear-confirmation packet to its DTE.

9. The clear-request packet emerges from the network as a clear-indication packet that is passed from the DCE to the DTE.

10. The DTE responds with a clear-confirmation packet to finish the call teardown process.

This process applies to an X.25 service type called a *switched virtual circuit* (SVC) service. Also available on most PSPDNs is another service type called *permanent virtual circuit* (PVC) service. With PVC service, the use of signaling packets to set up and clear calls is not required. The user simply contacts her PSPDN operator and asks for a PVC to be provisioned between two end points. Some time later (usually several days), the user is informed that her PVC is available and that she can begin using it. The advantage of PVC service is that often, a DTE needs to connect only to one other DTE on the network. PVC service offers a way to reduce repetitive call setup overhead under those circumstances. Of course, the disadvantage of PVC service is its obvious lack of flexibility.

X.25 Optional User Facilities

Twenty-eight optional user facilities are defined in Recommendation X.25. An *optional user facility* is best thought of as a feature of the network that may be useful in certain applications. It is not mandatory that a particular PSPDN operator support all or even some of these optional user facilities; however, these facilities can often help generate additional revenue because of their status as network features. X.25 optional user facilities may be offered on a per-port basis in which all virtual calls made over a given physical port have access to the facility. Some facilities are offered on a per-call basis. In the per-call case, the desire to use a particular facility is indicated by special facilities fields in call-establishment packets.

A complete description of each of the X.25 optional user facilities is well beyond the scope of this text. The interested reader is referred to Section 6 of the Recommendation for additional detail. Several examples of X.25 optional user facilities are offered in the following sections so that the reader can understand the usefulness of these network features.

Closed User Group

Closed user group (CUG) is a security feature. The members of a CUG are assigned a CUG identifier that must be included in call-establishment packets to place calls to other members of the CUG. Without the CUG identifier, calls will not be accepted by other members of the CUG.

CUGs have many variations. In one variation, only members of the CUG can place virtual calls to each other. Access outside of the CUG is barred. Furthermore, access by anyone outside the CUG is also prohibited. In another variation, members of a CUG can place calls to any destination in the network, regardless of whether or not the destination is part of the CUG. Calls to CUG members from outside of the CUG, however, are prohibited. Finally, a CUG can be set up so that anyone on the network can call into the CUG, but CUG members are restricted to destinations that belong to the CUG. Although somewhat less sophisticated, the concept of the CUG closely parallels that of the modern virtual private network (VPN).

Reverse Charging and Reverse Charging Acceptance

Everyone has had the experience of making a collect telephone call. The reverse charging optional user facility provides a mechanism whereby a collect virtual call can be made in a PSPDN. Such calls will be accepted, however, only by destinations that have subscribed to the reverse charging acceptance user facility. This facility is useful for providers of services who wish to bundle network costs into the cost of their service. In such cases, the calling users do not see a bill for the use of the PSPDN; the called party is billed for all usage.

Flow Control Parameter Negotiation

This X.25 optional user facility permits users to negotiate both maximum packet size and window size on a per-call basis. By requesting an increase in one or both of these parameters, users can increase throughput on a given virtual call. Requests for non-standard maximum packet size or non-standard window size may or may not be granted by the PSPDN operator. Generally, if traffic on the network is light, these requests will be granted, often at additional cost to the user. If, however, the network is congested, a request to negotiate flow-control parameters may be denied, or a set of "middle ground" parameters might be suggested by the network. Note that these "negotiations" are not arguments. The user makes a request at call setup time. The network will either grant the request, deny the request, or suggest other values. If the network grants the request, the call is established. If the network denies the request, the call is cleared. If the suggested parametric modifications are acceptable to the requestor, the call is established; if not, the call is cleared.

One can see from these examples that X.25 optional user facilities are powerful features that can be quite attractive to users of a PSPDN. PSPDN providers, therefore, try to support as many of these features as possible. In doing so, they differentiate their offerings from those of other providers and, at the same time, increase their revenues by charging for use of the facilities.

Special Operations at the X.25 PLP

When compared to the OSI Model's Network layer, the X.25 PLP exhibits several unusual characteristics. The astute reader will have already noticed that, because X.25 is a point-to-point interface, no routing is performed by the PLP. The PLP merely provides addressing information to the network. Based on that information, the network builds a route to the destination by whatever means it has at its disposal. Such operations are not a part of the X.25 standard.

The PLP is also atypical of a Network layer protocol in that it performs several functions that are usually considered to be Transport layer functions. These operations involve the use of several special bits in the headers of PLP packets. Specifically, X.25 data packets contain a D-bit and a Q-bit (both in the GFI) as well as an M-bit (in the data packet PTI). The use of these bits is discussed in the following sections.

End-to-End Acknowledgement: The D-Bit

D-bit modification is an X.25 optional user facility. When subscribed to, D-bit modification supports end-to-end acknowledgement of packets at the PLP. Under most circumstances, local acknowledgement is used across the DTE/DCE interface. A packet submitted to the DCE is usually acknowledged immediately upon receipt by that DCE. The meaning of such a local acknowledgement is "the DCE has received the packet and will employ the underlying reliability of the network to guarantee its passage to the destination DTE." Note that with local acknowledgement, the user is relying on network reliability to ensure accurate delivery of the packet. There is no way to know, in an absolute sense, whether the packet in fact arrived at its destination. In such cases, the D-bit is set to 0 in the packet header and the Transport layer is responsible for checking the incoming packet stream for sequence integrity.

When D-bit modification is supported by the network and subscribed to by the user, the D-bit can be set to 1 in the packet header by the originating DTE. When this happens, the local DCE withholds local acknowledgement of the packet. When the packet arrives at the destination DTE with the D-bit set to 1, that DTE is responsible for providing acknowledgement. That acknowledgement is then carried back across the network to the originating DTE, thus affecting end-to-end packet acknowledgement. In such cases, the Transport layer does not need to concern itself with the sequence integrity of the incoming packet stream.

D-bit modification therefore has advantages and disadvantages. Its greatest advantage is the assurance of the arrival of the packet at its destination. In mission-critical settings, this can be an important benefit. On the other hand, since the local DCE must wait for the remote DTE to acknowledge packets with the D-bit set to 1, throughput tends to suffer when D-bit modification is used.

PLP Packetizing and Reassembly: The M-Bit

According to the OSI Model, the Transport layer is responsible for segmenting messages in such a way as not to exceed the maximum packet size required by the Network layer. At the receiver, the peer Transport layer reverses the segmentation process to reconstitute the message. Through the use of the M-bit (in the PTI field of data packets), the PLP takes over the packetizing and reassembly function from the Transport layer.

When a message whose length exceeds the maximum packet size permitted by the PLP arrives at that layer, the PLP segments the message into packets of the correct length. The result is a string of related packets; when reassembled, they will form the original message submitted to the PLP. To indicate to the receiving PLP that this segmentation operation has occurred at the transmitter, the transmitter sets the M-bit in all but the last packet in the string to 1. In the final packet of the string, the M-bit is set to 0 by the transmitter. By examining the state of the M-bit, the receiver can thus reassemble the string of packets into the original message prior to delivering it to its upper layers.

Sending "Qualified" Data: The Q-Bit

The Q-bit (found in the GFI of data packets) is used to indicate an alternative destination for the contents of the user data field of a particular packet. Under normal circumstances, the Q-bit in a data packet is set to 0. This indicates that the data contained in the packet is destined for the end user (whatever entity rests at the top of the protocol stack at the receiver). When the Q-bit is set to 1, however, it is an indication that the recipient of the contents of the user data field is not the "typical" end user, but some other entity at the receiving location. For example, we may wish to control the configuration of a remote end user device during a virtual call. Perhaps we want to alter the value of a Data Link layer parameter (such as window size) in the middle of a call. Using the Q-bit, we can send the command, along with the new parameter setting, in the data field of an X.25 packet. When this packet arrives at the receiver, its contents will be directed to the Data Link layer entity, rather than to the actual end user of the virtual call. Thus, the Q-bit permits the choice of one or the other of two destinations for the contents of each X.25 data packet.

The recipient of "qualified" data (Q = 1) must be agreed upon at call setup time. As the Q-bit is a 1-bit field, a maximum of two destinations is supported. One is always the actual end user of the virtual call (Q = 0); the other is decided when the call is established (Q = 1) and cannot be changed during the call.

OTHER X. SERIES RECOMMENDATIONS USED IN PSPDNs

At one point in the mid-1980s, more than 125 PSPDNs were being operated around the world. Most of these networks were owned and operated by national governments, specifically by PTT administrations within those nations. In countries where significant deregulation of telecommunications services had already taken place (such as the United States, the U.K., and Canada), PSPDNs were run by private carriers. The requirement for

PSPDN standards beyond the X.25 Recommendation was recognized and acted upon by the CCITT (now ITU-T). This section introduces and discussed several of the important standards ratified for use in PSPDN environments.

Interconnecting PSPDNs: Recommendation X.75

To expand the global usefulness of national PSPDNs, a mechanism was needed by which networks owned and operated by different administrations could be interconnected seamlessly and transparently. To that end, the ITU-T ratified Recommendation X.75 as a provisional recommendation in 1978. Full ratification came in 1980 at the Geneva plenary session.

Recommendation X.75 is used to support an interface between two Signaling Terminal Equipments (STE), each of which belongs to a separate, autonomous PSPDN administration. An STE is best viewed in this context as a service gateway between two networks. X.75 bears much similarity to X.25. Thus, this treatment of X.75 will focus on differences between that standard and the X.25 Recommendation.

X.75 is a three-layer protocol stack that conforms roughly to the lower three layers of the OSI Model. At the Physical layer, two Physical layer protocols are supported. For interconnection of PSPDNs at 64Kbps, the Physical layer standard is as specified in ITU-T Recommendation G.703. For interconnection at E-1 rates (2.048Mbps), Recommendation G.704 is employed.

At the Data Link layer of Recommendation X.75, LAPB is used in precisely the same way it is used in X.25. For environments in which additional bandwidth is required across the gateway, an extension to LABP, called the *LAPB multilink procedure* (MLP) is defined. When the MLP is used, many individual LAPB links are made to appear as a single link to the X.75 PLP. This is accomplished by "sandwiching" a new protocol layer, the MLP layer, between multiple LAPB entities and a single PLP entity. The MLP basically uses a second set of sequence numbers to ensure that the sequence integrity of frames is preserved across the set of independent LAPB links. Through the use of the LAPB MLP, the bandwidth available over inter-PSPDN connections can best be tailored to the traffic requirement across a particular gateway interface.

In the X.75 PLP, packet formats and operational procedures across the interface are nearly identical to those of X.25. The most noteworthy difference between X.25 and X.75 concerns the format of call-establishment packets. In addition to carrying information relative to requested optional user facilities, X.75 establishment packets contain fields called *network utilities*. Many of these X.75 network utilities are identical to X.25 optional user facilities. These utilities are sometimes used across a gateway to permit one network to inform another about supported capabilities.

A few X.75 network utilities are unique to that implementation of the PLP. For example, a utility called *transit network identification* contains a code (a transit network identification code, or TNIC) that indicates that a virtual call traverses a particular PSPDN on its way from the calling party to the called party. For additional detail regarding X.75 network utilities, the reader is referred to that Recommendation.

Switched Access to a PSPDN: Recommendation X.32

ITU-T Recommendation X.32 defines procedures for switched access to a PSPDN by a packet-mode DTE. PSPDN access via a Public Switched Telephone Network (PSTN), an Integrated Services Digital Network (ISDN), or a Circuit Switched Public Data Network (CSPDN) is included in the standard. X.32 access is useful for those users of packet-mode DTEs that do not require continuous access to the PSPDN. Such sporadic users can cut access costs by deploying X.32 instead of X.25 because, unlike X.25, X.32 does not require a dedicated (and often costly) physical facility for access.

In almost every respect, X.32 is identical to X.25. Once connected to the PSPDN, an X.32 DTE uses LAPB and the PLP (as defined in X.25) for its layer 2 and layer 3 operations, respectively. X.32 must address two issues that are of no concern whatsoever in X.25 settings: the accommodation of a half-duplex Physical layer and procedures used for identification of the calling party DTE.

X.32 and Half-Duplex Physical Facilities

When an X.32 DTE accesses the PSPDN through either an ISDN or a CSPDN, Physical layer communication is duplex. In some cases, however, access via the PSTN may require support for half-duplex physical connections. To meet this requirement, X.32 defines a half-duplex transmission module (HDTM), which is a layer of protocol that is sandwiched between the half-duplex Physical layer and traditional LAPB. Using buffering techniques, the HDTM manages the half-duplex facility in a manner that is completely transparent to LAPB.

X.32 and Calling Party DTE Identification

Identification of the calling DTE is perhaps the most important function of X.32. In an X.25 setting, DTE identification is not required because it is permanently attached to the network by a dedicated facility. When the DTE accesses the network via a dial-in procedure, however, identification becomes important for both billing and security reasons. X.32 offers three alternatives for identification and authentication of the calling party DTE:

▼ Use of caller ID information supplied by the access network
  (a layer 1 approach)

■ Exchange of identification information in LAPB XID frames
  (a layer 2 approach)

▲ Identification of the DTE through use of the NUI optional user facility
  (a layer 3 approach)

A particular PSPDN may support one or more of these approaches to DTE identification. Additionally, X.32 provides procedures for dial-out by the PSPDN to a packet-mode DTE attached to the PSTN, an ISDN, or the CSPDN. In such cases, mechanisms by which the DTE can identify the calling DCE are also present in Recommendation X.32.

PSPDN and ISDN Interrelationships: Recommendation X.31

ITU-T Recommendation X.31 specifies two scenarios in which ISDN more or less plays a role in supporting X.25-based PSPDN service. The scenarios were originally dubbed minimum integration and maximum integration but are now known simply as case A and case B, respectively.

In the case A scenario, a single 64Kbps ISDN B-channel is used to provide transparent access for a packet-mode DTE to a DCE in a PSPDN. In this setting, DTE identification is handled by one of the alternatives specified in Recommendation X.32. In essence, X.31 merely addresses Physical layer compatibility with the ISDN circuit. Once established, the ISDN connection carries packets (according to X.25 PLP formats) embedded within LAPB data link frames.

The case B scenario is somewhat more complex. This scenario presumes that the X.25 DCE function is now supported in the ISDN network itself. The presence of a stand-alone PSPDN is optional for X.31 case B operations. In this scenario, X.25 packets can be sent over either the ISDN B-channel or the ISDN D-channel (in the case of the ISDN BRI). When sent over the B-channel, LAPB is used at the Data Link layer; when sent over the D-channel, LAPD is the Data Link protocol. If the destination user is attached to the ISDN, the packets never leave the ISDN network. If the destination user is attached to a PSPDN, an X.75 interface between the ISDN and the PSPDN is used to transfer the packets between networks.

PSPDN Access by Asynchronous Devices: Recommendations X.3/X.28/X.29

Until the boom in PC popularity in the mid-1980s, many users accessed remote resources using asynchronous terminal devices. These devices were often called "dumb" because they had no processor-based intelligence and little if any storage (buffer) capability. Nonetheless, literally millions of these devices were in common use. In recognition of the fact that PSPDN operators could serve the owners of these asynchronous devices if there were standards governing their attachment to PSPDNs, the ITU-T ratified a set of three recommendations concerned with the use of PSPDN services by "start-stop" (asynchronous) devices.

Definition of the PAD Function: Recommendation X.3

Recommendation X.3 defines the concept of the packet assembler/disassembler (PAD) function in a PSPDN. A PAD is a protocol converter. In the case of the X.3 PAD, the ports to which users attach support asynchronous data formats; the port that connects to the DCE in the PSPDN supports X.25 formats. Thus, the PAD is a specialized form of an X.25 DTE. At the transmitter, the role of the PAD is to receive asynchronous characters from the start-stop DTE, remove the start and stop bits (for additional detail, see Chapter 4), and place the resultant asynchronous characters into packets that conform to the X.25 PLP. Those packets are then embedded in LAPB frames and are shipped to the DCE over the physical circuit. At the receiver, the process is reversed. X.25 packets arriving at the PAD are processed by stripping off the PLP overhead, recovering the original asynchronous characters, and delivering them, with start and stop bits replaced, to the start-stop DTE.

Recommendation X.3 also defines a set of 29 PAD parameters that ensure compatibility between the asynchronous DTE and the PAD and that codify the details of their interaction. The parameters permit many different asynchronous DTEs to access the PSPDN. They are modifiable by the user of the asynchronous DTE, as well as by X.25 DTEs on the network. A thorough discussion of the operation of each of the X.3 PAD parameters is beyond the scope of this text. The reader is referred to Recommendation X.3 for additional detail.

PAD Commands and Service Signals: Recommendation X.28

ITU-T Recommendation X.28 defines a set of commands and responses used by the asynchronous DTE to access and use the transport services of a PSPDN. Remarkably, anything that a packet-mode user can do over the X.25 interface is available to the asynchronous user over the X.28 interface. Operations such as establishing virtual calls, transferring data, and clearing virtual calls are supported by the X.28 command set. During all phases of a virtual call, the asynchronous user is provided status information through a set of responses known as *PAD service signals*. These service signals appear on the screen of the asynchronous device to keep the user abreast of what is going on inside the network.

For example, an asynchronous user wishing to place a virtual call over a PSPDN might type the following X.28 command string on the keyboard of their device:

    C:8026550940,G21,R<enter>

The X.3 PAD will interpret the X.28 command string as, "I would like to place a call (C) to destination address 8026550940. I am a member of closed user group #21 (G21), and I would like to reverse charge the call to the destination DTE (R)." If the call is accepted by the called DTE, the user would see the PAD service signal "com" appear on his computer screen. This service signal indicates that the virtual call has been established according to the request of the user and that it has entered the data transfer phase. The user now simply begins to type messages on the keyboard of the asynchronous device. The PAD will packetize these messages and deliver them to the remote DTE.

Control of the PAD by an X.25 DTE: Recommendation X.29

In some cases, an X.25 DTE might wish to control the behavior of an X.3 PAD to which asynchronous users are attached. For example, under normal circumstances, the PAD is set up to provide local echo of each character typed. Thus, when a user presses a key on the keyboard, that character appears on the screen a short time later. Such "echoplex" operation is often used as an error-control mechanism in asynchronous environments. At certain times during a session, however, it is desirable to turn off this local echo feature. For example, when the user is asked to submit a password for access to an application, we generally don't want the password to appear on the CRT screen, lest its integrity be compromised by prying eyes. In these cases, the remote X.25 DTE must control the PAD to turn off the local echo feature momentarily and then turn it back on once the password has been submitted. Recommendation X.29 supports this type of operation.

Recommendation X.29 supports control of the X.3 PAD by the X.25 DTE by defining a message set to affect such control. The PAD control messages are carried across the X.25 interface in the user data field of X.25 packets. PAD control messages are distinguished from normal user data messages through the use of the Q-bit. When the contents of a data packet are destined for the asynchronous end user's CRT display, the Q-bit is set to 0. When, on the other hand, the packets contain X.29 PAD control messages, they are sent as "qualified" with their Q-bits set to 1. This is but one of the common uses of the Q-bit in PSPDN environments.

Other PAD Implementations

The ITU-T established a PAD standard for asynchronous devices because of their prevalence and because the asynchronous communications protocol is in the public domain. Unfortunately PAD standards for other types of non–packet-mode DTEs were never ratified by that body. The thinking was that other networking protocols were vendor-specific and proprietary and did not fall under the purview of the ITU-T. Nonetheless, the benefits that could be obtained by using the PSPDN to carry other types of traffic did not go unnoticed by users of those architectures.

As a result, a number of quasi-standardized approaches for converting other formats and protocols into X.25 were developed. Among the most important of these was a *de facto* standard for the transport of IBM's binary synchronous communications (BSC) formats using X.25. Referred to in the industry as *3270 display system protocol* or 3270 DSP, this approach segmented a BSC block into X.25 packets for transport across a PSPDN. Reassembly of the BSC block was affected by the PAD at the destination. 3270 DSP probably did more to extend the life of bisynchronous communications than any other development after its discontinuation by IBM in 1973.

PAD implementations have also been developed to support IBM SNA, Burrough's poll/select architecture, and Digital Equipment's digital network architecture (DNA), among others.

A Global PSPDN Addressing Scheme: Recommendation X.121

As mentioned previously, call-establishment packets carry addressing information. The address of the called party is always present in these packets; the address of the calling party may be included as well. Although it is not always the case, most PSPDN implementations use the addressing format specified in ITU-T Recommendation X.121. Along with X.75, X.121 supports the notion of a global, ubiquitous, transparent PSPDN.

The X.121 addressing format is shown in Figure 16-7. X.121 addresses are a maximum of 14 digits in length (and they can be as short as 5 digits) with each digit encoded in binary coded decimal (BCD) format.

The first four digits of the X.121 address are called the *data network identification code* (DNIC). The DNIC can be further subdivided into a one-digit zone (Z) code and a three-digit data country code (DCC). The zone code indicates where in the world the particular network resides (the ITU-T subdivides the world into seven zones); the DCC identified a particular network in that zone. DNICs are assigned by the ITU-T.

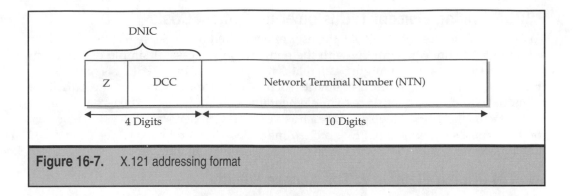

Figure 16-7. X.121 addressing format

The remaining 10 digits of the X.121 address make up the *network terminal number* (NTN). NTNs are typically assigned by some addressing authority within a country. For example, in the United States, area codes are assigned by the Federal Communications Commission (FCC). The remainder of the address is assigned by the administrator of the particular network (such as Verizon, SBC, and so on).

PSPDN ACCESS OPTIONS, PRICING MODELS, AND APPLICATIONS

Along with the hundreds of individual PSPDNs established and operated by telecommunications authorities throughout the world, a corresponding number of specific service delivery and pricing approaches have been used. Most of these, however, have certain elements in common. This section will examine those common elements with the goal of understanding how PSPDN pricing schemes were often directly responsible for the types of applications that were deployed over these networks.

Figure 16-8 illustrates the pricing elements common to almost all PSPDNs. Each element is numbered on the figure and will be discussed briefly in the following sections.

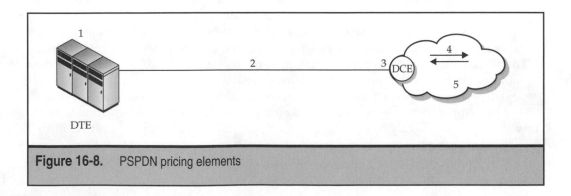

Figure 16-8. PSPDN pricing elements

PSPDN Pricing Element 1: Customer Equipment Costs

To utilize the services of a PSPDN, the user often needed to purchase hardware and/or software to ensure compatibility with the network. In the case of simple dial-in asynchronous access, a modem might be all that is necessary to achieve PSPDN connectivity. In the case of a more complex user device (such as a minicomputer or a router), the user might have to purchase software to implement the X.25 interface as well as hardware to connect to the network at high speed (such as a CSU/DSU). Generally, investment in customer premises equipment (CPE) is a one-time proposition and needs to be considered only when making the initial migration to a PSPDN environment.

PSPDN Pricing Element 2: The Access Facility

Although the cost of the access facility might be bundled with the cost of the actual access port (element 3), it is more typical to treat the access facility as a separate entity for pricing purposes. The actual charge for the access facility will depend entirely on the type of facility deployed. As we have seen, the type of facility is often a function of the specific protocol to be deployed for access.

Dedicated facilities (perhaps for X.25 access) are priced on a monthly basis. The actual cost will depend on the speed of the facility and the distance it traverses between the user and the entry point to the PSPDN. For short, low-speed analog facilities (such as 3002 circuits), these costs can be quite modest. But for lengthy, high-speed digital facilities, costs can escalate rapidly.

Switched (dial-up) facilities (perhaps for X.32 access) are generally charged on a usage-sensitive basis, where the usage is measured in minutes of occupancy. In addition, a fixed monthly cost component (how you probably pay for local telephone service) may be assessed in these settings. Switched access to a PSPDN is generally accommodated through a PSTN, an ISDN, or a CSPDN.

When a PSPDN provider designs its network, every attempt is made to locate access nodes as close to potential users as possible. In this way, the cost of access, either dedicated or dial-up, is minimized for the user.

PSPDN Pricing Element 3: Port Charges

When a customer connects to a PSPDN, the customer occupies a physical port on the network. This occupancy yields a charge commensurate with the speed and technology characteristics of the port in question. Generally, dedicated ports are charged on a monthly basis because the user occupies them all of the time. The actual monthly charge will depend on the speed and port technology used. As a general rule, the higher the data rate supported by a given port, the more expensive it will be. Additionally, digital ports are usually more expensive than analog ports, regardless of port speed.

For dial-in ports, users are typically charged by the minute for the time that they occupy the port. There may also be a speed component to the pricing formula, in that high-speed dial-in ports cost more per minute than their lower-speed counterparts. Some

PSPDN providers offer what can be termed a *private dial-in port*, whose access number is known only to members of a private group of users. When private dial-in service is supported, ports are priced as they would be in a dedicated scenario (on a flat, monthly basis).

PSPDN Pricing Element 4: Data Transport Charges

It is in this pricing element that PSPDNs differ from every other public network service type. Universally, PSPDN providers charge for usage by the actual amount of data transferred within a virtual call. Some charge by the character, some by the packet, and some by the segment. Regardless, the fact that one pays for the volume of traffic delivered by the network constrains the cost-effective use of these networks to applications that have low line utilization characteristics. As soon as the volume of traffic for a given application reaches a certain point, some other technology will emerge as a more cost-effective alternative to the PSPDN.

As mentioned at the beginning of the chapter, bursty, transaction-based applications like credit card validation, bank machine transactions, reservation system applications, and short database lookups are ideally suited to PSPDN technology and pricing models. In fact, one of the most difficult aspects of selling PSPDN services to end users is the degree of knowledge a salesperson must possess about the nature of the customer application. Unlike other network services, PSPDN services can be sold effectively only when one has a near-exact idea of the traffic volume generated by the particular application. Under some circumstances, this understanding can be hard to obtain. As a result, the sale of PSPDN services is not for the faint of heart!

PSPDN Pricing Element 5: Extra Cost Options

Providers of PSPDN services often charge for features over and above basic data transport. For example, the use of the X.25 optional user facilities may incur additional charges in certain PSPDN environments. Providers may also charge a premium for innovative billing features. Almost all providers charge extra for the establishment and maintenance of PVCs on their networks. An understanding of the features available on a given network and the cost of their use can be arrived at by carefully studying a provider's tariff or price list.

SUMMARY

Although the heyday of X.25-based PSPDNs was the mid-1980s or thereabouts, their use in many parts of the world remains strong to this day. This enduring popularity comes in part as a result of the efforts expended by standards bodies, most notably the ITU-T, to ensure compatibility across a myriad of equipment vendors and service providers in the PSPDN environment. This chapter has explored the most important of these standards. Hopefully, we have shown that the ITU-T's standardization efforts have resulted in a global, interconnected service environment in which providers can cost-effectively satisfy user requirements when those requirements meet certain criteria. Despite the fact that PSPDN offerings have been gradually disappearing in certain geographies (most notably North America), and that the Internet has taken the place of some of these offerings in recent times, the PSPDN remains as one of the three truly ubiquitous networks on the globe (along with the PSTN and the Internet).

Perhaps most important in the grand scheme of things, however, is the role that X.25 has played as a prototype for the development of standards and protocols that are of critical importance in today's telecommunications marketplace. Frame relay (see Chapter 17) is a direct descendent of X.25. In that ATM technology (see Chapter 18) has its roots in frame relay, one can safely say that, without X.25, we may not currently be enjoying the benefits that a shared, multiservice, converged network infrastructure can confer in terms of both cost and performance. In this regard, it may be wholly appropriate, years into the future, to look back and say, "X.25 is dead…long live X.25!"

CHAPTER 17

Frame Relay

Frame relay describes a standards-based interface between user equipment and a virtual circuit-based wide area network (WAN). Frame relay assumes the presence of robust higher layer protocols that operate in the hosts using the frame relay network. Frames with errors are discarded by the network; error correction (usually by retransmission) occurs under the control of the higher layers. Extreme network congestion could also result in discarded frames. While frame relay can transparently transport frames carrying information from any protocol, it depends on the presence of higher layer protocols to ensure end-to-end, error-free delivery of information.

Frame relay nodes perform a relay function only. That is, the node accepts a frame, does some minimal error checking, and then relays it towards its final destination. If the node has any trouble handling a frame, the frame is discarded.

Frame relay standards define the interface to a WAN providing unreliable virtual circuit service. In addition, it defines the interface among frame relay networks. Frame relay services have been enormously successful for service providers and users.

FRAME RELAY BASICS

Frame relay reduces network delay and improves nodal performance by reducing the amount of protocol processing performed at each node. To appreciate frame relay, recall that traditional packet-switching networks employ three levels of protocol: Physical, Data Link, and Network.

Since packet networks were originally built in the late 1960s, Physical layer transmission facilities have improved and error rates have declined dramatically. By assuming a largely error-free physical link, most of the error detection methods, and all error correction methods, can be removed from the Data Link layer. Because errors occur infrequently, the error correction mechanisms that are still employed on an end-to-end basis by the customer are sufficient.

This leaves the Data Link layer performing error detection and data delimiting functions only. Could we move Network layer functions to the Data Link layer? An Address field could be used to route information from multiple higher-layer sources and contain congestion control information. If we moved these functions to the Data Link layer, we could obviate the need for a Network layer. Our network now switches frames, not packets, and we have just developed frame relay.

Frame Relay Specifications

Frame relay products and implementations are based on standards. The International Telecommunication Union-Telecommunication Standardization Sector (ITU-T) provided the framework for frame relay in Recommendations I.122 and I.233. ANSI and other national standards groups have assimilated this work for further development. ANSI and the ITU-T have subsequently developed a series of standards completely defining the Data Link layer protocol and call control procedures. An industry group, the Frame Relay Forum (FRF) has developed a series of implementation agreements to help ensure equipment interoperability.

ANSI T1.606 provides a general description and framework of frame relay services and is roughly equivalent to Recommendations I.122 and I.233. These descriptions are based on the premise that frame relay would be an alternative to X.25 packet service in ISDN and, therefore, would rely on the standards developed to supplement existing ISDN specifications. In reality, frame relay has never been implemented as part of ISDN; instead it has been implemented as a stand-alone network using proprietary protocols between switches, or as a network service offered on an Asynchronous Transfer Mode (ATM) backbone. The FRF has also published a companion implementation agreement (FRF.1.1) for the User-to-Network Interface (UNI). Since the initial implementation agreement, the FRF has published the following:

▼ **FRF.1.1 User-to-Network Implementation Agreement** This agreement specifies the use of ITU-T Q.933 Annex A for signaling. It also defines the applicable physical interfaces and indicates the use of ITU-T Q.922 Annex A for data transfer. FRF.1 was revised in FRF.1.1 to include high-speed interfaces and optional loopback detection procedures.

■ **FRF.2.1 Network-to-Network Implementation Agreement** FRF.2 addressed the Phase 1 network-to-network implementations. The primary thrust of the document was defining when a multinetwork permanent virtual circuit (PVC) can be considered active. The second release of NNI IA is FRF.2.1.

■ **FRF.3.1 Multiprotocol Encapsulation Implementation Agreement** FRF.3 specifies how multiple protocols, when carried over the same frame relay PVC, can be identified in an agreed upon fashion.

■ **FRF.4 Switched Virtual Circuit Implementation Agreement** This agreement specifies a subset of Q.933 messages for the implementation of non-ISDN switched virtual circuits. Both the number of signaling messages and the content of the messages were considerably reduced.

■ **FRF.5 Frame Relay/ATM Network Interworking Implementation Agreement** This agreement pertains to running end-to-end frame relay, but interworking with an ATM network for transport at some point between the two end users.

■ **FRF.6 Frame Relay Service Customer Network Management Implementation Agreement** The FRF worked closely with the Internet Engineering Task Force (IETF) to develop a Simple Network Management Protocol (SNMP) management information base (MIB) for frame relay customer network management.

■ **FRF.7 Frame Relay PVC Multicast Service and Protocol Description Implementation Agreement** This agreement is based on ITU-T Recommendation X.6 ("Multicast Service Definition," 1993). The multicast supplementary services provide the ability for frame relay service providers to offer point-to-multipoint frame delivery services.

■ **FRF.8 Frame Relay/ATM PVC Service Interworking Implementation Agreement**   This specification is a joint implementation agreement between the ATM Forum and the FRF. It differs from FRF.5 in that it assumes the customer on one end of the connection has frame relay service and that the customer on the other end has ATM service. The two services, therefore, must interwork.

■ **FRF.9 Data Compression Over Frame Relay Implementation Agreement**   This IA describes the encapsulation of data compressed using the Data Compression Protocol within a frame relay network.

■ **FRF.10 Frame Relay Network-to-Network SVC Implementation Agreement**   This IA describes the use of SVCs in a multinetwork environment. It includes procedures for call establishment and management.

■ **FRF.11 Voice Over Frame Relay Implementation Agreement**   This IA extends frame relay application support to include the transport of digital voice payloads. It includes a variety of voice compression techniques and support for signaling necessary to establish voice and facsimile calls.

■ **FRF.12 Frame Relay Fragmentation Implementation Agreement**   This specifies techniques for fragmenting and reassembly of long frames. The main purpose for fragmenting frames is to control delay and delay variation when real-time traffic, such as voice and data traffic, share an interface.

■ **FRF.13 Service Level Definitions Implementation Agreement**   This IA describes a standard mechanism for defining service level agreements (SLAs) between customers and service providers.

■ **FRF.14 Physical Layer Interface Implementation Agreement**   This IA describes the Physical layer protocols currently used for frame relay, either at the User-to-Network Interface (UNI) or the Network-to-Network Interface (NNI). The IA covers options from 56Kbps to 600Mbps (OC-12).

■ **FRF.15 End-to-End Multilink Frame Relay Implementation Agreement**   This IA defines the mechanism for using multiple Physical layer interfaces to support a single frame relay data stream. The IA was created to provide compatibility between users who need more than single DS-1 but do not have the budget or bandwidth requirement for a DS-3.

▲ **FRF.16 Multilink Frame Relay UNI/NNI Implementation Agreement**   This is the latest in the series of IAs, and it specifies techniques for multilink operation between the user and the network as well as between networks. This IA supports NxDS1 and NxE1 interface configurations.

FRF Implementation Agreements can be downloaded from the Internet using a World Wide Web browser directed at **www.frforum.com/5000/5000index.html**.

Service Types

Frame relay was developed to augment the packet switching capabilities of X.25 for an ISDN. It provides a virtual circuit service, like X.25, and supports both switched virtual circuits (SVCs) and permanent virtual circuits (PVCs). A third type of service is being offered today as an enhancement to PVC service, the "soft" or "switched" PVC (SPVC).

Frame relay service was initially offered in early 1991. Since that time, access to the network has been over some sort of digital access facility, such as 56/64Kbps, DS-1/E-1, DS-3/E-3, OC-3c, or OC-12c. Initially, PVC services have been available, primarily relying on the LAPF-Core description in ANSI T1.618 and on FRF.1.1 and FRF.14.

SVCs have been slow to come to market, but users are beginning to demand SVC service because PVCs are somewhat inflexible. Most carriers have indicated plans to offer SVCs for several years; in late 1997, MCI (now WorldCom) became the first service provider to announce SVCs. These SVCs are based on the FRF's User-to-Network SVC Implementation Agreement (FRF.4). This agreement defines a subset of the ISDN call control procedures needed for equipment to set up and tear down switched frame relay calls. Since that time, SVCs have been offered internationally in Spain by Telefonica Transmission de Datos and domestically by Qwest Communications. It should be noted the demand for SVC service has been limited. Most frame rely network customers are happy with PVC connectivity.

The third type of frame relay service, the switched permanent virtual circuit (SPVC), offers the reliability of a PVC and the ease of use of the SVC. The *S* portion of the PVC is between switches in the frame relay network. If a switch-to-switch link fails, the switches automatically reroute and reconnect the PVC.

PVC frame relay service can be thought of as a logical or virtual replacement scheme for private line services. Consider a single private line being replaced by a single frame relay PVC. The user at the remote location continues to talk to the host computer with no apparent change in the normal routine. The application program running at the host has not changed—only the physical connection between the host computer and the remote terminal has changed. The dedicated private line between them has been replaced by a frame relay PVC.

Addressing Virtual Circuits

Figure 17-1 shows a frame relay network with a single customer with four locations. The user equipment attached to the frame relay node may be any data communications device, such as a bridge or router. Customer locations are interconnected over *virtual circuits*, which are logical, bidirectional, end-to-end connections that appear to the user as dedicated links. Virtual circuits are identified by the data link connection identifier (DLCI), which will be carried in the Address field of each frame relay frame.

It is important to note that in most public networks, DLCIs have *local* significance and are a local representation of an end-to-end virtual circuit connection. As an example, Figure 17-1 shows manufacturing's DLCI 17 associated with the virtual circuit connection

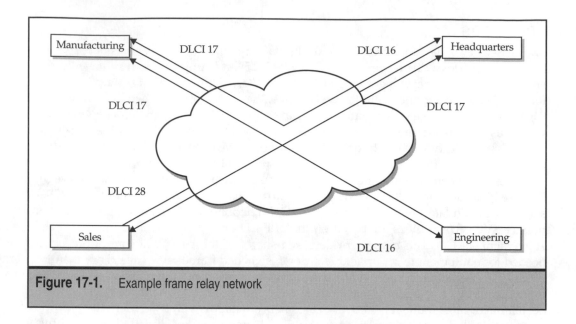

Figure 17-1.   Example frame relay network

to headquarters, where the same virtual circuit is referred to as DLCI 16. Headquarters also has a DLCI 17, which is the virtual circuit to sales (DLCI 28). There is no ambiguity here, because the DLCIs have meaning only at the local user network interface; it is the responsibility of the network to provide the necessary DLCI translation from node to node.

Two frame relay devices (such as router or frame relay access device, or FRAD) can communicate only if a virtual circuit is between them. Thus, in the example shown in the figure, data cannot be exchanged directly between Sales and Engineering. In the PVC environment, these virtual circuits are established at service subscription time and cannot be changed without contacting the service provider.

Switched Virtual Circuits

The PVC frame relay services offered today have been able to satisfy most customer needs. However, as customers on these networks mature, the need for SVC is likely to increase. PVC service works fine for fixed locations, but if users have only an occasional need to send data to a different location, the PVC service is of limited value. The user might have to seek a solution other than frame relay to fill this need. SVC service would serve this need very well.

Any-to-any connectivity is also being requested by customers. A switched service with a worldwide numbering plan would be ideal. However, with rapid advancements in switched Asynchronous Transfer Mode (ATM) services, it remains to be seen whether frame relay SVC services will evolve as the any-to-any connectivity service of choice.

Switched Permanent Virtual Circuits

The third and newest type of frame relay virtual circuit is the SPVC. The terminology for this type of circuit is still being debated; in some publications it is referred to as a *switched* PVC, in others a *soft* PVC. We will use *switched* to be consistent with the FRF.10 IA.

An SPVC is a PVC to the end devices. A customer is assigned permanent DLCIs at the UNI and the customer premises equipment (CPE) requires no additional signaling capability. The switched portion of the virtual circuit is established between the serving switches. When a PVC is provisioned, the interswitch portion of the circuit is established using routing tables and signaling messages rather than operational commands.

The benefit of the SPVCs is if a node or link failure occurs between the serving switches. In this scenario, the serving switch searches out another path through the network and reestablishes the PVC via these operational circuits.

LAPF-CORE FRAME FORMAT

Frame relay uses the transmission format of a classical HDLC frame. The fields of the frame and their functions are as follows:

▼ A flag is placed at the beginning and end of every transmission to provide frame synchronization. The bit pattern that makes up the flag (7E) is often used as an idle pattern as well.

■ The Address field carries the virtual circuit identifier, frame priority, and congestion notification information.

■ The Information field carries higher layer data.

▲ The frame check sequence (FCS) is used to detect bit errors.

DLCI Values

The default two-octet frame relay header contains a 10-bit DLCI field that will encode up to 1024 unique connection identifiers (0–1023). Because of various reserved values, only DLCI numbers ranging from 16 to 991 can be assigned to user data connections. This means that a given frame relay link can support only 976 user connections.

DLCI value zero (0) is reserved for frame relay signaling. Specifically, DLCI 0 will carry Q.933 SVC signaling and the ITU-T/ANSI PVC management protocols.

DLCIs in the range of 1 to 15 are currently reserved for future definition. Similarly, the DLCIs ranging from 992 to 1007 are also reserved. With regard to the latter DLCI range, DLCI 1007 is reserved for the defined but never implemented Consolidated Link Layer Management (CLLM) protocol.

Lastly, DLCI 1023 is reserved for the FRF's Local Management Interface (LMI) PVC management protocol.

Explicit Congestion Notification

Congestion control is usually a function of the Network layer protocol. Because frame relay does not have a Network layer, this function is moved to the Data Link layer. Furthermore, rather than have the network reduce the flow of information itself, notification of a congestion condition is passed to the end user.

Congestion notification is performed on each PVC by using the Forward Explicit Congestion Notification (FECN) and Backward Explicit Congestion Notification (BECN) bits in the Address field.

Congestion and Frame Discard

Frame relay supports an unreliable connection oriented service, and as such it operates under a simple rule: If there is any problem handling a frame, discard it. A number of protocol-related reasons might indicate that a frame be discarded:

- ▼ Invalid DLCI
- ■ FCS error
- ■ Frame too short or too long
- ■ Missing flag(s)
- ▲ Lack of octet alignment

A final reason for discarding frames is congestion. Severe congestion will cause network performance to degrade severely. It is the network's responsibility to maintain a service so it will take whatever steps are necessary to continue operation. To alleviate congestion, the amount of traffic must be reduced; if the end users do not reduce their traffic into the network, the network must discard some of the traffic.

FRAME RELAY CLASS OF SERVICE PARAMETERS

The standards define a set of Class of Service (COS) parameters for frame relay. Officially, frame relay can commit to throughput only related classes of service (such as the committed information rate and committed burst size). The lack of delay-related controls, both maximum delay and delay variation (sometimes called *jitter*), is the reason that frame relay is normally called a *data only solution*. Recognizing the need for voice/video support, many carriers offer service level agreements (SLAs) that, for a price, add delay and availability commitments to those supported by standards-based frame relay. Frame relay COS is based on the following parameters:

- ▼ **Access rate** The data rate of the access facility, in bits per second (bps).
- ■ **Committed rate measurement interval (Tc)** The interval of time over which information transfer rates are measured. The Tc is calculated as committed burst size divided by the committed information rate (Bc/CIR).
- ■ **Committed burst size (Bc)** The maximum number of bits the network agrees to transfer under normal conditions, during the time interval Tc.

- **Committed information rate (CIR)**   The information transfer rate, in bps, that the network commits to support under normal conditions (Bc/Tc).

- **Excess burst size (Be)**   The maximum number of bits, in excess of Bc, that the network will attempt to deliver over time interval Tc.

- **Excess information rate (EIR)**   Information transfer rate, in bps, the network will attempt to deliver under normal conditions (Be/Tc).

▲ **Discard eligibility indicator (DE)**   Indicates a frame that should be discarded in preference to other frames when frames must be discarded due to congestion.

Let's put these terms in their proper perspective. The access rate is simply the data transfer speed of the physical facility that connects the CPE to the frame relay network node. The CIR is the information transfer rate at which the network commits to deliver frames over some period of time. Any frames transmitted at or below the CIR are not marked eligible for discard (the DE bit is not set).

If the CPE transmits information at a rate greater than the CIR, the excess frames are marked discard eligible (the DE bit is set to 1) by the access node. There is, however, a limit to how much information the CPE can submit in a given time interval—that is provided by the excess burst size. If the CPE transmits more than Bc + Be bits in time Tc, the excess frames may be discarded immediately by the access node.

These COS parameters should be agreed upon between the user and the network provider *for each individual PVC on an access link*. Mathematical equations, as described, will be applied by the network access node to each PVC on the access link to determine whether frames are within the COS or are above the COS agreed to and should be marked eligible for discard.

Oversubscription of CIR

Many network providers allow customers to oversubscribe the amount of CIR they are allocated on an access link. Oversubscription is realized when the sum of the CIRs on a multi-PVC access link is greater than the rate of the access link. Of course, the customer can send no more than 64Kbps across a 64Kbps access facility, so how does this work? This only works because of the bursty nature of data traffic. The definition of "bursty" traffic indicates that there will be relatively long intervals of time between the bursts of data. Data traffic that meets this definition is also a good candidate for oversubscription of CIR. The longer the interval between bursts, the better the fit of the data to oversubscription and the less likely that multiple PVCs will reach their CIR at the same time.

Zero CIR

In 1994, a new twist was added to the CIR concept called *zero CIR*. A zero CIR provides a level of service consistent with its name. Specifically, users have a committed bandwidth of 0bps. So all frames transmitted to the network are at a rate in excess of the CIR and are, therefore, marked discard eligible.

Zero CIR was introduced for marketing purposes rather than technical ones, yet it provides customers with a cost-effective use of frame relay's bandwidth. When combined with a guarantee of delivery of 99 percent of all packets, it makes you wonder why anyone would bother with the complexity of the CIR.

LINK MANAGEMENT

Frame relay was initially designed to operate on the B-channels of an ISDN interface. As such, all management functions for the interface were carried out by the D-channel. When frame relay migrated from the ISDN environment to an independent existance, management functions had to be added to the protocol suite.

When multiple standards organizations developed link management protocols, one of the major differences between the protocols involved the choice of data link connection identifier (DLCI) on which to "signal" management information. One group felt that call control signaling should be done on DLCI 0 and management should be done on DLCI 1023; the other group felt that all signaling should occur on DLCI 0. The two groups could not agree.

Now three protocols provide PVC management, one from the FRF and the others from ANSI and the ITU-T. The ANSI and ITU-T methods are essentially the same—the few minor differences are in the field structure of the messages. This is because the ANSI standard is viewed as a national standard, which requires additional fields to indicate this in the message. The ITU-T is the international version and requires no additional identifying fields. For simplicity we will refer to these standards as ANSI/ITU-T, reflecting the common functionality.

Link Management Functions

The basic functions in link management protocols provide user equipment with the ability to track the status of its connections in the network as well as provide notification of the addition and deletion of PVCs.

Two messages are defined for basic operation: STATUS ENQUIRY and STATUS. The user sends a STATUS ENQUIRY message to the network to ask about the status of PVC connections; a STATUS message is returned by the network containing information about all PVCs on the link.

FRAME RELAY ACCESS OPTIONS

Access to a frame relay network is scalable from 56Kbps to 600Mbps. The choice depends on the customer requirements and the options available from the frame relay service provider. Two provisioning options are also available—*permanent leased* lines and *switched* lines.

For leased lines, the mainstay was for many years the 56Kbps digital data service (DDS) circuit; today T-1s are the primary access option. Many corporations are finding that Internet access and corporate frame relay access can be combined on the same T-1 using fractional service. High-speed dedicated access is provided starting at the DS-3 level and is currently supported with SONET standards through OC-3c (155Mbps). International frame relay service is supported on the Conference of European Postal and Telecommunications Administrations (CEPT) E-1 and E-3 standards as well as the European SONET equivalent, SDH.

Switched access to frame relay can be provided on the digital switched platforms Switched 56 and ISDN. Typically this type of access is used for backup.

THE NETWORK-TO-NETWORK INTERFACE

The initial implementations of frame relay did not require connections to other frame relay networks because so few of these networks existed. Relatively quickly, however, public and private frame relay networks sprang up in many different locations and the need for interconnection became obvious. A standard, the equivalent of packet switching's X.75, was needed for frame relay internetworking.

During 1992, ANSI worked on the development of T1.606b, the "Frame Relaying Network-to-Network Interface Requirements." Published in early 1993, T1.606b established protocol requirements for two phases of NNI protocol implementation and development. The first phase, for PVC-only service, stated the identical requirements that were applied to the UNI protocol with only one change—*bidirectional polling*, which was optional for the UNI, was made mandatory for the NNI. No changes were made to the data transfer specifications. Thus, a simple initial NNI implementation was made possible with no changes to the bits and bytes of the protocol. A procedural change was needed, however, in that each network would now determine the "active" or "inactive" status of each PVC in this new multinetwork environment. This kind of a procedural agreement could best be handled by an industry group rather than a standards group.

In the summer of 1992, the FRF published an implementation agreement for frame relay internetworking, defining how both public and private frame relay networks can effectively interwork with one another. Of particular importance, this agreement established the criteria under which a network could consider a multinetwork PVC to be "active" and thus report it as such across the customer UNI.

The major issue for the NNI development was to specify how link management and data transfer could be accomplished in a seamless fashion between two frame relay networks. In particular, users did not want to be bothered by the fact that more than one network connected two sites; users wanted to see a single virtual circuit between two points, much the same way X.25 networks provided a seamless interconnection.

As mentioned, the ANSI NNI requirements document also established the requirement for a second phase of development work. No new protocol work was needed for the first phase, but the second phase required protocol development. The second phase encompasses higher speed interface requirements and the support of both PVC service and SVC service.

The FRF has also kept pace with the industry for NNI. Three additional implementation agreements were added to the original NNI IA (FRF.2.1). These are the NNI Switched Virtual Circuit IA (FRF.10), the Frame Relay/ATM Interworking IA (FRF.5), and the Frame Relay/ATM Service Interworking IA (FRF.8). Together these documents specify a modern frame relay network.

FRAME RELAY AND ATM

Two basic forms of fast packet technologies could potentially form the basis for broadband services: frame-based and cell-based. ATM was designed for everything in the sense that ATM networks are meant to give adequate service for data, voice, video, and any other users on the ATM network alike—for details on ATM read Chapter 18.

Frame relay's goals were more modest to begin with: to speed up X.25 packet data service for possible broadband data use. However, if ATM is for everything, maybe frame relay is just another ATM service? Since frame relay defines only a UNI and an NNI, it can be be used for interswitch communications on a frame relay network. To this end, two frame relay/ATM interoperability standards exist, the first, Frame Relay/ATM PVC Network Interworking, provides for the transport of frame relay traffic over an ATM backbone, the second, Frame Relay/ATM PVC Service Interworking, provides for the interoperability between frame relay and ATM devices. In fact, most frame relay networks are implemented on an ATM backbone.

Frame Relay/ATM PVC Network Interworking Using FRF.5

Network interworking is the first step in providing interoperability between frame relay and ATM environments. Frame relay frames are encapsulated in ATM, thus maintaining the frame relay header and trailer. If an ATM device is to interact with a frame relay device, it must implement the interworking function. In other words, it must place information in a frame relay frame, which is, in turn, encapsulated in an ATM frame. Thus, the device is processing both frame relay and ATM information.

Frame Relay/ATM PVC Service Interworking Using FRF.8

Service interworking allows for transparent communication between frame relay and ATM environments. This allows network designers to select the best technology for their needs, without limiting their ability to interconnect locations using a different service. The interworking function handles the required processing between the frame relay and ATM environments, so any ATM end devices do not have to concern themselves with processing two sets of protocols.

FRAME RELAY CUSTOMER PROFILE

No single network service or technology is right for all customers—not even frame relay, despite some press to the contrary, can satisfy all customer needs. Frame relay is well-suited to customers with the following general characteristics.

▼   The customer should have multiple geographically dispersed business sites. Because frame relay is typically not billed on a distance-sensitive basis, greater economies are found as the sites are farther apart.

- There should be a requirement for a high degree of connectivity. If the sites are close together or only a few of the sites must be interconnected, a private line solution may be better. If a significant amount of site-to-site connectivity is required, however, frame relay's virtual circuit service may provide improved connectivity at a lower price than private line solutions.

▲ The customer should have a requirement for flexible communication between sites. Private lines lock a customer into a specific set of point-to-point connections; frame relay can provide a full mesh interconnection between many sites, either on a subscription basis (PVC) or dynamically (SVC).

While other network services may satisfy a customer with these needs, frame relay is particularly well-suited for a customer already using X.25 packet switching. A customer may need more connectivity and/or performance than its current network service, or its budget, provides. In these cases, frame relay provides a potential solution. While frame relay is more complex than "X.25 on steroids," it does provide a forward migration path from X.25. A customer using X.25 today may be happy with the service and the technology but might be concerned that a 64Kbps access rate is too slow. The frame relay alternative uses a similar technology but provides better performance at a relatively low upgrade cost.

The ideal customer configuration for frame relay has the following characteristics:

▼ Frame relay is well-suited for intracorporate communications, primarily because PVCs will be between two sites with the same owner *or* between a vendor site and a customer site.

- Frame relay is ideal for centralized applications, such as client/server or terminal-to-host, again because of today's requirement to establish PVCs.

▲ Frame relay PVCs will be defined in accordance with the traffic patterns of the user applications. Because PVCs cannot be dynamically reconfigured, they are well-suited to relatively static traffic patterns.

Frame relay has proved to be successful in replacing private line connections. The PVC simply replaces the individual circuit, usually at a much lower cost, yet with the same fixed cost pricing structure. This allows frame relay customers to simply replace their fixed cost private line networks with lower fixed cost frame relay networks.

Frame Relay in an ISP Network

Let's take a look at frame relay in an ISP network. Customers access the network using a variety of access technologies, including private line, frame relay, and ATM. Private line access is terminated at a router, or at the frame relay or ATM switch. Frame relay and ATM connections are terminated at the appropriate switch. It is worth mentioning that many switch manufacturers support frame relay and ATM on a single switch architecture; the service is determined by the line card.

Within the ISP network, frame relay and ATM are implemented to better utilize the Physical layer structure of their network. A higher degree of mesh connectivity can be achieved using frame relay and ATM, thus reducing the number of hops required, and delay incurred, in traversing the network. Many providers combine private frame relay and ATM networks with public services. In fact, many ISPs have substantial frame relay and ATM networks, perhaps larger than many frame relay and ATM service provider networks.

Some of the benefits of using frame relay in an ISP network are listed here:

▼ **Distance insensitive pricing** This allows an ISP to offer service nationally at the same rates.

■ **Customer side multiplexing** This reduces the number of physical ports and hardware at the ISP's points of presence (POPs).

■ **Protocol transparency** Frame relay can be used without altering the normal traffic flow.

■ **High-speed interfaces** An ISP can use a single technology in all locations without the worry of technical obsolescence.

▲ **High availability** Frame relay is available internationally.

Frame Relay and Other Protocols

It is often said that frame relay provides *transparent transport of higher layer protocol information.* From the network's perspective, this is quite true; the frame relay network does not care what type of payload it relays through the network, much less why the user generated it. However, the end user equipment needs to be able to differentiate between different higher layer protocols for several reasons, including these:

▼ To allow virtual circuits to employ any protocol to which both ends agree

▲ To allow the CPE to perform protocol-specific functions, such as routing, bridging, and various table update functions

Multiprotocol Standards

Several specifications describe the different aspects of multiprotocol operation in a frame relay environment, including the ones here.

▼ **RFC 2427, an update to RFC 1490 (IETF)** Describes how to transport many different types of protocols in the Information field of a LAPF-Core frame, including all IEEE 802 LANs, Fiber Distributed Data Interface (FDDI), IP, ISO's Connectionless Network Protocol (CLNP), LLC, and SNA/NetBIOS. It also describes bridging issues and the use of the Address Resolution Protocol (ARP) so that a frame relay device can associate a local DLCI value with another device's IP address.

- ■ **FRF.3.1 (Frame Relay Forum)**   Essentially a reprint and endorsement of RFC 2427, an update to RFC 1490 by the FRF.

- ▲ **Q.933 (ITU-T)**   Defines SVC signaling, including the information elements necessary to allow the calling party to specify the layer 2 and layer 3 protocols that will be used over an SVC once it is established.

Applications: LAN Interconnection

Connecting LANs has been the primary application from frame relay networks since the technology was introduced. The LAN provides a cost-effective mechanism for sharing local resources, frame relay provides equivalent capabilities for the wide area. The nature of the applications used in LAN environments call for more bandwidth than X.25 is able to provide. Also, the cost of a private line network can be prohibitive. Frame relay provides the perfect compromise—high bandwidth at a reasonable price—making frame relay the technology of choice for most LAN interconnection applications.

Applications: SNA

Despite all of the changes that have occurred in computing over the last two decades, applications running in an SNA environment still represent a large portion of networking traffic in large corporations today. Access has historically been provided using terminal devices connected to a cluster controller (CC), which is, in turn, connected via a private line to a communications controller (COMC), then on to the mainframe. Today the access can be provided by a PC connected to a LAN via an SNA gateway provided in the PC, or more often in a server, connected to the COMC; however, the end applications are still communicating using SNA.

The use of frame relay in these environments has been growing steadily in recent years. Financial service companies, banks, hospitals, travel agencies, retailers, package carriers, and governments are among the list of users that have chosen frame relay to provide their networking infrastructure. IBM has been a key player in the acceptance of frame relay as a private line replacement; in fact, IBM is a major user of frame relay within their own network.

Applications: Voice Over Frame Relay

Beyond integrating the data networks, many companies are adding voice and video support to their networks. By trading 9.6Kbps private lines for 56Kbps frame relay, new applications such as voice and fax transmission are possible, leading to even greater savings.

Initial voice over frame relay (VoFR) implementations were proprietary. In May 1997, the FRF approved a VoFR Implementation Agreement (IA), FRF.11. The purpose of this IA is to extend frame relay support to include voice and fax traffic. A variety of low bit rate voice compression techniques are provided for; fax traffic is carried on a data channel.

Several vendors now offer VFRADs, which are compatible with FRF.11. This is a common application today, enabling private telephone service over frame relay networks and resulting in cost savings.

All frame networks suffer from a network characteristic known as *serialization delay*. This occurs when a short, delay-sensitive frame gets stuck behind a long data frame on a serial port. Since bits are sent one at a time, and frames cannot be stopped once they have been started (by standard frame definition), there is nothing to do but hold back the voice frame until the data frame has finished. On frame relay networks with frame-based backbones (most common), this serialization jitter can destroy voice quality.

Cell-based frame relay backbones (based on Asynchronous Transfer Mode or ATM, for instance) minimize this jitter because small cells all but eliminate serialization delay (or at least stabilize it within strict parameters). In fact, merging voice and data is what ATM does best. Yet VoFR continues to be widely deployed. This is due in large part to the low cost of frame relay.

FRAME RELAY IMPLEMENTATION ISSUES

Once the standards are in place and the addressing is taken care of, one would think that frame relay would be ready to run. A number of other issues must be dealt with before the service is released to the users. The first of these is the management of the frame relay network. The service provider and the customer must reach some agreement about how the service is to be managed. Another is the type of service to be used by the subscribers; unicast and multicast services are now available from most vendors. Cost, service profiles, and guarantees are all issues that must be contracted. Finally, frame relay equipment decisions must be made. All these issues fall under the category of implementation. And the correct choice and final decision are just that, depending on the application.

Customer Network Management

One of the initial drawbacks of frame relay was the loss of control from a network management perspective. A network manager could not see into the frame relay network. The solution to this problem is frame relay service supported with Customer Network Management (CNM). If both the customer and the network provider implement a standard procedure for CNM, common equipment can be used and the customer can be given a set of tools for managing their portion of a service provider's frame relay network. Most of the current CNM systems offer performance monitoring, fault detection, and configuration information about the customer's virtual circuits.

If a customer's PVC spans multiple networks, another advantage of having a standard for CNM can be seen. Each network can monitor only its "segment" of a multinetwork PVC. However, if the customer has standard access to both networks, then the PVC can be monitored on an end-to-end basis. When combined with the existing management capabilities for frame relay CPE (such as routers), CNM gives the customer a powerful management capability for complete service management from a single platform.

The CNM defined in FRF.6 is based on SNMP. A PVC would provide the typical access between the customer's frame relay equipment and the service provider's CNM system. The service provider has responsibility to assure that each customer is given access only to their own information on the shared public network. The customers usually access the network management agent using SNMP encapsulated in UDP and IP over a frame relay PVC.

An option for customer network management is the use of a Web server and browser. The service provider maintains a Web server that communicates to the frame relay network components. Status information is retrieved by the server and posted on customer Web pages.

When compared to the CNM systems defined in FRF.6, the Web-based systems are easier to set up and simpler to use. The disadvantage is the absence of SNMP. If customers manage their internal network with an SNMP-based system, the Web-based systems are going to be harder to integrate.

Multicast Service

Multicast service can be described as a "one in, many out" service. FRF.7 describes multicast service for PVCs, and the multicast arrangement must be set up in advance. A multicast "group" is established with a number of "members" and a service description (i.e., how frames from/to this member are treated). The member that initiates the multicast is called the "root" while receiving members are called "leaves."

FRF.7 describes three options for multicast service:

▼ In One-Way multicast, frames sent by the root are broadcast to all of the leaves. No frames are sent from any of the leaves to the root. This is intended for a broadcast application, where the leaves would respond to the root on their individual PVCs.

■ In Two-Way multicast, frames sent by the root are broadcast to all the leaves. In addition, frames can be sent by a leaf to the root, but not to any of the other leaves. This form of multicast is well-suited to an SNA environment, where a primary station sends polling and selecting messages to secondary systems, but the secondaries do not communicate directly with each other.

▲ In N-Way multicast, all members are transmission peers. All frames sent from any source are sent to all members. This is intended for a full mesh broadcast environment and might be well-suited for routers exchanging tables; it simulates a connectionless service.

The DLCI of virtual connection from a root to the multicast server is called the Multicast Data Link Connection Identifier (MDLCI). In the FRF agreement, this can be any valid DLCI. (The original LMI specification contained an optional extension for multicast service that supported the use of DLCI 1019 for multicast service.)

The frame relay designer should be aware of the multicast options that a network supports. In some cases, using multicast can have significant performance advantages. Multicast service is currently being offered as a service differentiator by frame relay service providers.

Frame Relay Service Costs

It is often difficult to compare frame relay costs from one service offering to another because the tariffs differ. The local loop access to the frame relay service provider's point of presence (POP), for example, might or might not be included in a price quote.

Two basic pricing models are used for the actual transport of user data. Most service providers bill a flat monthly rate, usually based upon the port connection, CIR, and/or number of PVCs. WorldCom for example, bills its service based on the port connection speed and the sum of the CIR of all PVCs defined at a customer's site. The CIR is measured in a simplex fashion, reflecting only the outgoing traffic. CIRs are assigned to individual PVCs and may range from 16Kbps to the port speed, in increments of 16Kbps. The total aggregate CIR at an interface may total up to 200 percent of the port speed (up to 400 percent for 56/64Kbps ports). Local loop access is additional.

The second billing model is usage-based, possibly with a maximum monthly rate. As an example, usage charges range from 30 percent to 120 percent of the fixed rate plan, and usage charges are based upon CIR, distance, the amount of traffic within the CIR limit, and the amount of traffic that exceeded the CIR.

It is important to note that frame relay charges are usually based upon outgoing CIR, which implies that PVCs are simplex. Frame relay PVCs are, in fact, full duplex according to the standards. "Simplex PVCs" are merely a convenience for billing.

Frame Relay Equipment

Frame relay products are continually evolving to meet the demands of the network providers. The focus has been on developing mechanisms to support the needs generated by the increasingly sophisticated frame relay service offerings. Major development has taken place among value-added capabilities, including: network management, disaster recovery, scalability, and Quality of Service (QoS) provision.

Switches

Frame relay has created a huge market for switch manufacturers. They sell their wares to traditional full service telecommunications carriers, ISPs, specialized data service providers, and major corporations. Major switch manufacturers include Alcatel, Cisco, IBM, Lucent, Siemens, and Nortel Networks.

FRADs and Routers

We should first spend a moment discussing the difference between *FRADs* and *routers*—a seemingly simple task that can be complex.

FRADs are generally used to interconnect small branch locations in enterprise networks. They are low cost, yet powerful, devices, often with an integrated CSU/DSU. They support a wide variety of traffic types including legacy, LAN, and voice. Support for encapsulation for a wide variety of legacy protocols has been a key factor in the migration from private line networks to frame relay. FRADs offer support for a broad range of protocols, making them cost-effective as endpoints in a network. FRADs do not usually perform intermediate routing, nor do they offer as rich a set of security capabilities as routers.

Routers are functionally identical to FRADs as far as frame relay is concerned. The difference is that the routers support a greater set of routing and bridging protocols. The devices are optimized for LAN traffic and consider frame relay as only another type of wide area interface.

Today there are scores of manufacturers of FRADs and routers. The differences between the offerings from the major manufacturers are minimal; however, care must be taken to ensure the protocols you need are supported. Differentiation between the products is evident in the way each handles quality of service by protocol or application, congestion control, disaster recovery options, and management capabilities. Often the selection of vendor is simply driven by cost and/or need to integrate with the equipment found in the rest of the enterprise network.

CONCLUSION

This chapter provides the background and the implementation issues related to frame relay. This chapter opens with a definition of the network-based service, followed by the protocols and the parameters used in the network-based service. Finally, the applications, customers and issues related to the deployment of frame relay are covered. Frame relay has enjoyed one of the fastest growth characteristics in the industry. It remains one of the leading WAN protocols for LAN interconnection and Internet access.

CHAPTER 18

Asynchronous Transfer Mode

rior to the availability of inexpensive processing power and custom-designed integrated circuits, the networking world was a simple place. Real-time, delay-sensitive applications like voice were supported over circuit-switched networks that offered the quality of service characteristics required for the transport of such traffic. Data traffic, on the other hand, was supported most effectively on packet-switched networks that efficiently shared bandwidth but introduced significant delay into the equation. Delays in early statistically-multiplexed networks were measured in the tens to hundreds of milliseconds range. Delays of this length were unacceptable in real-time applications like voice and video, which had to operate properly within the constraints of human sensory systems. Voice that was unintelligible and video that was unwatchable were not particularly in demand in the marketplace, regardless of how low the cost might have been.

The introduction of Asynchronous Transfer Mode (ATM) technology in the early 1990s radically changed the picture. Here was a statistically multiplexed technology that was capable of handling all types of traffic on a single switching fabric. Voice, video, and all types of data are accommodated by ATM without degrading the performance of any particular application. How is this feat accomplished?

CELLS AND SPEED: THE KEYS TO ATM

ATM is a cell-relay technology. Unlike X.25 and frame relay in which packets and frames can be any size up to some maximum, all ATM cells are the exact same length: 53 octets (or bytes). Of those 53 octets, 5 are used for network processing overhead and the remaining 48 are used to carry user payloads (which might include additional overhead, depending on the traffic type). By fixing the data unit size to 53 octets, ATM exhibits a determinism that cannot be achieved with variable-length data units. Buffer management within nodes is greatly simplified as well.

ATM implementations typically occur in hardware. Specialized high-speed chipsets are responsible for ATM protocol processing. The use of these chip-based implementations reduces queuing delays in switches from the milliseconds range to the microseconds range. As such, real-time traffic suffers very little in the way of cumulative delay across even a large ATM network.

This combination of fixing the size of the ATM data unit and implementing the ATM protocols in high-speed hardware confers upon ATM the ability to converge all traffic types onto a single, highly-efficient switching platform.

ATM Traffic Classes

Figure 18-1 compares the four ATM service classes. These four service classes encompass every traffic type that one would encounter on a network now and in the future. Thus, ATM has truly been designed to be a multiservice networking platform.

Class A service is connection-oriented service. It supports constant bit rate traffic that requires that an end-to-end timing relationship be maintained. This service class is typically used for uncompressed streaming voice and video signals.

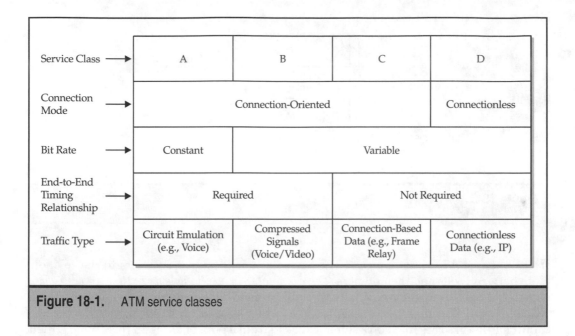

Figure 18-1.   ATM service classes

Class B service is connection-oriented and differs from Class A service only by its support for variable bit rate signals. End-to-end timing is still required for traffic that uses Class B service. Signals that require the service supported by Class B include compressed and packetized voice and video data.

Class C service is connection-oriented and is intended to support traffic types that exhibit variable bit rates but do not require that end-to-end timing be maintained across the network. Traffic that uses Class C service would include, but not be limited to, connection-oriented data such as frame relay frames.

Class D service is for support of data traffic that is connectionless in orientation. Such traffic is characterized by variability in its bit rate and the lack of a requirement for end-to-end timing. Perhaps the best example of such traffic is IP packet data.

ATM LAYERS AND PROTOCOLS

As illustrated in Figure 18-2, ATM is a *layered* architecture. The ATM layers correspond to layers 1 and 2 of the OSI Reference Model. Therefore, ATM is essentially a data-link implementation.

Prior to discussion of ATM protocols, we should reflect on the "asynchronous" nature of ATM technology. At the Physical layer, ATM utilizes bit-synchronous protocols (such as SONET/SDH) exclusively. Interestingly, ATM is also cell-synchronous at the Data Link layer. Examination of an ATM port under idle conditions (no user traffic) will reveal that a synchronized stream of empty cells is being generated and transmitted.

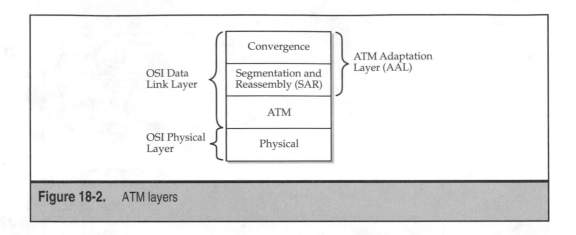

Figure 18-2.   ATM layers

How, then, is ATM *asynchronous*? The asynchronous designation refers to the relationship between cells and their owners. *There is absolutely no relationship in time between a particular cell and the current user of that cell.* Unlike time-division multiplexing environments where a particular time slot is assigned to a particular user for the duration of a connection, cells in ATM are filled on a statistical basis from a pool of users. In this context, then, *asynchronous* really means *statistical*. ATM is thus a departure from other data-link implementations that are truly frame asynchronous. In frame relay, for instance, when users are not sending data, no frames are being generated, and the only source of synchronization occurs at the Physical layer.

The ATM Physical Layer

Many Physical layer protocols have been standardized for use with ATM. Originally, the International Telecommunication Union-Telecommunication Standardization Sector (ITU-T) envisioned ATM as the Data Link layer protocol for a service offering called *broadband ISDN* (Integrated Services Digital Network). The concept of broadband ISDN builds on the narrowband ISDN idea of multiple services delivered over a single, standards-based physical interface.

A major difference, however, between the narrowband and broadband versions of ISDN is the data rate supported over the various interfaces. The broadband Basic Rate Interface (BRI) runs at 155.52Mbps (SONET OC-3c or SDH STM-1); the broadband Primary Rate Interface (PRI) runs at 622.08Mbps (SONET OC-12c or SDH STM-4). In the ITU-T view, then, the Physical layer standards for use in ATM environments are SONET and synchronous digital hierarchy (SDH). Procedures for mapping ATM cells into concatenated SONET/SDH frames have been developed (for a discussion of concatenated SONET frames, see Chapter 10). Currently, standards for mapping cells into SONET are defined at the OC-3c, OC-12c, and OC-48c levels. In SDH environments, standards for ATM cell mapping exist at the STM-1, STM-4c, and STM-16c levels.

To broaden the appeal and deployment of ATM technology, standards have been developed for mapping cells onto metallic carrier facilities at lower bit rates than those offered by SONET and SDH. For North American environments, DS-3 (44.736Mbps) can be pressed into service to transport ATM cell streams. Elsewhere in the world, the E-3 (34.386Mbps) and E-4 (139.264Mbps) carrier levels can be used to transport ATM traffic. Finally, for certain ATM network interfaces, DS-1 (1.544Mbps) can be employed to carry cell streams.

The Physical layer implementations apply to public ATM networks and have been sanctioned by standards bodies such as the ITU-T and the American National Standards Institute (ANSI). Of course, there is no reason why these Physical layer protocols cannot be employed in private ATM networks as well. For private ATM networks, however, other Physical layer alternatives have been delineated in ATM Forum implementation agreements. These include a specification for running ATM over Category 3 unshielded twisted pair (UTP) at either 51.84Mbps or 25.6Mbps. Another specification for running ATM over either fiber-optic cable or Category 5 UTP using signaling borrowed from the Fiber Distributed Data Interface (FDDI) standard at 100Mbps has been developed.

Finally, it is noteworthy that ATM's popularity in asymmetric digital subscriber line (ADSL) environments requires that ATM cells be delivered to the ADSL provider over the ADSL Physical layer.

In summary, ATM can be implemented using a wide variety of Physical layer protocols. This confers upon ATM a high degree of flexibility and scaleability in both public and private networking environments.

The ATM Layer

The ATM layer corresponds to the lower half of the OSI Data Link layer. Its basic responsibility is switching cells in a manner appropriate to moving them between a sender and a receiver. The basic data unit at the ATM layer is the *cell*. As mentioned previously, a cell is a 53-octet entity composed of 48 octets of payload and 5 octets of ATM layer overhead. The internal composition of the ATM cell header is shown in Figure 18-3. Each ATM header field and its function will be described briefly.

The Generic Flow Control Field

Occupying the most significant 4 bits of the first header octet, the generic flow control (GFC) field may be used in one of two ways. In an "uncontrolled" access environment, the GFC is not used and all bits are set to zero. This usage of the GFC field dominates in real-world ATM networks. In the "controlled" access mode, which has been defined by the ITU-T (but is rarely implemented in real networks), the GFC field can be used by a receiver to flow control a transmitter. In this mode, a "halt" command from the network stops cell flow from the user until such time as the congestion condition that caused the "halt" is remedied.

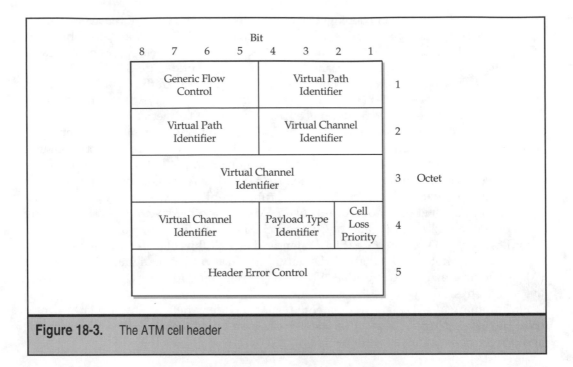

Figure 18-3. The ATM cell header

The Virtual Path and Virtual Channel Identifier Fields

The *virtual path identifier* (VPI) is found in the least significant 4 bits of the first ATM header octet and the most significant 4 bits of the second octet. This 8-bit field identifies a particular cell as belonging to one of 256 available virtual paths. Conceptually, the VPI (and for that matter, the VCI, or virtual channel identifier, as well) is akin to the *logical channel number* (LCN) in X.25 (see Chapter 16) and the data link connection identifier (DLCI) in frame relay (see Chapter 17). An ATM switching node that is capable of VP switching will interrogate the VPI field in each cell that it processes. If directed to do so by a table, the switch will swap one VPI for another, thus performing VP switching.

The VCI occupies the least significant 4 bits of the second ATM header octet, the entire third octet, and the most significant 4 bits of the fourth octet. This 16-bit field identifies a particular cell as belonging to 1 of 65,536 available virtual channels. A *virtual channel* (VC) is a simplex path. As with VP switching, a switching node that is capable of VC switching will interrogate the *virtual channel identifier* (VCI) field in each cell that it processes. If directed to do so by a table, the switch will swap one VCI for another, thus performing VC switching.

Because VCs are simplex in nature, duplex operations in ATM require two VCs, each operating in opposite directions. Often, the VCs that make up a duplex connection are bundled together in the same VP. This statement highlights an important aspect of ATM switching operations. Unlike X.25 and frame relay, which have a single logical identifier associated with each data unit, ATM cells are identified on two levels, the VP and the VC. A particular switch may perform VP switching, VC switching, or both. One can think of these logical identifiers as being "nested."

A bundle of VCs typically make up a single VP, and a bundle of VPs are typically found on a single physical facility. It is noteworthy that, with 24 bits dedicated to the VPI/VCI fields in the ATM cell header, one can support a maximum of 16,777,216 users on a single physical ATM link! Given the data rates encountered at the ATM Physical layer (such as OC-48c at 2.48832Gbps), it is not inconceivable that millions of users would share a single facility within a network. It must also be noted that certain VPI/VCI values are reserved for use in network management and resource management tasks such as fault reporting and signaling, among other things.

The Payload Type Identifier Field

The *payload type identifier* (PTI) field occupies bit positions 2, 3, and 4 of the fourth ATM header octet. As its name implies, the PTI permits differentiation of different payload types in an ATM network. Currently, two such payload types have been defined: the user data payload and the operations, administration, and maintenance (OAM) payload. User data payloads carry user information, whereas OAM payloads carry network management information.

The PTI field also plays a role in identifying whether or not a particular cell has encountered a congested state in its traversal of the network. Using a particular coding of the PTI field, each cell arrives at its destination with an indication either that congestion has been experienced somewhere in the network or not.

The Cell Loss Priority Bit

The least significant bit in the fourth octet of the ATM cell header is the *cell loss priority* (CLP) bit. The CLP bit can be set by an ingress switch whenever a user exceeds some contracted-for amount of throughput. When set to 1, the CLP bit indicates to ATM switches that a cell may be preferentially discarded if congestion conditions exist. In this regard, the use of the CLP bit in ATM is identical to the use of the discard eligible (DE) bit in frame relay (see Chapter 17). The parameter that when exceeded will typically cause the setting of the CLP bit by the ingress switch is called the *sustained call rate* (SCR).

The Header Error Control Field

In a study performed by AT&T in the early 1990s, it was discovered that, on fiber-optic systems, more than 95 percent of errors were of the single-bit variety. This is in marked contrast to metallic systems, which typically suffer burst errors where many bits are affected by some line impairment or noise. Because ATM was originally designed to run over fiber-based systems (SONET and SDH), a mechanism to detect and correct single-bit errors in the ATM cell header was included.

The header error control (HEC) field occupies the fifth octet of the ATM cell header. It uses a forward error correction scheme (for details, see Chapter 9) to detect and correct single-bit errors in the cell header. Note that HEC does nothing to provide error control for the payload portion of the cell. Protecting the integrity of the cell header is important to prevent such events as one valid VPI/VCI value being transmuted into another valid value by a bit error. In such cases, a cell disappears from one connection and appears on another unexpectedly. In a humorous (for the ITU-T!) assignment of nomenclature, the ITU-T calls such an event "unintended service."

The HEC field also supports cell delineation in some ATM implementations.

The ATM Adaptation Layer

Referring back to Figure 18-2, we see that the ATM adaptation layer (AAL) comprises the top half of the OSI Data Link layer. Furthermore, we see that the AAL is subdivided into two sublayers. The lower sublayer of the AAL is the segmentation and reassembly (SAR) sublayer; the upper sublayer of the AAL is the convergence sublayer (CS). The basic function of the AAL is to "converge" or "adapt" multiple traffic types into the ATM infrastructure. As such, AALs are traffic-type specific.

This far, four AALs have been defined. All use the same SAR sublayer, but each AAL type implements its own service-specific CS. The reader will note that the AAL definitions closely parallel the ATM service classes shown in Figure 18-1. This was the ITU-T's intent in developing the four service classes—each corresponds to its own AAL type. Care must be taken, however, not to limit thinking of the AALs in this way alone. In many real-world cases, an AAL of one type is used to support a service type that one might think required a different AAL type. After a brief discussion of the AAL sublayers, we will take a quick look at each AAL type and note the service for which it is intended.

The AAL SAR Sublayer

The function of the SAR sublayer is simple. SAR takes whatever is delivered from the CS and forms the 48-octet units that become the ATM cell payloads. The rule here is that *nothing* leaves the SAR that is not 48 octets in length. In some cases, SAR might add overhead of its own to the CS protocol data unit (PDU); in others it simply chops the CS PDU into 48-octet units and passes them down to the ATM layer.

The AAL Convergence Sublayer

The AAL convergence sublayer (CS) is responsible for receiving a PDU from the upper layers (such as an IP packet) and adapting it, typically by adding overhead, for presentation to SAR. It is at this sublayer that multiple traffic types converge onto the ATM protocol stack. Because each traffic type requires specific treatment in this operation, the four AAL types are differentiated by their CSs. Although a detailed study of each AAL type is beyond the scope of this text, a brief introduction to each AAL follows.

AAL Type 1   The type 1 AAL is intended to support constant bit rate traffic where a timing relationship is required on an end-to-end basis. Thus, it is typically used to support Class A ATM service.

AAL1 receives a constant rate bit stream from above and segments it into 47-octet pieces. SAR adds one octet of overhead to yield the 48-octet SAR PDU. The overhead is for sequencing and timing purposes. The SAR PDU is then passed to the ATM layer, where it becomes the payload portion of the cell. AAL1 is supported in many operational ATM networks.

AAL Type 2   The type 2 AAL is intended to support variable bit rate traffic where a timing relationship is required on an end-to-end basis. Its most typical use is in the support of compressed voice and packetized (such as MPEG) video streams—in other words, ATM Class B traffic.

Through a fairly complex process, an AAL2 CS submodule collects bytes from the upper layers and forms them into variable length CS *packets*. Each CS packet is given a 3-octet header and then passed to another CS submodule, where another octet is appended to implement additional protocol functions. The AAL2 SAR sublayer then chops the resultant stream of packets into 48-octet pieces for passage to the ATM layer as payload units. AAL2 is fairly new and is just now beginning to be implemented in operational ATM networks.

AAL Type 3/4   At one time, both AAL3 and AAL4 were defined in the ATM standards. They have subsequently been merged into AAL3/4. The AAL3/4 is intended to support either connection-oriented or connectionless variable bit rate data that does not require an end-to-end timing relationship. AAL3/4 can thus provide support for Classes C and D ATM service.

An upper layer PDU coming into the AAL3/4 CS (such as an IP packet) can be up to 65,536 octets in length. AAL3/4 CS appends a 4-octet header and a 4-octet trailer to the upper layer PDU to form the CS-PDU, which is passed along to SAR. AAL3/4 SAR chops the SAR PDU into 44-octet segments, and then adds its own 2-octet header and trailer to each segment to form the 48-octet SAR PDU. That PDU becomes the payload portion of the cell at the ATM layer.

AAL3/4 is rarely, if ever, implemented in operational ATM networks for several reasons. One issue is its complexity and high overhead. In fact, as we will see in the next section, the AAL5 was developed to provide a less complex alternative to the AAL3/4. Moreover, AAL3/4 formats are closely tied to a set of protocols developed by Telcordia for use in switched multimegabit data services (SMDS) environments. The spectacular failure of SMDS in the marketplace all but obviated the need to implement AAL3/4 in ATM networks. Of the AALs discussed herein, AAL3/4 is of only minor importance.

AAL Type 5   When the type 5 AAL was under development, it was referred to as the "simple and efficient AAL" (SEAL). In the realization that AAL3/4 was overly complex and overhead-intensive, AAL5 was adopted for functions originally intended to be supported by AAL3/4. Therefore, like AAL3/4, AAL5 is intended to support either connection-oriented or connectionless variable bit rate data that does not require an end-to-end timing relationship. As with AAL3/4, AAL5 is used in support of Classes C and D ATM service.

AAL5 can accept an upper layer data unit (such as a frame relay frame) up to 65,536 octets in length. To that PDU, AAL5 CS appends only a trailer. The trailer can be from 8 to 56 octets long, depending on the length of the original PDU. AAL5 ensures that the resultant CS-PDU is evenly divisible by 48, making SAR's job quite simple. The CS-PDU is then passed to AAL5 SAR, where it is chopped into 48-octet payloads for presentation to the ATM layer.

AAL5 is widely deployed in real-world ATM networks. Almost all variable bit rate, timing-insensitive data is handled by AAL5 in operational networks. In fact, on more than one occasion, this author has seen AAL5 pressed into service as a substitute for AAL2. In those instances, AAL5 did a fine job of supporting time-sensitive variable bit rate traffic. To reiterate, one must be careful in tying AAL types to ATM service classes. The ITU-T view is one way of interpreting such correspondence; the real world may hold some surprises.

Summary of ATM Layers and Protocol Operations

Figure 18-4 illustrates a generic summary of the operation of the ATM protocol stack. From the top down, an upper layer protocol data unit (PDU) arrives at the top of the ATM stack and enters the CS sublayer of the AAL. The CS-PDU is formed by appending overhead to the upper layer PDU.

As we have discussed, each AAL type has its own service-specific approach to this overhead. Once formed by CS, the CS-PDU is passed down to the SAR sublayer of the AAL. SAR may or may not add its own overhead. The major function of SAR, however, is to segment the CS-PDU into 48-octet pieces. These are the SAR-PDUs shown in the figure. SAR-PDUs are passed down into the ATM layer where the 5-octet cell headers are appended to form ATM cells. The cells are then mapped onto the appropriate Physical layer protocol for passage between ATM nodes in the network. At the receiver, of course, the processes shown in Figure 18-4 are reversed. The result is the ability to integrate multiple traffic types on a single switching fabric without compromising the integrity of any particular application data stream.

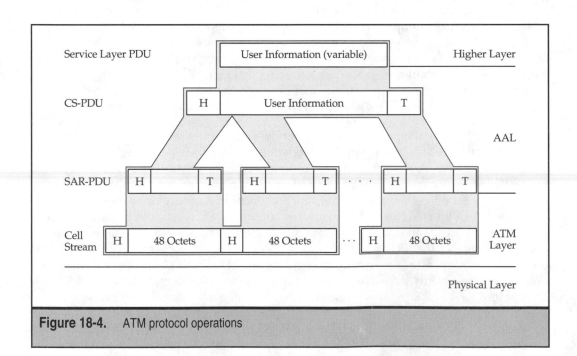

Figure 18-4. ATM protocol operations

ATM CONNECTIONS AND OPERATIONS

Like X.25 (see Chapter 16) and frame relay (see Chapter 17), ATM networks support permanent virtual circuits (PVCs) and switched virtual circuits (SVCs). In the ATM world, the latter are properly referred to as *on-demand* connections. Unlike X.25 and frame relay, however, connections in ATM networks will be used to support applications with widely varying Quality of Service (QoS) requirements. This complicates matters considerably, especially in the case of on-demand connections.

Prior to establishment of an on-demand connection, it must be determined whether or not the network can provide the requested class of service for the user. A technique called *connection admission control* (CAC) is used to make such a determination. During the call setup process, a *traffic contract* is negotiated with the network for a particular class of service. Various QoS parameters are agreed upon between the user and the network. During the data transfer phase of the call, the network must "police" the connection to ensure that the service contract is being adhered to by the user. Finally, a technique called *traffic shaping* is employed in ATM switching nodes to ensure that the user's service contract is being maintained by the network.

Figure 18-5 illustrates the ATM on-demand connection establishment and release process. All signaling messages referred to in this discussion are from ITU-T Recommendation Q.2931.

Here's how it works:

1. The user on the left of Figure 18-5 initiates the call with a SETUP message to the network across the user-network interface (UNI). The SETUP message indicates the AAL type requesting the connection, the various QoS parameters desired for the connection, and the endpoint address of the called party.

2. The SETUP message arrives at the destination UNI (right side of the figure). It contains the information in the sender's SETUP message but also includes a network-assigned VPI/VCI to apply to the call.

3. Meanwhile, the calling party receives a CALL PROCEEDING message that contains a network-assigned VPI/VCI to apply to the connection.

4. The called party then responds to the SETUP message with a CONNECT message, which is acknowledged by the network across the UNI. The network informs the calling party of the success of the connection by sending a CONNECT message across the UNI.

5. Upon acknowledgement of that message by the calling party, the connection is established and enters the data transfer phase.

Termination of the ATM connection occurs when one of communicating parties sends a RELEASE message across its UNI (right side of the diagram in the figure). A cause code is typically provided by the clearing party in the RELEASE message. The RELEASE message appears at the other UNI, and, when both sides respond with RELEASE COMPLETE messages, the connection is released.

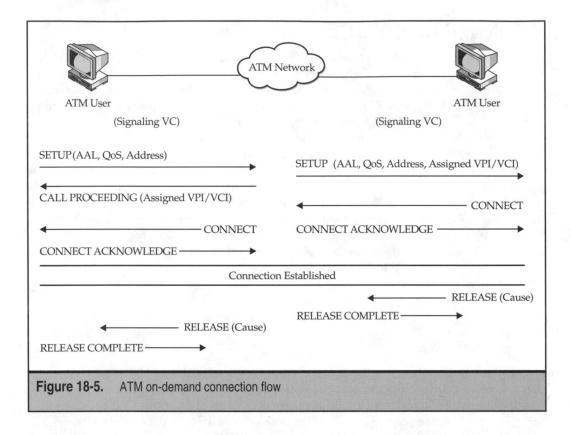

Figure 18-5.   ATM on-demand connection flow

As shown in Figure 18-5, SETUP messages indicate the AAL requesting the connection as well as a set of QoS parameters to be applied to the connection. The following sections will look in some detail at five established ATM traffic classes and the QoS parameters applicable to each.

ATM Traffic Classes and Quality of Service Parameters

Figure 18-6 shows five ATM traffic classes as established by the ATM Forum. As shown in the table, each traffic class is associated with a number of QoS parameters. When an on-demand connection is established, values for these parameters are negotiated on a per-connection basis. The following discussion will introduce each ATM Forum traffic class, the QoS parameters associated with that class, and the consequences of violating the network service contract for each traffic class.

Constant Bit Rate (CBR) Service

When a CBR connection is established, the user negotiates a peak cell rate (PCR) with the network. In addition, the user specifies a maximum cell delay variation tolerance (CDVT) to ensure that jitter is held within acceptable limits for a particular application. If the connection is

Service	Descriptors	Guarantees			Feedback
		Loss	Delay	Bandwidth	
CBR	PCR, CDVT	Yes	Yes	Yes	No
rt-VBR	PCR, CDVT, SCR, MBS	Yes	Yes	Yes	No
nrt-VBR	PCR, CDVT, SCR, MBS	Yes	No	Yes	No
UBR	PCR, CDVT*	No	No	No	No
ABR	PCR, CDVT, MCR	Yes	No	Yes	Yes

*Not necessarily subject to enforcement

Figure 18-6.   ATM traffic classes (ATM Forum)

accepted, the network must provide a guaranteed level of bandwidth, delay performance, and loss performance. The network in not obligated to provide feedback regarding performance to the user in CBR settings. If the user exceeds the PCR negotiated at call set-up time, cells submitted in excess of the PCR are discarded upon ingress.

Typical applications for CBR service include telephony, television, radio, audio feeds, video-on-demand, and circuit emulation services (such as those emulating a T-1 circuit).

Real-Time Variable Bit Rate (rt-VBR) Service

When a rt-VBR connection is established, the user negotiates both a PCR value and a sustained cell rate (SCR) value with the network. The SCR value may be thought of as an average cell rate over time. A maximum burst size (MBS) value is also negotiated. Finally, maximum CDVT is specified for the connection. When the network accepts the rt-VBR connection, it must guarantee bandwidth, delay, and loss performance. Like with CBR connections, the network is not obligated to provide feedback regarding performance to the user in rt-VBR environments. If the user exceeds the PCR value, cells in excess are discarded upon ingress. If the SCR or MBS values are exceeded, cells in excess are typically marked as eligible for discard by the network (that is, CLP is set to 1).

Typical applications for rt-VBR service include compressed and packetized real-time signals such as compressed voice, audio feeds, and video.

Non-Real-Time Variable Bit Rate (nrt-VBR) Service

When a nrt-VBR connection is established, the user negotiates the same set of QoS parameters with the network as in the rt-VBR scenario (PCR, SCR, MBS, and CDVT). In fact, the only difference between the real-time and non-real-time variants of VBR service is that the network offers no guarantee of delay performance in the latter case. The astute reader will recognize that nrt-VBR ATM service is almost identical to frame relay service in concept.

Typical applications for nrt-VBR service include frame relay support and support for mission-critical business applications (such as banking transactions, reservation systems, and factory automation).

Unspecified Bit Rate (UBR) Service

When a UBR connection is established across an ATM network, the user negotiates only a PCR bandwidth value. The user may specify CDVT as well, but the network is under no obligation to enforce this parameter in UBR settings. For UBR connections, the network guarantees nothing; UBR is a best-effort cell delivery service. If the user exceeds the PCR value, however, the network will discard violating cells upon ingress. Unfortunately, when a UBR connection is established across a congested network, high levels of cell loss are probable. The less congested the network, the higher the probability of cell delivery.

Despite the high probability of cell loss under certain circumstances, UBR service is employed in non–mission-critical application settings. These include credit card validation (the user will try again if an attempt fails), e-mail, facsimile, news feeds, general file transfers, and LAN-to-LAN connectivity.

Available Bit Rate (ABR) Service

In an attempt to avoid the high levels of cell loss associated with UBR service in congested networks, the ATM Forum introduced ABR service. When an ABR connection is established in an ATM network, the user negotiates a PCR value as well as a minimum cell rate (MCR) value. In addition, a maximum CDVT value is specified by the user at call setup time. For an ABR connection, the network guarantees delivery of cells sent at or below the MCR. Cell loss performance is also guaranteed by the network for ABR connections. Cells submitted to the network in excess of the PCR are discarded upon ingress.

Users in ABR environments begin the data transfer process by sending at the MCR. Over time, the rate of cell delivery to the network is increased until feedback from the network (on a special VP/VC) indicates that further transmission at this rate will result in cell loss. At that point, the user throttles back its sending rate until feedback from the network indicates that no cell loss is likely to occur. The cycle then repeats, with the user gradually increasing transmission rate until feedback indicates imminent cell loss and then backing off to avoid such loss. By way of this rather elegant scheme, the user and the network dynamically establish an acceptable transmission rate while, at the same time, avoiding cell loss.

ABR service is typically used for the same applications as UBR service. ABR is especially useful when cell loss concerns are high. Other applications that might make use of ABR service include military applications, supercomputer applications, and loss-critical client/server applications.

Summary of ATM Connections and Operations

The previous discussions have focused on the complexity of establishing on-demand connections in an ATM multiservice environment. Although the actual procedures for establishing and releasing connections in this environment are similar in concept to those used in X.25 environments (see Chapter 16), additional complexity comes from the traffic-dependent nature of ATM connections. A lot of work needs to be done to set up an

ATM on-demand connection because the connection must function properly for the type of traffic it is intended to support.

Although we did not go into detail about the data transfer phase of the typical ATM connection, it is complex as well. Two separate entities are responsible for monitoring a user's traffic contract when data is being exchanged. One process, the generic cell rate algorithm (GCRA), determines whether an incoming cell conforms to the traffic contract negotiated at setup time. Another, the usage parameter control (UPC), is responsible for taking actions on nonconforming cells. Additionally, traffic shaping processes may be present to alter the sender's traffic characteristics so that the resulting cell flow will conform to the connection's traffic contract. Given the sheer numbers of connections that may need to be handled by a single ATM switch port at any given time, it is truly amazing that ATM networks function as advertised. Perhaps this accounts for the origin of the anthem of the ATM cynic, "ATM handles all traffic types equally poorly."

ATM APPLICATIONS

While those in the industry were arguing ATM's merits in the mid-1990s, the technology was quietly becoming entrenched in public as well as private networks. As this book goes to press, ATM is a mature, widely-installed technology that continues to show growth in the near term. We will now look briefly at some applications of ATM in public and private networks.

ATM in Carrier Backbones

Among the first organizations to deploy ATM technology in their networks were carriers, both local exchange and interexchange. For these companies, ATM offers efficiencies not available in traditional circuit-switched environments. As such, the deployment of ATM in carrier backbones provides an opportunity to realize substantial cost savings and, in some cases, increased profit margins. The following sections make no attempt to provide an exhaustive treatment of every carrier application of ATM. Rather, they are intended to give the reader a sense of how carriers use ATM in their backbone and access networks.

Cell Relay Services

The deployment of ATM in carrier backbones permits these organizations to offer a new set of services to their customers. Referred to by most carriers simply as *cell relay services* (CRS), this offering supports transport of information for customers who have deployed native-mode ATM equipment in their networks. As the name implies, this offering collects cell-based traffic from the customer at one location and delivers that traffic to another location transparently. The user interface to CRS varies depending on the particular carrier, but generally, they are high-speed. DS-3 and SONET interfaces (such as OC-3c and OC-12c) are fairly typical in these settings.

Frame Relay Support Services

It is estimated that well over 85 percent of frame relay service sold by carriers today is actually carried across the backbone using ATM. In this scenario, the carrier switching node is an ATM node with a frame relay interface. Frames received from customers are converted to ATM cells using AAL5 at the ingress node, and then at the egress node the process is reversed. From a customer viewpoint, the carrier's use of ATM is invisible. From the carrier's perspective, supporting frame relay customers in this fashion permits them to achieve economies of scale by mixing frame relay with other traffic types (such as CRS) on the backbone. For a more detailed description of this use of ATM, the reader is referred to FRF-5, an Implementation Agreement from the Frame Relay Forum.

Another use of ATM in frame relay environments is discussed in depth in FRF-8. In FRF-8 environments, frames received from users on one side of the network are actually converted to ATM cells for delivery to native-mode ATM equipment on the other side of the network. For example, a large corporate user with a centralized headquarters location and a number of small branch offices may be running ATM at headquarters, but not at the branches. In this case, each branch can be served by an economical, low-speed (such as 64Kbps) frame relay interface. Traffic from the branches, once converted to ATM, can be delivered in that format to the headquarters location.

Circuit Emulation Services

In circuit emulation services (CES) environments, carriers use ATM to deliver private line services to their customers. Many customers require point-to-point leased lines (such as T-1s or T-3s) to construct router-based LAN internetworks or PBX-based private voice networks. Rather than using traditional cross-connect technology to provision such services, carriers often use ATM to emulate private lines across their ATM backbones. AAL1 is typically pressed into service in these settings. Once again, the use of ATM is invisible to the customer, but the carrier attains the cost benefit associated with mixing traffic types on a single backbone fabric.

ADSL Support Services

A recent use of ATM technology is in the provisioning of ADSL services. Carriers often offer ATM-based connections to Internet service providers (ISPs). These ISPs essentially buy CRS for their connections to serving carriers. To deliver traffic in ATM formats to the ISP location, the conversion to ATM cells is typically performed by the ADSL modem at the customer's premises.

Using a technology called point-to-point protocol over Ethernet (PPPoE), IP packets are first put into Point-to-Point Protocol (PPP) frames. These frames are then encapsulated into Ethernet frames inside the user's PC. The Ethernet frames are then delivered to the remote ADSL modem where, using AAL5, they are *cellified* into ATM formats for delivery to the ADSL line. At the carrier central office, the cells are switched to the appropriate ISP based on the VPI/VCI value in the cell header. Not only does this mechanism satisfy the ISP's desire to receive traffic in ATM formats, it allows the carrier to support "equal access" to any number of ISPs without costly point-to-point connections to every one of them.

ATM in Private Networks

Although not as widely deployed in private networks as in public carrier backbones, ATM has made inroads into certain private networking environments. The use of ATM in campus environments, usually to interconnect LANs, is one such environment. The following sections will examine two private network applications of ATM technology. Again, the intent is not to be exhaustive but to give the reader a sense of ATM's value in these settings.

LAN Emulation

LAN emulation (LANE) is defined by the ATM Forum as a way of creating a virtual LAN (VLAN) with an ATM network in the middle. As one might expect, LANE is a complex environment. Shared medium LANs (see Chapter 15) are typically broadcast oriented, whereas ATM is typically a point-to-point technology. Therefore, LANE defines procedures whereby ATM is used in the multicast mode to emulate a shared medium LAN. These procedures involve the deployment of a number of specialized server functions in the network, as well as modification to existing ATM switches or IP routers. LANE uses AAL5 to cellify LAN frames for transport across the ATM campus network. In summary, LANE is a form of LAN bridging that requires both Q.2931 call control functions and ATM multicast ability to be implemented in the network. For companies that deploy LANE, cost reductions can be realized by combining LAN traffic with other traffic types on a single switching fabric.

Multiprotocol Over ATM

Because LANE environments are bridged (that is, they operate at layer 2 of the OSI Model), routers must be retained to support traffic between emulated LANs. If ATM switches, however, could be made to perform the tasks of routing nodes, routers could be eliminated and substantial cost savings realized. This is the goal of the ATM Forum's multiprotocol over ATM (MPoA) specification. With MPoA, layer 3 routing is performed by an ATM switch without the need to reassemble cells into layer 3 packets. MPoA requires a LANE implementation with Q.2931 support present. MPoA represents not only cost savings for the user, but it offers a significant performance boost by eliminating the need for layer 3 processing of data units. When most traffic on a network consists of packets within cells, as is the case for many corporate networks and ISP networks, such enhanced performance becomes a valid reason in itself for the deployment of MPoA. MPoA is typically seen in IP environments, but it will work with any layer 3 protocol (such as IPX).

SUMMARY

With its roots deep into X.25 and other statistically multiplexed technologies, ATM is the first architecture designed from the ground up as a multiservice convergence vehicle. With its built-in Quality of Service (QoS) capabilities, ATM is capable of transporting voice, video, compressed signals, and all types of data traffic over a unified switching fabric. Although somewhat complex, the ability of ATM to deliver efficiency and economies of scale has made it popular in carrier backbones, ISP environments, and some private networking settings. Despite indications to the contrary from some industry pundits, this author believes that ATM technology will continue to play a central role in many networking environments for years to come.

PART V

The Internet

CHAPTER 19

The Internet Protocol Suite

In the late 1980s, many large corporations found themselves thrust into a multivendor computing and communications environment. Such an environment was created by workgroups unilaterally purchasing LANs and associated applications, and by mergers and acquisitions. As users began to need access to the resources of other workgroups, the requirement to interconnect LANs on a company-wide basis emerged.

None of the major computer vendors (such as IBM and DEC) offered a model for communications that supported connectivity and interoperability among many different vendors' equipment. Corporate information systems managers began to search for a model for multivendor interconnection and soon discovered the Internet model and its foundation—the TCP/IP suite (Transmission Control Protocol/Internet Protocol).

Initially defined in 1974 as the networking protocol for UNIX systems, TCP/IP enjoyed an early foothold in academic computing networks. When the U.S. Department of Defense standardized TCP/IP as the networking protocol to be used on the ARPANet (Advanced Research Projects Agency Network—the huge WAN used by the DOD in the 1960s), TCP/IP's future was assured. Since that time, TCP/IP has become the most widely used protocol suite in the world. Fighting off several challengers over the years, TCP/IP has emerged as a strong and flexible protocol that supports a variety of services. While TCP/IP will adjust in coming years to support new applications, TCP/IP will be the mainstay of networking for some time. This chapter will outline the major concepts of the TCP/IP suite starting with LAN standards, moving to addressing issues, and finally discussing how TCP interacts with some of the most popular applications on the Internet.

THE NETWORK INTERFACE LAYER

While not officially defined by the Internet Engineering Task Force (IETF), the bottommost layer of the IP suite, the Network Interface layer, is commonly included. It defines the software and hardware used to carry TCP/IP traffic across a LAN or WAN. In the past few years, the number of options available for this process has grown significantly.

One of the primary advantages of a layered protocol suite is that one layer can change without affecting the others. In the IP suite, this is demonstrated dramatically. TCP/IP can utilize virtually any network interface. While TCP/IP has been demonstrated with smoke signals and via carrier pigeon, some of the higher bandwidth options such as ISDN (Integrated Services Digital Network), frame relay, ATM (Asynchronous Transfer Mode), ADSL (asymmetric digital subscriber line), and PPP (Point-to-Point Protocol) have been used to transport IP packets.

This wide implementation of network interfaces has been one of the reasons that the IP suite has been so successful. Networks generally end up employing any number of network types on different segments. Likewise, when global connectivity is the goal, there is no way to guarantee what the data link of any given remote layer will be. This adaptability ensures the ubiquity of TCP/IP.

THE INTERNET PROTOCOL

IP was first defined in 1974 and was the ARPANet Network layer protocol of choice by 1982. Today, IP version 4 (IPv4) is the cornerstone of the Internet and the most widely implemented protocol in private internetworks as well. IP provides a connectionless Network layer service to the end-to-end layers. Its primary responsibility is to enable the routing of packets through interconnected networks.

The use of a connectionless protocol evolved from the earlier use of a connection-oriented service provided by ARPANet. When the ARPANet grew, encompassing many networks (some unreliable), it was decided that it would be appropriate to provide a more reliable connection-oriented service on an end-to-end basis over a less-than-reliable network. Therefore, the Network layer became connectionless and the connection-oriented services migrated up to the Transport layer, reflecting a "why do it twice" philosophy.

It is appropriate to take a moment and reflect on the word "unreliable." In an Open Systems Interconnection (OSI) context, the term refers to the fact that IP does nothing to ensure reliability; there is no guarantee of delivery, no guarantee of sequential delivery, no acknowledgments, no error control, and no flow control by the network. If a packet *is* lost, there is no indication to the end host and the upper layers must resolve the problem. Hence, IP is unreliable in that it cannot ensure orderly delivery. Unreliable then, is simply another way of saying that the reliability of the service is directly tied to the reliability of the underlying network.

IP implements a *datagram* network. Packets are routed on a packet-by-packet basis from IP node to IP node. For each link and each node, IP does its best to deliver the packet. To do this, IP defines mechanisms for packet fragmentation and reassembly, hierarchically structuring the IP address, controlling packet lifetime in the network, and verifying the integrity of the IP header. As a connectionless protocol, IP handles each datagram individually (each packet is individually routed, which means that packets can be dynamically routed around network trouble spots). Hosts do not need to establish a logical connection before exchanging IP datagrams.

IP Addressing

Each device attached to a TCP/IP-based network must be given a unique address. These addresses are carried in the IP packet to identify the source and destination hosts. IP defines 32-bit (four octet) addresses, which permit more than 4 billion hosts to attach to a single internetwork. IP addresses are represented in a format known as *dotted decimal*, where each octet is represented by its decimal equivalent and the four numbers are separated by a dot (.). With 8 bits, you can represent numbers between 0 and 255. Therefore, the IP address represented in binary as 10000000 01000011 00100110 00010111 is written (and spoken) in dotted decimal as 128.67.38.23.

Each IP address has two components: a *network identifier* (NETID) and a *host identifier* (HOSTID). The NETID identifies the specific network to which the host is attached. The HOSTID uniquely identifies a host within that network. This distinction is important because routers route to a given NETID and don't care about the HOSTID.

IP does not permit the NETID or HOSTID to be all 1s or all 0s. All 1s mean *broadcast* and can be used for all networks or all hosts. For example, the IP address 128.17.255.255 has a NETID 128.17 and a HOSTID 255.255. It means *all hosts on the network with NETID 128.17*. To any IP device, 128.17.0.0 refers to the entire 128.17 network, regardless of HOSTID.

Address Classes

The relationship between the sizes of the NETID and the HOSTID is important. If we presume that the address is divided in half, a 16-bit HOSTID allows up to 65,534 hosts (bit values of all 0s and all 1s are not allowed) to attach to a single subnetwork. A 16-bit NETID allows a total of 65,534 such networks to exist in the same internetwork. Although this might seem like an ideal balance at first glance, it begins to look seriously flawed when we consider that the average LAN has fewer than 100 hosts. This means less than 100 of the possible 65,534 addresses for each network are being used; the rest are wasted. At 100 hosts per subnetwork, fewer than 6,554,000 hosts will probably attach to the internetwork, wasting over 99.8 percent of the available addresses—certainly not an efficient mechanism. As proof, consider that the entire address space would have been exhausted by mid-1995 had addresses been assigned this way.

To strike a balance between efficient use of the address space and the needs of different networks, five classes of address were defined, designated Class A through Class E. The three main classes of interest to us are A, B, and C. The essential difference between these classes, as depicted in Figure 19-1, is the relative size of the NETID and the HOSTID fields.

Class A addresses have an 8-bit NETID and a 24-bit HOSTID, with only 126 possible Class A addresses. With 8 bits in the NETID and the requirement that the first bit be set to 0, only the numbers from 0 to 127 can be represented. Zero is not allowed and 127 is reserved, leaving the numbers from 1 to 126. Thus, a Class A address has a number between 1 and 126 in the first octet (for example, 85.13.6.12 and 126.52.37.120). Class A addresses are intended for use in very large networks since the 24-bit HOSTID can uniquely identify more than 16 million hosts. Approximately 90 Class A addresses are in use.

Class B addresses have a 16-bit NETID and a 16-bit HOSTID. Subnetworks can support up to 65,534 hosts with 16,382 possible network addresses. The value of the first octet represents the range 128 to 191. Thus, IP addresses 128.5.12.87 and 153.200.59.254 are examples of Class B addresses. Class B addresses are intended for moderate sized networks. Many large corporations have a block of Class B addresses assigned to them, rather than a single Class A address.

Class C addresses are intended for small subnetworks. Class C addresses have a 24-bit NETID and an 8-bit HOSTID, permitting more than 2 million possible network addresses. The first number of a Class C address always falls in the range 192 through 223 (for example, 199.182.20.17 and 220.16.39.52) and each subnetwork can support up to 254 hosts (0 and 255 are not permitted).

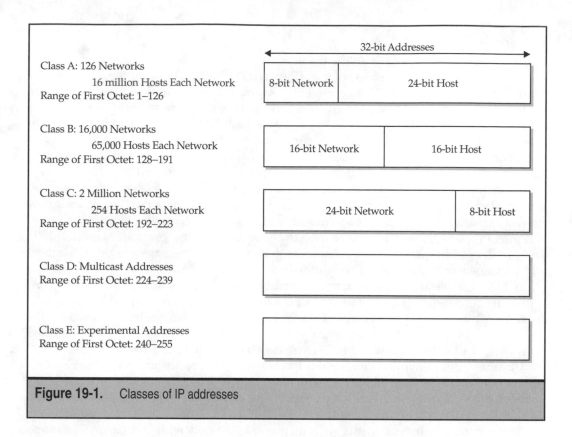

Figure 19-1. Classes of IP addresses

Class D addresses begin with a digit between 224 and 239 and are used for multicast applications such as the Open Shortest Path First (OSPF) routing protocol or multicast voice and video services.

Class E addresses begin with a number between 240 and 255 and are used for experimental purposes. This address class should not be used in a network unless you are inordinately fond of troubleshooting.

The concept of address classes has traditionally been important in the discussion of IP addressing; yet it is declining. IP modifications such as Variable Length Subnet Masking (VLSM) and Classless Interdomain Routing (CIDR), discussed a little later, have made the organization of IP addresses into classes obsolete. They are still important, however, as they are constantly being referred to. For example, an IT manager may ask an ISP for a Class C address to number the company LAN with. Today, she is not asking for a NETID of between 192 and 224, she is asking for 254 IP addresses!

Subnet Masking

In computer science, a *mask* is an electronic filter used to extract particular portions of a unit of data. Likewise, a mask is used in IP to extract the NETID from the address. The mask is the same length as the address (32 bits) and is usually depicted using the same dotted-decimal notation used for addresses. In the mask, every bit that is part of the NETID is set (to 1) and every bit that is part of the HOSTID is unset (to 0). It is also common to see the mask represented using the *slash* notation. In this case, the number of NETID bits is explicitly stated following a slash (/) after the IP address. For example, 150.128.110.1/16 means that the first 16 bits of this address is the NETID—150.128.0.0.

By manipulating the bits in a subnet mask, a group of IP addresses can be broken up into smaller networks. For example, by adding 2 bits to a 24-bit subnet mask, four new networks, each supporting a smaller number of host IDs is created. The explanation of this phenomenon is that 2 bits can be used to create four possible combinations: 00, 01,10, and 11. Thus adding 3 bits to a 24-bit mask would create eight new networks. This important concept is best illustrated through the following example.

In this example an ISP has assigned company A one Class C address, 200.6.2.0, to use within its corporate network. The 200.6.2.0 address is for use on the LAN side of the router, not the WAN side that connects to the ISP. The addresses on the WAN side of the router come from the ISP address space. Using the 200.6.2.0 address with a default Class C mask of 255.255.255.0 allows up to 254 hosts to exist within the one corporate network.

Company A's network administrator has identified the requirements of the new corporate network to be four subnets with no more than 60 hosts attached to any one of the subnets. How can these requirements be met since company A's ISP has provided them with only one Class C address?

The solution is to extend the default mask associated with the Class C address (255.255.255.0 or /24) so that more bits are allocated to the network portion of the address and fewer bits are allocated to the host portion. With a default mask of /24 we have 8 hosts' bits to work with. The question becomes this: how many of the 8 bits are required to create four subnets, and can the remaining hosts' bits support 60 hosts on each subnet?

By extending the subnet mask by 2 bits, four smaller networks are created. The remaining 6 bits are used as the host IDs for each of these new networks. Since 6 bits allows 64 possible combinations, the network administrator is able to meet the addressing needs of her company while still using only the single address block provided by the ISP. Figure 19-2 shows how this works.

VLSM

Traditional subnetting is also referred to as *classful* addressing in reference to the well-known address classes. One limitation of this classful subnetting had been that all subnet masks in a network needed to be the same length. This was wasteful for several reasons. Consider the preceding example. What if the network administrator of company A was to decide that one section of the network would never need more than 12 hosts? Under classful addressing, each network would still require 60 IP addresses. The 12-host network would waste 48 IP addresses.

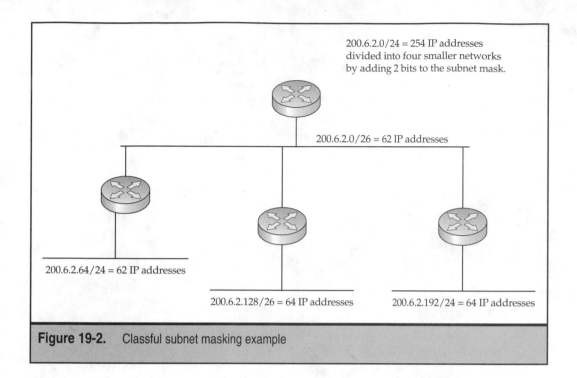

200.6.2.0/24 = 254 IP addresses
divided into four smaller networks
by adding 2 bits to the subnet mask.

200.6.2.0/26 = 62 IP addresses

200.6.2.64/24 = 62 IP addresses

200.6.2.128/26 = 64 IP addresses

200.6.2.192/24 = 64 IP addresses

Figure 19-2.   Classful subnet masking example

VLSM allows you to subnet Class A, B, or C addresses using different length masks. It is beneficial because it allows more efficient use of an assigned address space.

Referring again to our preceding example, company A is using the network 200.6.2.0/24. Without VLSM, company A would have to use the same mask for all subnets. In this example, the network administrator has determined that four networks are still necessary, but only two of them would ever grow to support 60 hosts. The other two networks will never grow larger than 25 hosts. In this case, the number of bits that the network administrator borrows from the NETID can vary. In the case of the 60-host networks, 2 bits can be borrowed making the subnet mask 26 bits long. In the case of the 25 host networks, 3 bits can be borrowed, making the subnet mask 27 bits long. Not only does each subnet mask more accurately reflect the number of hosts on the network, but it will also save more IP addresses. In this case, the network administrator of company A has saved 64 IP addresses for future growth, a significant improvement over classful subnetting which reserved no IP addresses. While the math behind this operation is not complex, it is more detail than intended for a book on basic telecommunications (Figure 19-3).

Care must be taken when using VLSM. Using a range of subnet masks within a single classful address requires that the routing protocol supports VLSM. Support for VLSM requires that the protocol not only advertise the available network number, but also the mask (as any less information is ambiguous). Furthermore, this is a relatively new addition to IP. Some older routing protocols such as Routing Information Protocol (RIP) or Cisco's Interior Gateway Routing Protocol (IGRP) do not understand VLSM and cannot be used in these environments.

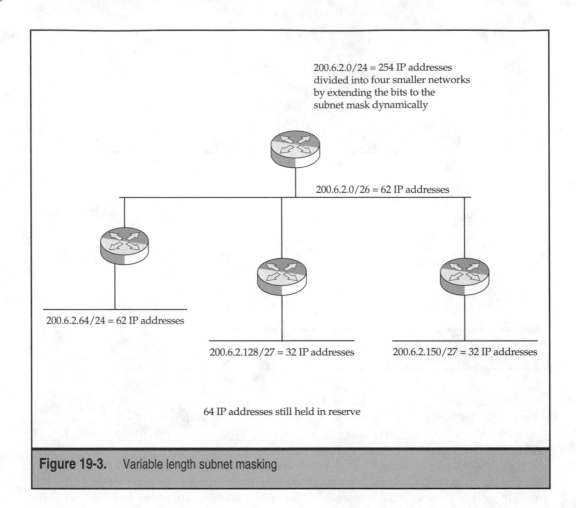

200.6.2.0/24 = 254 IP addresses
divided into four smaller networks
by extending the bits to the
subnet mask dynamically

200.6.2.0/26 = 62 IP addresses

200.6.2.64/24 = 62 IP addresses

200.6.2.128/27 = 32 IP addresses

200.6.2.150/27 = 32 IP addresses

64 IP addresses still held in reserve

Figure 19-3.   Variable length subnet masking

CIDR

The IP address space is slowly but surely being depleted, with Class B addresses potentially in short supply. Using Class C addresses for small networks (those with fewer than 254 hosts) is greatly relieving the strain on the IP address space, but is causing another problem. IP routing is based on tables that must be populated by the routing algorithms. However, every known IP network requires a routing table entry somewhere. CIDR, defined in RFCs 1518 and 1519, provides a way to limit the explosive growth in the size of routing tables within the Internet by building routing tables independent of the IP address class.

The basic idea behind CIDR is to group blocks of addresses together so that a single routing table entry can point to several addresses. While VLSM can be considered a way to dynamically make networks smaller, CIDR allows smaller networks to be aggregated into larger blocks. Suppose, for example, that a single organization is assigned eight Class C addresses. If we are clever in our allocation scheme, we will assign the eight addresses as a block so that all eight can be *summarized* in a single routing table entry. To provide for this summarization, CIDR requires that multiple IP addresses that are to be summarized share the same high-order bits in their address. In addition, routing tables and routing algorithms must be modified so that routing decisions can be based on a 32-bit address and a 32-bit mask. Finally, routing protocols must be extended to carry the 32-bit mask along with the 32-bit address. (OSPFv2 and RIP version 2 both have this capability.)

Consider the following example. RFC 1466 recommends that new Class C addresses for Europe be assigned in the range from 194.0.0.0 to 195.255.255.255. This represents some 130,000 possible Class C addresses, but all share the same value in the seven high-order bit positions. This means that all European addresses require but a single table entry (194.0.0.0) and a 7-bit mask (254.0.0.0). Without CIDR, these 130,000 networks would require the same number of routes in all backbone routers around the world, a number that is larger than the current routing tables!

CIDR can work as described in the preceding example because it follows the rule that the best match is always the longest match—that is, the match with the highest number of 1 bits in the 32-bit mask. Let us suppose that one European network provider (whom we will call EurNet) is assigned a block of 16 Class C addresses in the range 195.10.16.0 through 195.10.31.255. EurNet, then, would have the routing table entry 195.10.16.0 along with a 20-bit mask, 255.255.240.0, written as 195.10.16.0/20. Now, suppose a datagram needs to be routed from Japan to a host within the EurNet IP address space, such as 195.10.27.5. Between Japan and Europe, intermediate routers need know only the summarized European CIDR block. Once the packet reached Europe, EurNet's more specific route would begin to appear in the routing tables. This would allow the packets to ultimately reach EurNet despite the presence of two routes in the router, one for EurNet specifically and the other for Europe's CIDR block.

NAT

A number of products are available for the Internet that perform *network address translation* (NAT). While their primary function is explained in the name, it is useful to review what the products do and why. NAT helps work around several addressing problems that are common to the Internet today.

The first problem arises from the fact that the Internet uses centrally administered IP addresses. The NETID portion of the IP address space is administered by the Internet Registries, while the HOSTID portion is administered by the organization assigned a particular NETID. Many companies adopted use of IP in their internal networks and were either sure that they would never connect to the Internet or were unaware of its existence; in either case, they just grabbed a convenient NETID and used that IP address.

What happens when this company attempts to connect its network to the Internet? There is an excellent chance that they will be using someone else's assigned NETID. After getting an assigned IP network address, what's next? One option is for the "offending" organization to renumber their network. This obviously will solve the problem, but it is not a viable solution for very large private networks, as it is impossible to quickly change to a new address scheme, and "illegal" addresses would continue to be used for at least a little while. It also requires that the company obtain as many official IP addresses as it has hosts, since each host will be directly connected to the public Internet.

A second alternative is to use network translation software in the router or special gateway that maps the "public" IP addresses to the "private" IP addresses. In a way, the NAT device is acting as a "proxy" on behalf of the internal private hosts. Address conservation is accomplished by the fact that the NAT device needs only a small number of official IP addresses assigned to it. It can map the official addresses to the internal hosts as needed and does not have to have a permanent one-to-one mapping between the private internal addresses and the public external addresses.

IPv4 is the most commonly deployed version of IP and it uses 32-bit addresses. IP version 6 (IPv6), currently in the early stages of implementation, will use 128-bit addresses and accommodate traditional IPv4 and IPX address formats, as well as a new IPv6 format. While some networks will quickly make the transition to IPv6, most will make the transition slowly because they will incur some high costs and not realize a significant benefit. Address translation software will provide an easier migration to this new addressing scheme and provide compatibility between IPv4-only and IPv6-only hosts.

More than just translating an IP address, NAT also involves modifying the IP checksum and the TCP checksum. In fact, NAT must modify any occurrence of an IP address above the IP layer. Here are some examples:

▼ The Internet Control Message Protocol (ICMP) embeds the IP header of the control message. Hence, when an ICMP message is sent through a NAT device, the device must change the contents of this embedded header.

■ The File Transfer Protocol (FTP) includes instances of IP addresses in some commands.

▲ Domain name system (DNS) queries contain IP addresses in the DNS header.

Most NAT implementations support ICMP, FTP, and DNS. Simple Network Management Protocol (SNMP), IP multicast, and DNS zone transfers are not handled by most NAT implementations, however. Furthermore, because of the use of cryptographic checksums when using IP security, NAT has particular problems with these protocols. Changing the IP address as it would for an ICMP would render the checksum and therefore the security invalid! Newer releases of NAT software support TCP/UDP applications that carry the IP address in the application data. Examples of these applications include H.323, RealAudio, and NetMeeting. For protocols that are not supported, careful placement of the NAT device can usually overcome most translation problems.

Perhaps the largest benefit of NAT, however, is that it has dramatically reduced the speed at which public IP addresses have been allocated. With NAT, an entire company can represent its internal network with a single IP address.

DHCP

As a network administrator, it takes about 5 minutes to configure a desktop computer with the parameters needed to access the Internet. This is not a chore when a company has only 10 computers. When this number is multiplied by 100 however, 5 minutes at each machine suddenly becomes an administrative nightmare. This is not a one-time configuration, either. As the network grows and evolves, the desktop computers need to be reconfigured. Any network administrator who has had to deal with these changes has pondered a better way of administering this network. The answer is to let a computer server configure the machines dynamically every time a user turns on.

The Dynamic Host Configuration Protocol (DHCP) provides a mechanism to automatically and dynamically assign an IP address to a host. DHCP is based on the Bootstrap Protocol (BOOTP). However, the IP address assigned to a given host by BOOTP is always done statically, with the host always being assigned the same IP address.

DHCP defines three address allocation mechanisms:

▼ **Dynamic allocation**   The server assigns an IP address to a host on a temporary basis. The host can keep this IP address for some period of time—called the lease—or until it explicitly releases the address.

■ **Automatic allocation**   The server assigns a permanent IP address to a host. In this case, the host has asked for an infinite lease.

▲ **Manual allocation**   A host's IP address is assigned by a network administrator and the DHCP server is used simply to convey this address to the host.

Dynamic address allocation is the only one of the three mechanisms that allows automatic reuse of IP addresses. It is ideally suited to networks with many roaming systems; a laptop, for example, can be assigned an IP address from one location of the corporation's address space on one day, and it can receive an IP address from another location's address space on the next day.

As can be imagined, the ability to administer IP addresses centrally without having to touch potentially thousands of client computers is a great boon to efficient network administration. In addition to assigning IP addresses, DHCP can also automatically configure a client computer with a host of other parameters commonly used on TCP/IP networks such as the subnet mask, default gateways, DNS servers, WINs servers, and more.

DHCP is not perfect for all scenarios however. While DHCP does have the ability to statically assign IP addresses to the same device each time it connects to the network, in practice this is rare. Configuring IP addresses on a handful of servers is easy enough, and why introduce another point of failure for a server in the event of the DHCP server going down?

DHCP is ideal for client devices that consume rather than provide network services. Being assigned a new IP with each power cycle does not affect reach ability to the client since they typically initiate the connections. Servers and routers on the other hand do suffer if IP addresses keep changing. It will be difficult for a client to contact a server if the IP address of that server is changing from time to time!

Network address translation (NAT) and the Dynamic Host Control Protocol (DHCP) are often used together. In the visual, the DHCP server has a static IP address of 192.168.0.5, from the private Class C range, and is configured to assign private addresses to clients from a pool, whose addresses range from 192.168.0.32–192.168.0.254. The File Transfer Protocol (FTP), the World Wide Web (WWW), and the Simple Mail Transfer Protocol (SMTP)/Post Office Protocol version 3 (POP3) servers have static addresses ranging from 192.168.0.2–192.168.0.4. Lastly, the router has a static address of 192.168.0.1 on the internal network and a public address of 208.106.162.1 on the external network. (The external network number is from the ISP address space.)

The ISP assigns its customer a public prefix of 208.132.106.0/28. In this case the customer has addresses 208.162.106.0–208.162.106.7 as public addresses. Given that all 1s and 0s are reserved subnetwork addresses, the customer has six host addresses that range from 208.162.106.1 to 208.162.106.6. This is a common scenario for customers with a business DSL service.

The router performing the NAT function in Figure 19-4 is configured to statically map the private server addresses to public addresses (e.g., 192.168.0.2–208.162.106.2, 192.168.0.3–208.162.106.3, etc.). Relative to the clients, the router is configured to map all other private addresses (i.e., 192.168.0.32–192.168.0.254) to one public address (i.e., 208.162.106.6). From the Internet, it looks as though the host whose address is 208.162.106.6 is very active since all the clients access Internet resources from this one address.

IPv6

In the late 1980s, the IETF recognized that the useful life of the IPv4 would come to an end. Address space, in particular, was cited as IPv4's biggest weakness; although a 32-bit address should allow users to identify up to 4 billion hosts, the hierarchical address structure of IP combined with poor early allocation of addresses have inefficiencies that would have resulted in early address depletion. In 1991, work began on the next generation of IP, dubbed *IP Next Generation,* or *IPng*. The goal of the IPng work was to define a successor to IPv4, which was defined in RFC 791 in September, 1981. While still quite usable and implemented on the current Internet, IPv4 has been showing its age in a number of areas, and a replacement protocol is needed. Through many iterations and proposals, IPv6 was officially entered into the Internet Standards Track in December, 1995.

The most significant change made in IPv6 is the vastly increased address space. While taking the time to increase addresses, however, the IETF also decided to improve upon IPv4 in other ways. The IP packet header was streamlined to allow more efficient packet processing. Header extensions and options that had been clunky in IPv4 were modified to allow easier integration of new options as we make more demands of IP in the future.

Finally, recognizing that the Internet was not the small, closed community it was when IPv4 was developed, security options are integrated directly into the IPv6 header. In addition, because it would be some time before IPv6 was widely deployed, the IPSec standards used for security on today's networks were borrowed directly from IPv6.

At first glance, the 128-bit address scheme appears to be flat in structure, with no concept of class, or even NETID or HOSTID. IPv6 does not even use the a subnet mask! How

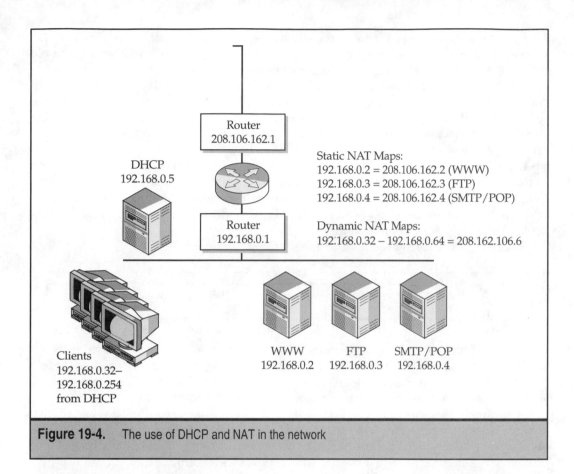

Figure 19-4.   The use of DHCP and NAT in the network

is routing supposed to be accomplished? Delving a little deeper into the allocation of bits in the address, an approach that better suits the commercial, provider-centric Internet becomes apparent. RFC 2073 describes this "global provider-based unicast address."

Unlike IPv4, next generation addresses will have an explicit geographic scope as part of their structure. The 5-bit Registry ID is the first part of the routing hierarchy, indicating which regional registry is ultimately responsible for a particular network's address allocation—the Internet Assigned Numbers Authority (IANA) will assign address space to these entities. Today three such organizations are identified: the American Registry for Internet Numbers (ARIN) in the U.S., Réseaux IP Européen Network Coordination Centre (RIPE NCC) in Europe, and Asia Pacific Network Information Centre (APNIC) in Asia.

As the majority of Internet users have network access via an ISP of some sort, routing will also be based on the provider of network access, using the provider ID. The regional registries will determine how these identifiers are allocated. Since each regional registry could conceivably decide upon a different assignment scheme, the length of the field is not specified; instead, it is simply expressed as *n* bits in length.

Every purchaser of service will require a unique identifier called the *subscriber ID*, whose structure will be decided by the service provider. This is somewhat like the old NETID, as it indicates the specific subscriber network. Some subscribers might want to obtain their addresses directly from a registry, which, in effect, will give them a unique provider ID independent of their ISP's provider ID. The field's length will vary based on the provider ID's length.

Finally, much like the HOSTID of IPv4, the intrasubscriber ID identifies individual interfaces on the network, as well as to which subnet the devices belong. This allocation is up to the local administrator. The intrasubscriber ID is made up of the 64 remaining bits in the Address field, which gives administrators a great deal of flexibility in assignment. One attractive approach is to use the IPv6 autoconfiguration capabilities to combine a 16-bit Subnet ID with a 48-bit Media Access Control (MAC) address, giving each subscriber over 65,535 enormous subnets.

The entire structure of the IPv6 address is illustrated here:

←		128 bits	→
Prefix 3 bits	Registry ID 5 bits	Provider ID/Subscriber ID 56 bits	Subnet and Host (MAC) Address 64 bits

Some would call 128-bit addresses overkill, but when the new structure of the Internet is considered, the allocation of the bits does make a great deal of sense. There is plenty of room for growth, as well as excellent support for existing addressing schemes. And remember that once upon a time, we all thought 32-bit addresses would last forever.

Although it is not certain if and when the Internet will move to IPv6, it is certain that when it does the Internet will slowly migrate to IPv6 and IPv6 addressing. The IPv6 transition mechanisms include a technique for hosts and routers to dynamically tunnel IPv6 packets over the IPv4 routing infrastructure. IPv6 nodes that utilize this technique are assigned special IPv6 unicast addresses that carry an IPv4 address in the low-order 32 bits.

An IPv6 mapped address is an IPv4 address transported over an IPv6 network. If a user with a machine configured for IPv4 wants to communicate on the Internet with a server that is using IPv6, the packet reaching the server contains an IPv6 mapped address. A compatible address, on the other hand, is an IPv6 address that is meant to be used when communicating with an IPv4 network. In this example, the user uses an IPv6 protocol stack trying to communicate with a server running IPv4. In this case, the IPv6 address must either be compatible from the start or be translated at some point into an IPv6-compatible address.

Despite the necessity of IPv6 development and its almost guaranteed success, the transition from IPv4 to IPv6 will not occur quickly nor painlessly. The single fact that tens of millions of hosts exist on the public Internet alone that need to be upgraded suggests that almost nothing (even a new World Wide Web application) can move *that* quickly. The major concern about IPv4 was the depletion of address space, but CIDR, DHCP, and NAT have all played a part in reducing the rate of address depletion. IPv6 offers several other advantages over IPv4, many of which can be migrated into IPv4. While this might be costly, it will likely be less costly than changing to IPv6. Fred Baker, chairman of the IETF, once indicated there could be another (better) replacement for IPv4 by the time address depletion becomes a real problem.

The transition to IPv6 will start with the network infrastructure, meaning that routers must be the first to be upgraded. In addition, the DNS, routing protocols, subnet masks, and any protocol carrying an IP address must be modified for the new IPv6 128-bit addresses. Later, hosts and end applications will transition to the new protocol.

It is clear that there will be IPv6/IPv4 coexistence for the foreseeable future. This will be the case for at least two reasons. First, the transition will take years. Second, some network administrations will not convert to IPv6 because they will derive no benefit from IPv6 and will, therefore, be reluctant to expend the time, energy, and cost on such an upgrade.

IP Routing

IP routing is the process of moving an IP packet from its source network to a destination network. As mentioned in previous chapters, the instructions that tell a routing device how to direct these packets could be manually or statically entered. While straightforward, this solution suffers from an inability to scale. As networks become more complex, we would ideally like a way to dynamically create the IP routing tables. This requires each router to learn the network topology and then to identify the *best* path to the destination. To top it all off, we would like these routers to change their tables automatically to reflect any changes in the network so that traffic forwarding is not disrupted.

While this is a tall order, network engineers are a resourceful lot, and dynamic routing protocols were developed to fill a need. By using messages (protocols) between the participating routers, a router is able to learn the network topology including destinations to all networks and redundant paths, insert them into a routing table for the router to use when forwarding packets, and finally react to a change in the network caused by either failures in the network or the addition of new hardware.

The process of learning the network can be done in a number of different ways; thus there are many examples of dynamic routing protocols with perhaps a half dozen or so in common use. Despite the variety of dynamic routing protocols, only two major philosophies exist for learning about the network on modern networks.

The first philosophy for enabling a dynamic routing protocol to learn about the network is called *distance vector*. When networks became large enough to make it clear that static routing alone would not be viable, the family of distance vector routing protocols was developed. These are simple protocols that essentially have routers connected to the same network share their routing tables with each other. So, for example, if router 1 has an entry in it that router 2 does not have, router 2 will update itself with the new information that it has learned from router 1.

In a large network, each router is constantly sharing its routing table with directly attached neighbors and updating it as the network changes. If another router were added to the sample distance vector–enabled network illustrated in Figure 19-5, it would learn about network 10 from router 2 since router 2 has inserted that entry into the routing table it shares with all of its neighbors.

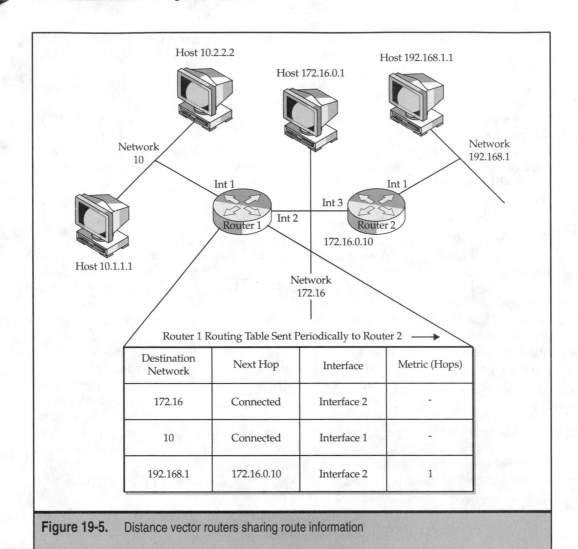

Figure 19-5. Distance vector routers sharing route information

Some consideration of distance vector routing protocols reveals that they are fairly simple protocols. The only real information that a router knows about the rest of the network is what it hears from its neighbors. Routers separated by several networks never hear from each other directly. Also, what they do hear about the rest of the network is fairly simple—essentially the information in a routing table. Thus, router 3 knows that router 1 is one network away from router 3 in the direction of interface one, but it has no idea what the conditions of the network are, what the bandwidth of the links are, and so on. These two problems lead to inefficiencies in the network as a whole.

The other major philosophy of dynamic routing protocols is known as *link state*. Being a protocol family that was developed some time after distance vector routing protocols, link state tries to address the deficiencies of distance vector routing. To do this, all routers share information with each other in a process known as *flooding*, illustrated in Figure 19-6. Thus, in our sample network, instead of router 3 hearing about router 1 from router 2, router 3 will hear about router 1 straight from the source.

This process of flooding allows all routers in the network to build their routing tables from information learned firsthand from all other routers in the network. This eliminates many of the most troublesome problems of distance vector routing protocols.

One of the most fundamental rules of routing is that if a network does not show up in a routing table, the packet is dropped. Every router must have a route to every destination on the Internet to make it possible for us to surf the Internet and check our e-mail. This makes sense until you learn that there are hundreds of thousands of networks on the Internet and no one router can look through that entire list for every packet!

The default route solves this problem. It tells the router: "If you don't know what to do with this packet after checking your routing tables, forward it this way!" Hopefully the router at the other end of the default route will have either another default route or will be able to forward the packet to another router that does have an entry in the IP routing table for the destination network.

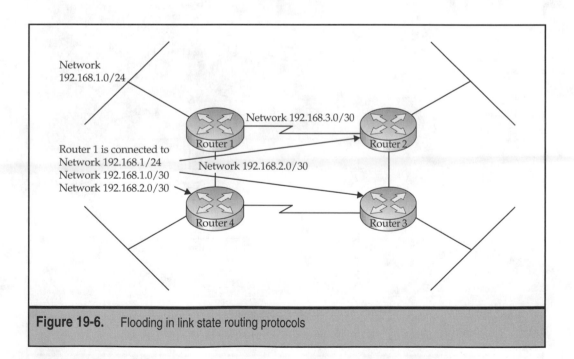

Figure 19-6.   Flooding in link state routing protocols

The three types of routes used to build an IP routing table—static, dynamic, and default—are often used simultaneously on one network or on one router. Let us examine the most common applications of these approaches.

Figure 19-7 shows the relationship between a typical ISP and a customer that has contracted with the ISP for Internet access. During the course of buying the access link to the Internet, the ISP will have also provided the customer with the other essential element to Internet access: an IP network address and host addresses.

While some sort of dynamic routing protocol could be used to connect the customer network to the ISP network, there is generally no need to do so, and this is discouraged for two reasons. The first is that running a distance vector or link state routing protocol over a WAN link uses too much of the already scarce bandwidth for routing table updates. The second reason is that the ISP generally is not going to trust that the customer has configured his dynamic routing protocols correctly. Since there is only one place in the world that the customer network should be found—off a particular port on a particular router belonging

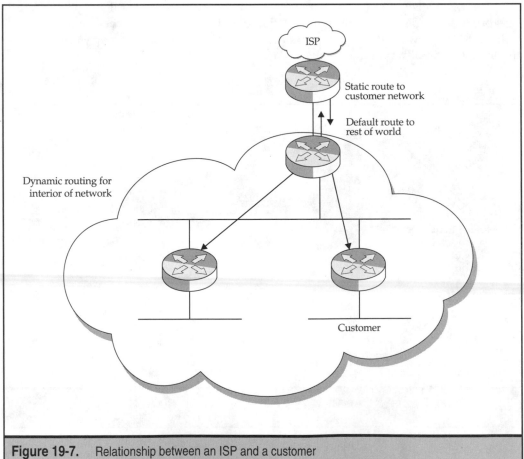

Figure 19-7. Relationship between an ISP and a customer

to the ISP—the ISP will set a static route to the customer network. Redundancy is not an issue because if the link between the customer and the ISP were to fail, there is no other way to reach either network.

From the customer point of view, there exists its network and the rest of the world. The router between the customer and the ISP WAN link has a pretty easy decision to make. If the destination of a packet is not on this local network, it has to be someplace "out there." A default route on the customer network edge router pointing to the ISP is enough to route traffic destined for the outside world correctly.

Internal to the customer network may be any number of subnetworks and routers. To assure connectivity between these internal networks, some sort of dynamic routing will generally be used.

Routing Information Protocol

The version of the Routing Information Protocol (RIP) used by IP routers is a distance vector protocol specified in RFC 1058. RIP uses *hop count* as its sole routing metric. Per the specification, a router extracts subnetwork and distance information from its routing table, broadcasting it to all neighboring routers and hosts every 30 seconds. RIP, sometimes called RIP version 1, is now referenced as "historic" by the IETF. Despite its historic classification, RIP is found on quite a few production networks.

RIP has a number of advantages. First, it is an exceedingly simple protocol to implement. Simplicity usually means "inexpensive to develop and low in computational overhead." Second, RIP has the advantage of being widely implemented and supported. Historically, RIP has been the most popular routing protocol for use in TCP/IP internetworks, largely due to its incorporation in Berkeley Software Distribution (BSD) 4.2 UNIX.

RIP also suffers from some serious shortcomings. Designed in a simpler time for simpler networks, RIP has begun to show its age as corporate networks have increased in size. Some of the problems are discussed in the following paragraphs.

In modern networks, use of hop count as a routing metric is not always efficient. It does not take load into consideration. Worse still, in internetworks with links of varying bandwidth, hop count ignores the differences. RIP would prefer a two-hop path over 9.6Kbps facilities rather than a three-hop path over 1.54Mbps facilities. A network manager can arbitrarily assign high hop counts to low-speed links, but the 15-hop limit (discussed in a minute) places severe restrictions on this approach.

Table exchanges, how RIP exchanges routing information, in large networks can consume significant bandwidth. This is because the routing tables grow as the internetwork grows, and the exchanges between neighbors consist of a significant portion of the routing table.

RIP also has a problem with *convergence.* A network has converged when all the devices are routing based on common information. A router running RIP typically detects local failures right away, but the routers one hop away do not detect the problem until they receive an update; the next closest routers will require yet another exchange, and so forth. This problem is addressed using *triggered* or *event driven* updates (routers generate updates upon detecting a change, rather than waiting for the next update interval). Unfortunately, this introduces the potential for oscillating routing tables and a flood of routing updates. To prevent this, RIP defines a *holddown* time. Once a router has updated its tables with a new route, it does not issue another update for a defined period of time (typically three times the update interval). The combination of triggered updates and the holddown time is that bad news travels fast (routers learn quickly that a route is lost), but good news travels slowly (learning new routes can take a significant amount of time). To put a bound on this time, the diameter of the network is limited to 15 hops.

Finally, the RIP update does not include the subnet mask assigned to particular subnetworks. This means that addressing strategies (such as Classless Interdomain Routing and noncontiguous subnetting) cannot be implemented in a network using RIP.

OSPF

Open Shortest Path First (OSPF) is a link-state routing protocol for use with the IP Network layer. Like RIP, it is intended to be used within the confines of a single network, also known as an *autonomous system* (AS). An AS is any network under the administrative control of a single entity. While technically any type of routing protocol creates an AS, OSPF was one of the first to formalize the term. Today many routing protocols make use of the term AS to define the area in which the routing protocol is operating. Unlike RIP, OSPF has been expressly designed to accommodate today's networks, which can grow to be very large, with dozens of hops, many different link types, and multiple paths through the network to any particular destination. Work was begun on it as early as the mid 1980s, but OSPF is now in its second version, which is described most recently in RFC 2328.

As a link state protocol, OSPF shares information about the AS topology so that all routers can build a link state database containing all topology information for the AS. Each router then calculates its own routing table by running a shortest path first (SPF) algorithm with itself as root.

One of OSPF's primary benefits is its speed. OSPF converges rapidly when a network topology change occurs. This speed is inherent to link-state routing protocols that flood link-state updates (small packets), rather than an entire routing table, when a change occurs. OSPF also allows for the division of autonomous systems into areas, which limits the amount of update traffic that must travel throughout the AS and keeps router memory requirements reasonable. This also helps contribute to faster convergence when network changes occur.

OSPF uses a single dimensionless metric that is assigned by the network administrator on a per-link basis. Accordingly, that number can represent speed, actual cost, reliability, or whatever combination of these attributes is desired. This allows network administrators quite a bit of flexibility in engineering traffic on their networks compared to earlier routing protocols.

IP subnetting, classless interdomain routing and variable-length subnet masking are all accommodated by OSPF because it defines a database and update structure that uses mask information, as well as address information, to identify subnetworks.

Finally, OSPF allows for a number of different authentication schemes for use between routers, thereby making it difficult for accident or malice to corrupt routing tables. RIP lacks such features and would allow a malicious user or program to change routing tables easily, disrupting network operations.

TCP/UDP TRANSPORT LAYER

The primary aim of the Transport layer is to reliably move an entire message from one end system (i.e., host) to another. In addition, the Transport layer can make decisions about how that message is transported, to save money or maximize speed.

The Transport layer must use the services of the Network layer to achieve this goal. This quite directly implies that the nature of the Transport layer's job is strongly related to the nature of the underlying network.

The information that peer Transport layers exchange is officially called the *transport protocol data unit*. In TCP/IP, the end-to-end unit is called a *segment*. Many other protocols simply call this unit a *message*.

A lot can go wrong with a message sent across a network. If sent as a series of packets on a packet-switched network, the messages must be divided into packets and reassembled across the network. This is true for newer frame- and cell-based networks as well as traditional packet networks. With connectionless packet (datagram) services, packets might arrive out of sequence or be lost. In all cases, the Transport layer must be able to detect and correct end-to-end errors.

Classes of Service

Different users (actually the higher layers) require different qualities of service. Transport protocols must provide a means for a user to specify what Quality of Service (QoS) is required on a particular connection. A connection carrying voice data through a packet network requires low and stable delays, for example, but it can tolerate an incredible number of bit errors; a connection carrying fund transfer information between banks can never tolerate a single bit error, but it might be able to proceed more slowly.

The transport user in a connection request specifies the QoS parameters. For each of these parameters, it is typical to include desired values and minimum accepted values. If the Transport layer cannot provide the minimum values, it sends a disconnect message to inform the user that the call request failed. If the Transport layer can meet all the parameter requirements, the connection request passes through the lower layers to the receiving Transport layer. This Transport layer must also examine the parameter values to determine whether it can accept them and set up a connection (whether all values can be met) or must disconnect (it cannot meet all of the minimum values).

To meet these specific QoS parameters, the Transport layer of the IP suite employs the use of two protocols: TCP and UDP.

TCP is designed to operate over a wide variety of network topologies and provide connection-oriented service with guaranteed delivery and sequentially. TCP is necessary to provide reliable interprocess communication to higher layers over possibly unreliable subnetworks.

The services provided by TCP can be characterized by at least four well-known features.

First, *TCP is a stream-orientated protocol.* Upper layers provide a data stream for transmission across the network (i.e., TCP inserts no record markers). The stream is a sequence of octets (or bytes). The destination TCP entity passes the stream to the destination application in the same sequence. In effect, TCP provides an ordered pipe that allows one application entity to pass information to another.

TCP is a three-phase communication protocol. The first phase involves the establishment of a connection with the destination application. This process means that TCP is connection orientated. This is analogous to the process of placing a telephone call. We must first establish a connection to the destination host. Once the connection has been established, data can be transferred. In our phone call analogy, this is equivalent to the act of having a conversation. When data transfer is complete the connection must be terminated (i.e., the phone must be returned to its cradle). This frees network resources for other connections.

TCP implements a buffered transmission stream. When a connection is established, two buffers are allocated at both ends of the connection. One is a transmit buffer and the other is a receive buffer. When the application submits data to be transmitted by TCP, TCP buffers it in the transmit buffer until it has accumulated some pre-established minimum amount or until a timer expires, whichever comes first. When TCP is ready to transmit, it assembles a unit of transfer called a *segment* and sends it across the network using the services of the IP. When the data arrives at the other end of the connection, it is buffered in the receive buffer. The application can remove it from this receive buffer as needed, which is not necessarily in the same size units that were transmitted.

Connections provided by TCP allow concurrent transfer of data in both directions, forming a full-duplex stream. From the perspective of two communicating entities, the full-duplex stream appears as two parallel connections with no interaction between them. This gives peer TCP entities the ability to piggyback control information concerning one direction of communication on segments traveling in the opposite direction.

User Datagram Protocol

UDP provides a connectionless, best effort delivery service—that is, UDP provides no connection establishment, acknowledgment, sequencing, flow control, or retransmission services. This lack of UDP functionality should not be seen as a weakness but a purposeful design decision. By eliminating these complexities, the resulting protocol is much simpler for a host to implement and it typically consumes less network bandwidth than TCP.

Applications that use UDP must provide their own forms of error recovery, due to UDP's lack of error control. For example, UDP-based data communications applications such as the SNMP or Sun's Network File Sharing (NFS) implement application level–based acknowledgment and error correction functions. Since real-time traffic such as interactive voice cannot withstand the delays associated with retransmissions, UDP as a transport for these types of applications makes a great deal of sense. After all, the complexities of TCP provide little benefit to real-time applications. UDP also makes sense for short-lived transactions, such as DNS queries, which do not warrant the overhead associated with the connection setup, tear down, and acknowledgment function associated with the TCP protocol.

So what does the UDP protocol do? In a nutshell, UDP provides for application level multiplexing (via ports) and segment error detection based on a simple checksum. UDP can also provide padding functions via its Length field.

UDP plays a significant role in the voice over IP (VoIP) environment as it is the primary vehicle by which real-time media is transported. Since corrupt UDP segments are "silently discarded" by the receiving UDP layer, the functionality associated with both the detection of, and the recovery from, sequencing errors/media loss falls upon the shoulders of the UDP application itself (e.g., the Real-time Transport Protocol, RTP). Since RTP (which supports voice and video in VoIP applications) implements sequence numbers, it can detect these anomalous events if they occur. In this case, a Protocol layer higher in the stack than UDP preforms the recovery from errors. In the RTP example, recovery from the loss of real-time media normally involves a simple "playback" of the last media block.

Well-Known Ports

If there were only one process ever running at any given moment during a TCP connection, the concept of a socket might not be critical. Imagine, however, a typical Internet device running Web and e-mail server software (and possibly more). The socket, as a combination of IP address and Transport layer port, guarantees that data not only reaches the right machine, but that it is also processed by the appropriate application.

This arrangement is much like the different layers of addressing in the physical world—apartment buildings are an excellent example. We first need a unique street address, like 467 Colchester Ave., for mail to be sent to the right building. Since apartment buildings comprise many different apartments, we also have to indicate an individual apartment number, such as Apartment 4. Our socket would therefore be 467 Colchester Ave., Apartment 4, combining street (IP) address with apartment (port) number.

TCP/IP applications are client/server by nature, so each side of a connection requires a unique socket. The client software is assigned an ephemeral, or dynamic, port greater than 1023 and the server processes are assigned a well-known port. For one HTTP connection, a browser might be assigned port 1024 for the duration of the session, while the server will be using port 80. An e-mail connection could also be running, which would receive a different, unique port number to ensure that the Web software wouldn't accidentally process its data.

The client has two sockets for its two application connections, expressed as *200.1.1.1:1024* for the browser and *200.1.1.1:1025* for e-mail. Similarly, the server currently has two corresponding sockets, the Web server at *130.1.1.0:80* and the e-mail server at *130.1.1.0:25*. When these connections are closed, the sockets are temporarily held in reserve to ensure that any late data is still handled correctly; subsequent connections receive new, incremented port numbers and, thus, new socket numbers.

Table 19-1 shows some of the common TCP/IP applications and their corresponding port numbers. TCP/IP applications commonly operate in a client/server mode, where the requester is the client and the system providing the service is the server. The port numbers shown here are the well-known ports associated with the server. The client randomly selects the client ports.

Port Number	Protocol (TCP/UDP)	Commonly Known As:
12, 20	TCP	FTP
23	TCP	Telnet
25	TCP	SNMP
53	UDP/TCP	DNS
80	TCP	HTTP
110	TCP	POP
161	UDP	SNMP
220	TCP	IMAP

Table 19-1.   Some Well-Known Ports

The Internet Assigned Numbers Authority (IANA) assigns port numbers. In the past, these numbers were documented through the RFC document series; the last of these documents was RFC 1700, which is now outdated. Now, the assignments are listed on the IANA server (**www.iana.org**), constantly updated, and revised when new information is available and new assignments are made.

Note that most of the applications specifically indicate use of TCP or UDP. A given function or task will be assigned either TCP or UDP by the protocol; neither the implementer nor user has a choice as to which transport protocol to use.

APPLICATION SERVICES

Most network users are not that interested in the particulars of networks, LANs, and Transport layer protocols. It is the benefits that this technology can provide that are important to the end users. In this section, we describe some of the most commonly used network applications and their function.

DNS

A 32-bit, dotted decimal IP address is used to identify host systems on the Internet. Although this is convenient for the network and for the routers, it is easier for users to remember names. It is also better for the network if users use names instead of numeric addresses; users will more accurately remember names, and system administrators can move systems around without having to notify everyone about a new address. The DNS describes the distributed name service on the Internet that allows hosts to learn IP addresses from Internet names.

The importance of the DNS cannot be overstated. When ARPANet was small, every host had to know the name and address of every other host. With the DNS, the capability is in place so that a host can learn another host's IP address from its name (or vice versa).

Consider that when a user wishes to connect to a given site, such as *www.hill.com*, the application must send out a request to the local name server to find out that this host's IP address is 208.162.106.4. In fact, the scenario is usually a bit more complicated than a simple query-response. When an application needs to learn the address of some host system, it sends an appropriate message to the local name server. If the requested name is not known by the local name server, it is the local name server's responsibility to forward the request to other name servers until the request is satisfied. Only when the name has been translated to an address can the local host try to determine a route to the destination host.

All Internet host names follow the same general hierarchy that can be read from right to left. The rightmost part of the address represents the top-level domain (TLD). A non-inclusive list of the generic top-level domain (gTLD) names used today include *.com* (commercial organizations), *.edu* (educational sites, generally colleges and universities), *.gov* (U.S. government agencies), *.org* (nonprofit organizations), *.net* (network operations hosts), *.biz* (businesses), and *.coop* (cooperative organizations). Other top-level domains include two character country codes as defined in International Organization for Standardization (ISO) Standard 3166.

The next-to-last portion of the address is generally called the *second-level* domain name, or simply domain name, and is assigned by the naming authority associated with each top-level domain. The fully qualified domain name *hill.com*, for example, is associated with the Class C IP address. When country codes are used, the appropriate country's Internet naming authority assigns names. The local Internet administrator, who is probably responsible for the assignment of individual host identifiers as well, assigns the other portions of the address.

The process of obtaining a domain name registration in the Internet has changed substantially over the past few years. Until mid-1998, Network Solutions, Inc. (NSI) was the sole accredited registrar for domain names. Through an organization called InterNIC, NSI would register a domain name for a fee of $100.

NSI's exclusive contract as a domain registration authority expired in 1998, and the IAB, seeking to lower prices and improve service issued a "Memorandum of Understanding" that opened the domain name registration function to competition.

Today, multiple domain name registrars exist within the U.S. To become a registrar, one must be accredited by the Internet Corporation for Assigned Names and Numbers (ICANN) (**http://www.icann.org/registrars/accredited-list.html**). As an accredited registrar registers new domain names, they are passed along to Network Solutions, Inc., for inclusion into the database. As overseer of the database, NSI charges a royalty of $6 for each domain name accepted from another registrar. It's important to note that domains were free before InterNIC started charging $100 for two years, $50/year thereafter. The initial fee for two-year registration is $70. Many registrars are offering longer terms (5, 10 years) for reduced rates. Although it is still possible to register a domain name directly through NSI, users are typically encouraged to use their ISPs to mediate registration. An entity wishing to register a domain name must fill out the appropriate paperwork, submit it to their ISP along with a payment of $70, and wait. A short time later, they receive indication that their domain name has been registered and is recorded in the NSI database.

FTP

The File Transfer Protocol (FTP) allows users to access remote file servers, list remote directories, and move files to or from remote hosts. FTP understands basic file formats and can transfer files in ASCII character or binary format. Defined in RFC 959, FTP provides a standard UNIX-like user interface, regardless of the actual underlying operating system. FTP allows a user to upload a file to a remote host, download a file from a remote host, or move files between two remote hosts. FTP was defined to allow file transfers between two systems without giving the user all of the capabilities that a Telnet session would (e.g., the ability to execute programs at the remote host).

FTP maintains separate TCP connections for control and data transfer. FTP commands and responses are exchanged on the control connection, and a new data connection is established by the server for each directory listing or file transfer.

FTP is one of the most useful TCP/IP applications for the general user. Anonymous FTP, in particular, is commonly available at file archive sites and allows users to access files without having an account already established at the host.

The age of the World Wide Web might appear to doom FTP, but in reality many people are still using FTP, though they might not realize it. Companies like Netscape, makers of a popular browser in the Internet, use FTP to distribute the latest version of their software. Rather than using a command-line interface, Netscape takes advantage of an easy-to-use graphical user interface (GUI) with which people are familiar. When a user clicks the file she wants, her browser is directed to an FTP server instead of a Web server. In fact, anonymous FTP is being used, with the browser automatically supplying the *anonymous* user name and e-mail address for the password. By shunting file transfers to another server, connections are freed on the Web server to distribute information that is better suited for Web access.

ICMP

In the connectionless, packet environment of IP, each host and router acts autonomously. Packet delivery is on a best-effort basis. Everything functions just fine as long as the network is working correctly, but what happens when something goes wrong within the subnet? As a connectionless service, IP has no direct mechanism to tell higher layer protocols that something has gone awry.

Furthermore, IP does not have a method for peer IP entities to exchange information; if an IP host receives a packet, it attempts to hand it off to a higher layer protocol. ICMP has been defined for exactly this purpose—IP-to-IP communication, usually about some abnormal event within the network.

There are several types of ICMP messages. Some of the most common are listed here:

▼ **Destination Unreachable**   Indicates that a packet cannot be delivered because the destination host cannot be reached; the reason is also provided, such as the host or network is unreachable or unknown, the protocol or port is unknown or unusable, or the network or host is unreachable for this type of service.

- ■ **Time Exceeded** The packet has been discarded because the Time to Live (TTL) field reached 0 or because all fragments of a packet were not received before the fragmentation timer expired. All IP packets carry a TTL field to prevent rogue packets from endlessly circling a network.

- ■ **Redirect** If a router receives a packet that should have been sent to another router, the router will forward the packet appropriately and let the sending host know the address of the appropriate router for the next packet.

- ▲ **Echo and Echo Reply** Used to check whether systems are active. One host sends an Echo message to the other, optionally containing some data, and the destination must respond with an Echo Reply with the same data that it received. (These messages are the basis for the TCP/IP PING command.)

SNMP

As its name suggests, the Simple Network Management Protocol was designed to ease the network management burden of TCP/IP-based networks. Originally, SNMP was to be the short-term solution for managing TCP/IP networks, but the widespread market acceptance for TCP/IP networks in general, and SNMP in particular, ensures SNMP's long-term success.

SNMP was conceived as a protocol for network performance monitoring, fault detection and analysis, and reconfiguration across TCP/IP networks. Because of the wide acceptance of the TCP/IP protocols within commercial sectors, nearly all networking products, such as workstations, bridges, routers, and other internetworking components, are being designed to use SNMP.

A network utilizing SNMP has two primary components: the management stations and the client agents. The management station implements the manager process, which might involve a GUI, and is the interface to the management information base (MIB), which is a database that contains information collected by the client agents. Whenever the manager needs information from the managed devices, it initiates SNMP messages to query the appropriate agent(s) for the data objects. The agents respond with the requested information, again using SNMP messages. Since SNMP relies upon short query/response messages, the User Datagram Protocol (UDP) is the connectionless transport service employed by SNMP.

The success of SNMP has been remarkable. It has grown from obscurity to become the de facto standard in a relatively short time. SNMP implementations now span more than just computer networks. SNMP may be implemented in almost anything that needs some range of control. Examples of SNMP use, or potential use, are power and gas telemetry and control, elevator control, and plumbing control.

SMTP/POP/IMAP/MIME

E-mail is one of the most widely used applications on the Internet. In fact, e-mail has been viewed by corporate America as an aid to worker productivity since the early 1980s and has been an important part of any host or LAN application suite or operating system for the last 15 years. Commercial e-mail services from companies such as AT&T, Sprint, and MCI WorldCom have been available for many years, well before the increased commercialization of the Internet. In almost all cases, these e-mail systems originally were proprietary, incompatible, and not based on SMTP.

The use of e-mail varies widely. Most major companies today have e-mail for at least internal communication. Networking a corporation e-mail system with a network such as the Internet adds tremendous benefit since it allows employees to communicate with professional colleagues from around the world and allows customers to communicate with the company. In this section, we discuss the protocols that enable the functionality of e-mail on a TCP network.

In SMTP, an item of mail is sent after the sender and one or more recipients are identified. When the SMTP server process accepts mail for a particular recipient, it has the responsibility to post the mail to the user if the user is local, or to forward the mail to the appropriate destination host if the user is not local. SMTP determines the appropriate mail server for a given user using the DNS's Mail Exchange (MX) records. As a message traverses the network, it carries a reverse path route so the sender can be notified of a failure.

SMTP's main functions are to identify the sender, identify the intended receiver(s), transmit the message, and provide confirmation that the message has been sent. SMTP defines a small set of commands to invoke its procedures, such as HELO (sic), MAIL FROM, and DATA. Responses are even more cryptic, consisting of a set of response codes, such as 250 (OK) and 550 (user unknown). Most SMTP implementations provide a better human interface or provide a gateway to another e-mail package like Lotus cc:Mail, Outlook, or Eudora.

SMTP defines a protocol for the exchange of mail messages between e-mail servers. In addition, SMTP does not provide a clean mechanism for users accessing and managing their e-mail. For example, SMTP provides no commands so that a user can reply to a message, forward a message to another user, or send carbon copies (CC) of messages. To fill this gap, the Post Office Protocol (POP) was developed. POP is intended to be used across a local network. The POP server maintains users' mailboxes. Each user's system has POP client software, which provides a rudimentary interface so that users can access and manage their e-mail.

The Internet Message Access Protocol (IMAP) is another method of reading e-mail kept on a mail server by permitting a client to access remote message stores as if they are local. Messages stored on an IMAP server can be manipulated from multiple computers without transferring messages between the different machines. In other words, a telecommuter with a laptop computer at home and a desktop computer at the office need not worry about synchronizing each computer since mail read at either machine continues to be maintained on the server. This "online" message access contrasts with the "offline" POP, which works best when you only have a single computer; mail messages are downloaded to the POP client and then deleted from the mail server.

While IMAP has more commands than POP, making it arguably more difficult to implement and use, the existence of development libraries and extensions give IMAP much flexibility and room for growth. Many management functions, sorting capabilities, and editing tools can be invoked on the server to improve performance and organization. Many POP clients now have similar options, but POP's main advantage is the maturity of applications using the protocol, with a wide array of products already available.

Just as the type of data that we employ over the network has changed over the last two decades, so has the type of information that we would like to exchange in e-mail. The Multipurpose Internet Mail Extensions (MIME) were defined specifically for this purpose. MIME supports the exchange of multimedia messages across the Internet by defining procedures that allow a sender to provide attachments to e-mail that employ a variety of non–text-based formats. The types of contents in parts of an e-mail message body are indicated by use of a subtype header in the message. Possible MIME content types are text, images, audio, video, and applications.

The use of mail-enabled applications is the subject of some controversy by security-minded users. Traditionally, a computer virus cannot be sent via e-mail and cause harm to a user unless the virus is sent as executable code to the user and the user runs the program. If the virus is an attachment, the user has the opportunity to scan the program before running it. But what happens when the virus is an attachment that may be automatically executed by a mail-enabled application? This situation is potentially worsened by the fact that MIME messages can contain external pointers to files that reside on other computers. Among the potentially dangerous mail-enabled applications that MIME can launch are FTP (both anonymous and nonanonymous) and SMTP-based e-mail.

HTTP

Tim Berners-Lee is considered by many to be the Father of the WWW. A physicist by training, he worked in the computer and telecommunications industries following graduation from Oxford and eventually accepted a consulting position as a software engineer with the European Center for Nuclear Research (CERN) during the late 1970s.

During his stint in Geneva, Berners-Lee observed that CERN suffered from the problems that plague most major organizations: information location, management, and retrieval. CERN is a research organization with large numbers of simultaneous ongoing projects, a plethora of internally published documentation, and significant turnover of people. As a consequence, Berners-Lee found that his ability to quickly locate and retrieve a particular piece of information was seriously impaired by the lack of a single common search capability. To satisfy this need, in 1980 he wrote a search and archive program called Enquire. Enquire was never published as a product, although both Berners-Lee and the CERN staff used it extensively. In fact, it proved to be the foundation for the World Wide Web.

In May 1990, Berners-Lee published *Information Management: A Proposal,* in which he described his experiences with hypertext systems and the rationale for Enquire. He described the system's layout, feel, and function as being similar to Apple's Hypercard, or the old "Adventure" game where players went from "page to page" as they moved through the game. Enquire had no graphics, and was rudimentary compared to modern Web browsers, but it did run on a multiuser system and could therefore be accessed simultaneously by multiple users.

In November 1990, Berners-Lee wrote and published, with Robert Cailliau, *Worldwide Web: A Proposal for a Hypertext Project*. In it, the authors describe an information retrieval system in which large and diverse compendiums of information could be searched, accessed, and reviewed freely using a standard user interface based on an open, platform-independent design. This paper relied heavily on Berners-Lee's earlier publication, *Information Management: A Proposal*. In this second paper, Berners-Lee proposed the creation of a "worldwide web" of information that would allow the dispersed CERN entities to access the information they need, based on a common and universal set of protocols, file exchange formats, and keyword indices. The system would also serve as a central (although architecturally distributed) repository of information, and it would be totally platform-independent. Furthermore, the software would be available to all and distributed free of charge.

After Berners-Lee's second paper had been circulated for a time, the development of what we know today as the World Wide Web occurred with remarkable speed. The first system was developed on a NeXT platform. The first general release of the WWW inside CERN occurred in May 1991, and in December the world was notified of the existence of the World Wide Web (known then as W3) thanks to an article in the CERN computer newsletter. Over the course of the next few months, browsers began to emerge. Erwise, a GUI client, was announced in Finland, and Viola was released in 1992 by Pei Wei of O'Reilly & Associates. NCSA joined the W3 consortium but didn't announce its Mosaic browser until February 1993.

Today's WWW consists of four principal components: Uniform Resource Locators (URL); HTML; HTTP and HTTP client/server transactions; and resource processing, which is generally carried out on the server using the Common Gateway Interface (CGI). URLs represent the standard addressing scheme utilized on the WWW to specify the location of information on the Web. An example showing standard URL format is *http://www.hill.com*. Here, *http://* indicates that the Hypertext Transfer Protocol will be used to transfer information from a *www* server located at a site called *hill*, that happens to be in the commercial (*.com*) domain.

HTML is the programming language used to format documents in a universal, standardized format so they can be downloaded from a Web server as hypertext documents. The markup language consists of formatting commands that allow a user to specify the location, size, color, alignment, and spacing of headings, full-text paragraphs, animation sequences, quotations, and images on a page, as well as functional buttons that lead to other hypertext documents.

The extensible markup language (XML) is considered the next generation of HTML. This language uses formatting commands like HTML but allows the user to define what these commands are. This allows closer integration with databases and user applications and will eventually expand the capabilities of the WWW.

Like all resources on the WWW, the location of these documents is specified through a URL. These URLs can point not only to text, but also movie clips, image files, music, and so on. If the information can be digitized and stored on a hard drive or CD-ROM, it can be pointed to with a URL using HTML formatting and downloaded to the requesting system using the HTTP.

The HTTP lies at the heart of the Web. It represents the command syntax used to transfer documents from a Web server to a Web client, similar to the File Transfer Protocol (FTP). In fact, HTTP is a client/server environment. Transactions are carried out between the client and the server, moving the requested file from the server to the client.

SUMMARY

Many protocols have been used in networks since the evolution of distributed computing. While originally developed as an interim solution, TCP/IP has grown to become the de facto standard of the Internet. This deployment alone requires that students of data communication have an understanding of its parts and role in modern networks. The current popularity of TCP/IP has been double edged. While its wide use ensures adoption and understanding by many, its wide use has also limited its ability to change and adjust to growing demands on the Internet as a whole.

CHAPTER 20

Internet Access
Options

Access to remote computing services by means of a telecommunications network is certainly not a new phenomenon. As far back as the 1960s, modems have been employed to render the telephone network capable of carrying data. In those early remote access applications, terminal-to-computer scenarios were typical. Whereas both asynchronous and synchronous Physical-layer protocols were employed, the applications basically involved sending keystrokes (or blocks of characters) from the terminal to the remote computer, and receiving character-oriented data in return to be presented to the user on a CRT screen.

The telephone network was used in two distinct ways in these remote timesharing settings. For higher-speed, business-oriented applications (e.g., IBM's Binary Synchronous Communications), leased telephone company facilities (often called "3002 circuits") were the norm. These 4-wire facilities were either point-to-point or multipoint, and depending on the modems employed, they operated at speeds from 1200 bps to 9600 bps. For remote timesharing in engineering and scientific applications, dial-up modems were used in conjunction with telephone company switching facilities to interconnect asynchronous terminals to minicomputers. These environments made use of 2-wire local loop facilities and modems that operated at speeds from 300bps to 1200bps.

From those early days, it was recognized that, all other things being equal, the faster the data rate afforded by a particular technology, the better. High access rates permitted operators to work more quickly and efficiently. In addition, operation at higher data rates made possible the transfer of more complex information. Unfortunately, narrowband networks like the public switched telephone network (PSTN) offered only so much in the way of access rates. Even when customers were willing to spend a lot of money (the most rudimentary dial-up modems cost hundreds of dollars and high-end leased line modems cost in the thousands), access rates were limited.

Today, the remote access problem is experienced most often in the context of the Internet. With the numbers of people with online capability growing at an exponential rate, and the increasing complexity of the data types to which they wish to connect (e.g., audio, video, and complex image graphics), the local access bottleneck has become a major obstacle to continued growth of the Internet itself.

This chapter discusses several alternatives for high-speed Internet access. The reader should note that these technologies can apply, in most cases, to corporate networking environments where high-speed access is desirable as well. Generally, technologies for high-speed Internet access fall into three categories. They are:

▼ Technologies based on the telephone network (modems and ADSL)

■ Technologies based on community antenna TV (CATV) networks

▲ Technologies based on wireless networks (fixed-base and satellite)

We will cover each of these major categories in its turn.

HIGH-SPEED INTERNET ACCESS AND THE PSTN

The PSTN is currently used in three ways to access the Internet. Most common by far is its use as a narrowband network to support modem-based access at speeds up to 56Kbps. The vast majority of residential users access the Internet with inexpensive (under about $100) modems.

To date, some users have ISDN basic rate (see Chapter 13) connections to the PSTN. For these users, access speeds of up to 128Kbps are possible. Unfortunately, sales of ISDN basic rate service into the residential marketplace have been lackluster to say the least. Thus, ISDN connectivity to the Internet accounts for fairly few users. In some markets, ISDN has seen an increase in sales due to the lack of availability of other high-speed options. When other options become available, however, ISDN service is usually terminated.

Over the last several years, an attempt to leverage the PSTN infrastructure to support truly high-speed access rates has been underway. The result is a technology called asymmetric digital subscriber line (ADSL). Using ADSL technology, users can achieve access rates of up to 7Mbps or thereabouts.

The next sections will focus on two PSTN-based Internet access technologies. They are analog dial-up modems and ADSL. In each case, we shall briefly describe the technology and then highlight the advantages and disadvantages of each.

Dial-Up Internet Access—The V.90 Modem

The majority of people who access the Internet by dialing through the PSTN use a modem described in ITU-T Recommendation V.90. Although the V.92 Recommendation has been ratified, and products are available, this newer modem has not yet seen widespread deployment. In any case, the fundamentals of V.92 closely mirror those of V.90, so we will concentrate on the latter specification.

The reader will recall (see Chapter 4) that a modem operates by modulating a digital bit stream onto a broadband carrier for transport over the PSTN. The use of a modem therefore requires no special loop conditioning. Depending on the quality of the connection obtained at any particular time, modems may deliver variable data rates. In this discussion, we will always refer to data rates as those maximally attainable under perfect conditions.

A V.90 modem operates in duplex mode but in asymmetric fashion. The maximum upstream (i.e., user-to-network) rate is 33.6Kbps, and the maximum downstream (network-to-user) rate is 56Kbps. This asymmetric operation comes as a result of the fact that V.90 is not a single modem specification, but a specification that defines two separate modems, one analog and the other digital. The so-called "analog" modem is the one associated with the remote access user. The "digital" modem is the one associated with the Internet service provider (ISP).

The use of the terms "analog" and "digital" is somewhat misleading in this context. Both modems, of course, perform digital-to-analog (D/A) conversion. The analog modem, however, is more like a conventional modem in that it impresses ones and zeros on an analog carrier for transmission. In the reverse direction, the analog modem performs an analog-to-digital (A/D) conversion much like a conventional modem. Therefore, the use of the analog modem mandates that two A/D conversions occur between the transmitter and the receiver. One A/D conversion occurs at the telephone company central office, where the local loop is digitized at a channel bank. The other occurs in the receiving modem, where the digital information is recovered through demodulation of the analog carrier signal. It is this requirement for two separate A/D conversions that limits the upstream rate in V.90 to 33.6Kbps.

The digital modem, on the other hand, attaches to the network using digital technology. The output of the digital modem consists of codes taken from the μ-255 encoding scheme used in the digitization of voice (see Chapter 10). Only seven of the eight bits for each PCM codeword are used, however, limiting the V.90 modem's downstream data rate to 56Kbps, instead of the 64Kbps associated with a digital voice channel. Attaching the modem to the network in this fashion eliminates one A/D conversion and delivers the 56Kbps data rate.

V.90 Modems—Pros and Cons

The most obvious drawback to the use of the V.90 modem is low speed. The data rates supported by V.90 are not considered "high speed." Table 20-1 shows the amount of time required to download a 3MB file from a server using various access technologies. The data in the table assumes that access rate is the limiting factor in receiving the file, and that the file is delivered as a continuous stream of bits. As can be seen in the table, the V.90 modem requires a little over seven minutes to transfer the file. Moreover, if line conditions aren't perfect, V.90 modems fall back to a lower speed, making the problem worse.

The single most important advantage to the use of the V.90 modem is cost. The modem itself is quite inexpensive (under about $100), and if a local call is all that's required to access your ISP, additional cost benefits accrue. Finally, the ubiquity of the PSTN is another advantage to V.90. For users that travel, V.90 facilitates connectivity from just about anywhere.

Because of the prevalence of V.90 technology in the Internet access world, we will use the V.90 data rates shown in Table 20-1 as a benchmark against which we can compare other access technologies.

Always-On Internet Access—ADSL

ADSL was originally developed by telephone carriers for delivery of high-quality, on-demand video services to residential users over the in-place local loop. This intention explains its asymmetric nature. To request a video or to control it while it is playing requires very little bandwidth in the upstream direction. To receive such a streaming video in real time from a server requires substantially more bandwidth in the downstream direction. Interestingly, this same traffic characteristic applies in most cases to Internet access.

Technology	Network	Maximum Data Rate*	Time to Download 3MB File
V.90 Modem	PSTN	56Kbps	7.1 Minutes
ISDN BRI	PSTN	128Kbps	3.1 Minutes
T-1/ADSL	PSTN	1.5Mbps	15.5 Seconds
ADSL	PSTN	7Mbps	3.4 Seconds
CATV Modem	CATV	10Mbps	2.4 Seconds
MMDS	Wireless	50Mbps	0.48 Seconds
LEO Satellite	Wireless	64Mbps	0.38 Seconds
LMDS	Wireless	155Mbps	155 Milliseconds

*Maximum is downstream where applicable.

Table 20-1.   The Worldwide Wait

Requests to servers (upstream transfer) are typically characterized by low volumes of data. Responses to browsers (in the downstream direction) may be considerably more voluminous.

Current ADSL services, therefore, operate upstream at maximum rates of about 384Kbps. Downstream data rates are very dependent on the length of the local loop, with 7Mbps available only to those who live in close proximity to their central offices (that is, nearer than 9000 feet). Once loop length exceeds about 15,000 feet, maximum downstream data rates diminish to about 1.5Mbps. The reader will observe on Table 20-1, however, that ADSL at 1.5Mbps is substantially faster than the V.90 modem. At maximum downstream rates, ADSL can deliver a 3MB file in less than five seconds!

ADSL Architecture

Prior to discussing the components of the ADSL architecture, we will look at how ADSL uses the bandwidth available on the local loop to support voice and data simultaneously. Figure 20-1 is a graph of signal energy as a function of frequency. As you can see, voice signals occupy the bandwidth of 0–4000Hz, whereas data signals utilize the remaining bandwidth up to 1.1MHz. The ADSL upstream channel operates at 25–200KHz; the ADSL downstream channel operates at 200KHz–1.1MHz. Some implementations use echo cancellation, where the upstream and downstream data channels can actually overlap to provide additional bandwidth for the downstream channel.

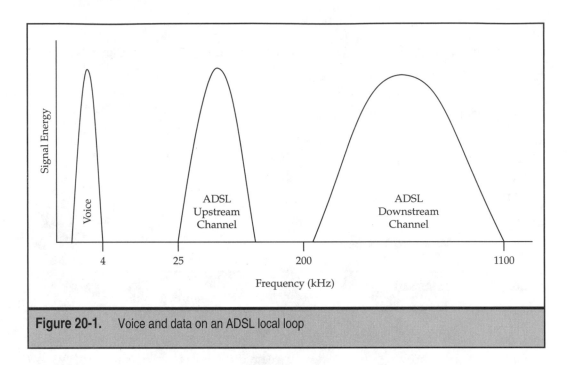

Figure 20-1. Voice and data on an ADSL local loop

Figure 20-2 illustrates the architecture and components of an ADSL connection. At the customer's premises, the local loop is terminated on a splitter. The function of the splitter is to separate voice signals from data signals on the local loop. Beyond the splitter,

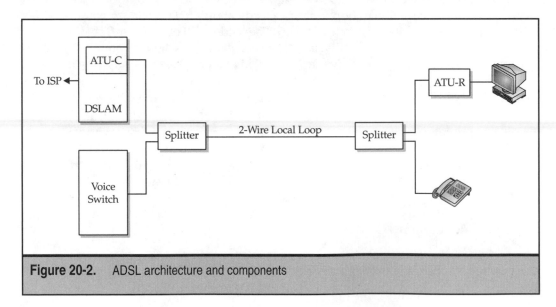

Figure 20-2. ADSL architecture and components

voice signals are directed to a standard analog telephone set and data signals are directed to a device called an "ADSL Terminal Unit – Remote" (ATU-R, sometimes called an "ADSL modem"). The ATU-R typically connects to the user's PC via a 10Base-T Ethernet port.

At the telephone company central office, the local loop terminates on another splitter. This central splitter performs the same function as the splitter on the customer's premises; it separates the voice signals from the data signals on the loop. From the splitter, the voice signals are delivered to a voice switch that supports traditional POTS service for the customer. The data signals are terminated on an equipment bay referred to as a "DSL access multiplexer" (DSLAM). Contained within the DSLAM are line card modules that perform the same function as the ATU-R does at the customer's location. These ADSL Terminal Unit – Central (ATU-C) cards demodulate the digital data stream to recover the customer data traffic. In typical ADSL implementations, customer data (typically IP packets) are broken into ATM cells by the ATU-R. Demodulation by the ATU-C, therefore, yields a cell stream. For this reason, most ADSL providers use an ATM switching fabric to interconnect the ISP to the ADSL user. The DSLAM itself may perform functions related to delivering the data to an ISP, such as ATM switching or IP routing. Alternatively, these functions may reside outside of the DSLAM in other equipment (that is, an ATM switch or an IP router).

ADSL Advantages

As mentioned previously, the most significant advantage of ADSL when compared to other methods for Internet access using the PSTN is speed. Referring back to Table 20-1, we see that neither ISDN nor the V.90 modem can begin to offer the data transfer rates available with ADSL. Even at its lowest speeds, an ADSL connection offers users the bandwidth of a T-1 connection in the downstream direction.

An often-overlooked advantage of ADSL is its "always-on" nature. Unlike a connection through the analog PSTN or an ISDN, the data portion of the ADSL circuit behaves like a leased line. No dial-up procedure is required for connection to the Internet. As a result, users can receive information from the Internet at any time. This enhances convenience and supports services known as "push" services. A "push" Internet service, as its name implies, delivers information to a subscriber at whatever time it becomes available. For example, a brokerage service with push capability allows a user to set a threshold value for a stock price. When that threshold is reached, the service informs the user immediately of the current price. Other push services include news services that filter information according to some user criteria and then deliver relevant information automatically to the always-on subscriber.

Another advantage in having a permanent connection to the Internet is the ability to host Web content. Although ISPs vary widely in their attitude toward individual subscribers hosting content, users that wish to do so require, at minimum, some form of always-on connection to the ISP. The astute reader will recognize that always-on Internet connectivity has its downside. An always-on user is more vulnerable to security problems than the occasional dial-up user. For this reason, enhanced security measures (e.g., a firewall) are often implemented by users connected to the Internet via ADSL.

The Trouble with ADSL

As a friend of this author remarked recently, "ADSL is the best high-speed Internet access service that you can't get." This lament sums up the basic problem with ADSL. There are a number of reasons why ADSL cannot be delivered to a large percentage of residential users. The following sections will summarize the major provisioning difficulties experienced by ADSL providers.

Distance Limitations   Proper operation of ADSL is a function of the length of the customer's local loop facility. Beyond about 18,000 feet, the service is simply not available. It is estimated that 20 percent of local loop facilities in the U.S. exceed 12,000 feet. For these customers, ADSL will either be unavailable or severely rate-limited. For customers whose loops are less than 12,000 feet in length, ADSL service may be available at speeds ranging from 1.5Mbps to 7Mbps.

Loop Impairments   In the early days of outside plant construction, telephone companies used pieces of lump inductance, called "loading coils," to bring the attenuation characteristics of long loops into line with the notion of "toll quality" voice transmission. That is to say that, on a loop that exceeds about 18,000 feet, attenuation of high-frequency (i.e., 2500–4000Hz) voice signals becomes unacceptable by telephone company standards. Because, in a particular binder group, some loops would turn out to be longer than others depending on their points of termination, loading coils were used freely in the loop plant. It is estimated that approximately 50 percent of all loops in the U.S. are loaded in this fashion. Unfortunately, the loading coils make the loop behave like a low-pass filter. Attenuation is improved across the entire voice frequency band (i.e., 0–4000Hz), but a severe "rolloff" phenomenon occurs just above 4000Hz. In that ADSL's upstream and downstream data bands occupy the region above 4000Hz (see Figure 20-1), ADSL signals do not pass over a loaded loop. Thus, to provision ADSL service to such a customer, the loading coils must be removed from the loop manually in a process that is expensive for the ADSL provider.

Another line impairment that bears upon a carrier's ability to offer ADSL services is an unterminated bridged tap. When telephone companies install loop wiring, for example into a new housing development, they run a binder group with all of its associated pairs down an entire street. Then, as houses become occupied, they select a single pair from the binder group and terminate it at a particular residence. Rather than make each loop the exact length required to reach each property, they leave all of the pairs in the binder group the same length. When service is required, the installer simply "bridges" on to an existing pair. Between this "bridged tap" and the end of the binder group, there remains a length of unterminated loop. Depending on the length of this unterminated bridged tap, ADSL service distance can be reduced. In addition, signal reflections from the unterminated portion of the loop can degrade service quality and lower effective data rates for ADSL.

Finally, a host of other loop impairments can reduce the quality of ADSL service. Dampness in binder groups, poor splices, and mixed wire gauges can all contribute to less-than-optimum ADSL performance.

Digital Loop Carrier (DLC) Systems   Approximately 30 percent of U.S. households are served by a telephone company digital loop carrier (DLC) arrangement. In Chapter 10, we explained that a DLC system uses high-speed digital facilities (e.g., SONET) to connect a central office switch to a remote terminal usually located fairly close to the customer. DLC systems thus extend the digital capability of the switch to a location remote from the central office. The job of a DLC is to digitize voice at the remote terminal location. You will recall (see Chapter 10) that the first step in the voice digitization process is to filter the analog signal to remove all frequency components above 4000Hz. Once again, ADSL's upstream and downstream data bands occupy the region above 4000Hz (see Figure 20-1), and so ADSL signals do not pass over a DLC system.

The solution to this problem is more complex that it looks at first glance. It would appear simple to retrofit the DLC remote terminal with ATU-C line cards, in effect turning it into a remote DSLAM. Unfortunately, this is a costly proposition for systems that were never designed to support ADSL. In addition, unless sufficient bandwidth is available between the DLC remote and its central office terminal, this approach cannot be considered. Serving a population of users with ADSL at 1.5Mbps each is a much different proposition than serving that same population with 64Kbps each.

Another option in these settings is simply to install a DSLAM at the remote location, perhaps alongside the existing DLC remote terminal. Although this approach is being taken by some ADSL providers, various difficulties have arisen. The issue of real estate at the remote location is an obvious one. If no room exists for the DSLAM, substantial construction costs may be incurred by the provider. In addition, under the Telecommunications Act of 1996, incumbent local exchange carriers (ILECs) must allow competitors to colocate equipment in their facilities. It is fairly easy to permit a competitor to colocate within the confines of a central office; at a remote terminal location, such colocation may be impossible. In fact, one large incumbent carrier has withdrawn its ADSL service from the market due to problems with colocation by a competitor!

Installation Woes and the Dreaded Truck Roll   Referring back to Figure 20-2, you can see that ADSL service requires two pieces of equipment on the customer's premises. There is the ATU-R, which is typically colocated with the user's PC, and there is the splitter. The presence of the splitter at the customer site has proved problematic for several reasons. It has been shown that the average customer has a great deal of difficulty installing the splitter. The problem is twofold. If the splitter is installed at the point where the loop enters the home, two sets of wiring (one for voice and one for data) are required from that point forward. Although most dwellings have 4-wire capability, most users cannot deal with continuity issues arising when trying to make the second pair operational. For homes without 4-wire capability, new wiring must be installed from scratch. If the splitter is installed at the PC location along with the ATU-R, other wiring problems arise. How are the telephones in the house to be hooked up to wiring that carries both voice and data signals?

These difficulties with the remote splitter have forced telephone companies to send installation technicians to subscriber's homes to install ADSL service. These "truck rolls" are expensive and can quickly erode any profit margin that the provider expects to make on the service. Finally, inside wiring is usually the responsibility of the homeowner. Even when the telco technician installs the splitter, the aforementioned wiring problems are not necessarily taken care of.

G.lite to the Rescue   In order to alleviate the problems associated with the remote splitter in ADSL environments, a new "flavor" of ADSL has been introduced. Called "universal ADSL" (UADSL) or simply "G.lite," this form of ADSL is a compromise solution.

On the upside, G.lite eliminates the splitter on the customer's premises. It is, therefore, fairly simple for the average homeowner to install by himself or herself. Instead of a splitter, G.lite uses passive filters. A high-pass filter in the ATU-R keeps voice from interfering with data, and low-pass microfilters installed at each telephone set keep data from interfering with voice. A splitter is retained at the central office with G.lite.

G.lite extends the reliable working distance of ADSL to a full 18,000 feet. Although full ADSL was intended to support loops of that approximate length, field trials proved that ADSL operates reliably out to distances of only 12,000–15,000 feet.

The price paid for eliminating the remote splitter in G.lite is fixed speed operation. No matter the distance from the central office, G.lite customers get a maximum downstream rate of 1.5Mbps. Upstream maximum data rate is 512Kbps, but providers routinely reduce that to 128Kbps or 256Kbps. Table 20-1 shows that these rates are clearly adequate for most Internet applications, but the promise of video-on-demand is severely constrained in the G.lite environment.

Finally, there is some question regarding service quality with G.lite. Passive filtration is simply not as effective as active splitting in ensuring that voice and data signals do not interfere with one another on the loop. As a consequence, when a phone goes off-hook while the user is on the Internet, data rates tend to suffer. Conversely, when a voice call is attempted when the loop is being used for data transfer, an annoying noise may be heard on the phone.

ADSL Summary

ADSL seems like a good idea on the surface. After all, here is a technology that can leverage the huge installed base of local loops to bring cost-effective, high-speed Internet access to the masses. Unfortunately, carriers that have tried to deploy ADSL have experienced a host of problems, many of which are discussed in this section. Current market estimates indicate that approximately 1.5 million ADSL lines have been installed in the U.S. to date. Given this installed base and the current rate of growth in ADSL lines, it is doubtful whether this technology will ever dominate the high-speed Internet access market.

HIGH-SPEED INTERNET ACCESS AND CATV NETWORKS

Currently, the dominant technology for providing high-speed Internet access to residential users is the cable modem. It is estimated that approximately five million cable modems are in operation. As its name suggests, the cable modem uses the infrastructure of the community antenna TV (CATV) network to provide high-speed, asymmetrical Internet access services to users. Although not as ubiquitous as the PSTN, the CATV networks in the U.S. pass most of the residences where high-speed Internet access is likely to be desirable. The next several sections will examine the use of CATV networks for data transport. We will examine the anatomy of a CATV network and discuss the advantages and disadvantages of cable modem service for Internet access.

Anatomy of a CATV Network

CATV networks can be broken into two categories, depending, for the most part, on their age. Older CATV networks consist purely of coaxial (COAX) cable from the service provider's "headend" location (think of the headend as the CATV provider's central office). Originally, all CATV networks were of this variety, and their intent was solely to support the delivery of analog broadcast video services to their users.

Newer CATV networks include a fiber optic component and utilize an architecture called "hybrid fiber/COAX" (HFC). From the headend, a fiber optic connection extends to a remote terminal location. At the remote terminal, the fiber gives way to COAX bus connections to individual residences. The CATV HFC configuration is similar to the telco DLC architecture, except that, beyond the remote terminal, HFC uses COAX in a bridged topology whereas DLC uses twisted pair in a point-to-point topology. HFC networks are better suited to the transport of digital information than their all-COAX counterparts. As such, HFC networks are often found in environments where the service provider desires entry into data-oriented markets like Internet access. Moreover, with an HFC architecture, high-quality digital video services are more easily deployed.

All-COAX CATV Networks

The top diagram in Figure 20-3 shows an all-COAX CATV network. From the headend, a randomly branching bus of COAX cable extends toward users' residences. The problem of attenuation is solved by the use of amplifiers in an all-COAX CATV network. At intervals, amplifiers are deployed to boost attenuating signals back to some acceptable level of amplitude. Also at intervals, the cable branches at locations where a passive splitter device

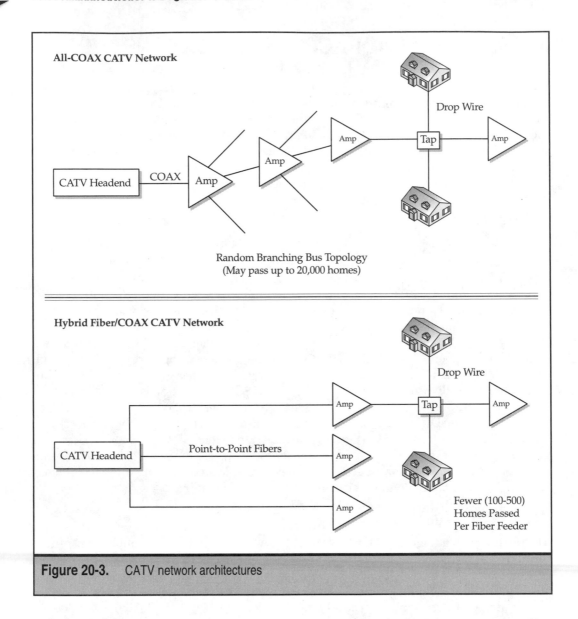

Figure 20-3. CATV network architectures

is installed. Finally, at the residence, the cable is tapped and a "drop wire," a thin section of coaxial cable, is used to deliver the signal into the home, where it terminates on a set-top box. The gain characteristics of the amplifiers used in an all-COAX CATV network vary from very large to small. As the cable branches, the amplifiers deployed contribute less and less gain to the signals.

It should be noted that the numbers of residences served by a single port on a headend can be quite large, on the order of 20,000 or so. This fact makes CATV networks notoriously prone to catastrophic failure. Failure of any component along the bus equates to an outage for any subscriber "below" that level on the bus. Failure of the headend can render 20,000 users without service simultaneously!

The total bandwidth available on an all-COAX CATV network is 750MHz. For delivery of analog video services, this bandwidth is divided into 6MHz channels. Thus, the typical CATV network can deliver about 125 channels of broadcast television. Some of these channels, however, are reserved for upstream transmissions (i.e., transmissions from subscribers to the CATV headend). These upstream channels were originally intended to support interactive TV services to public schools, but they are being used today to support upstream traffic in high-speed Internet access scenarios. The reader can see that it is a fairly simple matter to reserve one or more of the 6MHz channels on the cable for support of high-speed downstream data services.

Hybrid Fiber/COAX (HFC) CATV Networks

The bottom diagram in Figure 20-3 shows a CATV network that has been upgraded by the addition of fiber optic cable. From the headend, fiber facilities extend outward and terminate on remote terminal devices. At the remote terminal, the signals are removed from the fiber and placed on a random-branching shared bus implemented over coaxial cable. This hybrid fiber/COAX (HFC) architecture has several advantages over its all-COAX counterpart. The use of fiber in the backbone substantially reduces the number of amplifiers between the headend and the residence, resulting in improved signal quality. Perhaps most importantly, fiber in the backbone necessitates a point-to-point topology between the headend and the remote terminals. Rather than supporting a single COAX port that might serve 20,000 users, the headend now has multiple fiber ports, each of which serves far fewer users. In cases where fiber is deployed close to the user population, as few as 100–500 homes may be served by a single fiber strand. Failure of a single fiber or remote terminal affects far fewer users than a similar failure in an all-COAX environment. Moreover, with fewer households served by any one physical facility, there is less sharing of bandwidth, a factor that becomes critical when the network is used for high-speed Internet access.

CATV Standards and Protocols

Unlike the telephone industry with its legacy of regulation at the federal and state levels, the CATV industry is mostly regulated on a municipal basis. As a result, there has never been a mandate for equal access in the cable industry. In fact, CATV providers are not considered "common carriers" in the U.S. This situation has led to a number of proprietary implementations of HFC around the country. As long as the service provider supplies the customer premises equipment (i.e., the cable modem), the lack of standards in the CATV sector is not really a problem. From a user perspective, however, it would be nice to have some choice in the selection of premises equipment. As cable modems become more popular, enhanced features and functions will likely find their way into CATV Internet access environments. In order for every user to take advantage of such enhancements, standards are required for their definition and implementation. Furthermore, standards tend to unify markets and drive equipment costs down. Cost savings realized by CATV providers in this area will likely be passed along to users in the form of reduced service prices. In the competitive world of high-speed Internet access, such price reductions are important to ensure customer retention.

Data over Cable System Interface Specifications (DOCSIS)

The Data over Cable System Interface Specifications (DOCSIS) standard is a product of the OpenCable project. It is intended to support interoperability between headend equipment and cable modems from different vendors regardless of the details of the HFC implementation that interconnects them. The current version, DOCSIS 1.0, supports downstream data rates of up to 10Mbps. DOCSIS 2.0 will increase that rate to 30Mbps shortly. As CTAV providers embrace DOCSIS and equipment vendors begin to offer compliant equipment, the result is likely to be substantial growth in the numbers of users that rely on their CATV provider for high-speed Internet access.

Cable Modems for High-Speed Internet Access—Pros and Cons

Judging by the dominant position of cable modems in the high-speed Internet access market, there are advantages to this technology not shared by other technologies in this application. Perhaps the most important advantage of cable modems is service availability. The probability is high that if the customer is passed by the CATV infrastructure, he or she is a candidate for CATV-based Internet access service. No special conditioning of the physical facility is required to support data services, as is often the case with ADSL service. Installation of the cable modem is simple and straightforward. No modifications to inside wiring are required. In HFC-based systems, service quality tends to be quite good as well. In some older CATV systems, upstream channels are not supported on the cable infrastructure. In these cases, upstream connections are supported by the use of a modem and a telephone line.

Another important, and often overlooked, advantage to high-speed Internet access over the CATV network is the synergy between the Internet and broadcast television. Especially in cases where TV signals have been digitized to ride the HFC backbone, the possibilities for service integration are numerous and intriguing. Interactive applications in sports programming (alternative camera angles, scores, and statistics), home shopping (Web-based e-commerce, interactive help desk), and movie viewing (enhanced programming guides) come to mind immediately. Exploration of the synergy between the Web and traditional TV is in its infancy. CATV providers looking for innovative service packages to lure customers from other forms of Internet access (and television) will likely be quick to explore the possibilities offered by integration of the Web with broadcast television.

Cable modem service does exhibit several disadvantages. CATV networks suffer many of the same line impairments seen with other forms of high-speed Internet access technology. Drop cables are often installed poorly. Splices may be substandard, and connectors may wear out over time. Inside wiring may be suspect, from the standpoints of both cable type and routing within the home. These physical impairments are typically not a problem when broadcast video is the only application on the cable. The addition of data services, however, may bring to light problems with the cable infrastructure that were unnoticeable prior to their introduction. Other sources of noise such as radio frequency interference (RFI) may be present on the cable as well, and they will accumulate as signals are boosted by chains of analog amplifiers. The effect of physical and electrical impairments is typically reduced by the introduction of HFC into the CATV infrastructure.

The upstream channels in a CATV network are also a cause for concern about service quality. Originally seeking to curry favor with municipal regulators (e.g., the promise of interactive educational TV in schools), most CATV providers never intended to actually use the upstream channels in the network. This is blatantly obvious when one considers that all of the amplifiers deployed in early CATV networks pointed in the downstream direction only! The bandwidth reserved for the upstream channels was located in the 5–55 MHz range, an area of the spectrum affectionately referred to by CATV operators as the "garbage band." In this area of the cable spectrum, signals are susceptible to electromagnetic interference from such things as household dimmer switches, refrigerator motors, garage door openers, and the like. Because little consideration is typically given to the routing of the cable within the home, such sources of interference are likely to be encountered in some cases. Finally, because CATV networks are shared-medium networks, data rates tend to fluctuate as a function of the numbers of users on the cable at any given point in time. Providers of ADSL service are quick to point out that their service is superior to cable modem service because an ADSL facility is dedicated to a single user. Despite the misleading nature of such a claim (ADSL involves quite a lot of shared resources in current implementations), the slowdown experienced by cable modem users when their neighbors go online simultaneously has been perceived as a cause for concern for users of this technology.

Cable Modem Summary

CATV companies have experienced a high degree of success in selling high-speed Internet access services to their clientele. By offering cost-effective, easy-to-install, and relatively reliable high-speed data services, these providers have captured the lion's share of their target market. The integration of Web applications with traditional TV may present an opportunity for further, perhaps spectacular, growth in the years to come. Success breeds success, and in this instance, 5,000,000 users speaks for itself.

WIRELESS ALTERNATIVES FOR HIGH-SPEED INTERNET ACCESS

Wireless alternatives for high-speed Internet access fall into two basic categories. Fixed-base wireless technologies like local multipoint distribution services (LMDS) and multichannel multipoint distribution services (MMDS) are emerging as alternatives to wireline technologies like ADSL and cable modems. In addition, satellite-based services are competing for their share of the Internet access market. Satellite-based offerings include those that use satellites in geosynchronous orbit (e.g., DirecTV) as well as those that use satellites parked in low-earth orbit (e.g., Teledesic). The next several sections will briefly explore these wireless technologies, emphasizing the advantages and drawbacks associated with each.

Fixed-Base Wireless Alternatives

Two fixed-base wireless technologies are making inroads into the high-speed Internet access market. They are local multipoint distribution services (LMDS) and multichannel multipoint distribution services (MMDS). Both are radio-frequency systems that operate in the microwave band, but the similarities end there. Let's take a brief look at each of these wireless Internet access technologies.

Local Multipoint Distribution Services (LMDS)

LMDS began life as "cellular CATV." Using base stations with a coverage diameter of about three miles, LMDS underwent early trials as an alternative to CATV infrastructures for the delivery of broadcast video services. Because of the small antenna footprint and attendant expense of covering a large geography, LMDS never caught on as a video delivery mechanism. It has, however, garnered a lot of interest as a high-speed Internet access technology in densely populated areas.

LMDS systems in the U.S. are licensed to operate in the 21–38GHz range of the spectrum. They implement a wide variety of channel counts, data rates, and modulation techniques. Recent efforts by the Institute for Electrical and Electronic Engineers (IEEE) to standardize LMDS systems (for further information, see the IEEE 802.16 standard) are likely to pay dividends in terms of the near-term growth of the technology. Currently, symmetric and asymmetric LMDS services are available at data rates that can reach 155Mbps in the downstream direction. Teligent is one provider of such services.

LMDS providers are targeting urban areas and multiple dwelling units for their services. There is still much doubt about LMDS's viability in residential settings. As a general rule, the more dense the user population, the more cost-effective LMDS becomes.

The advantages of LMDS include potentially high data rates, ease of infrastructure deployment, and flexibility of service offerings. On the other hand, small antenna footprints offset some of the cost-to-deploy advantages of LMDS. Beyond cost to deploy, however, is cost of customer premises equipment, which in the case of LMDS is about $4000 per household at this writing. In addition, operation at very high frequencies renders

LMDS susceptible to impairments experienced in other microwave-based systems (e.g., rain fade). Finally, LMDS services for Internet access are not widely available at this point. Only time will tell whether LMDS can carve out a substantial niche in the high-speed Internet access market.

Multichannel Multipoint Distribution Services (MMDS)

MMDS was introduced as a 32-channel video delivery service in the 1980s. Due to lack of spectrum (i.e., low channel counts) and intense competition from wireline video providers (i.e., CATV companies), MMDS has been repositioned as a technology for high-speed Internet access.

MMDS systems operate in the region of the spectrum just above 2.5GHz. Each base station has a coverage diameter of about 30 miles, making MMDS cost-effective when serving less populous areas. It is estimated that a single base station needs to serve about 10,000 subscribers for a provider to break even on a capital expenditure basis. From a customer's point of view, equipment cost for MMDS service is about $500 per household. This is reasonable and in line with the cost of equipment for other approaches to high-speed Internet access. Offsetting these cost advantages somewhat is the line-of-sight requirement associated with MMDS. It is sometimes difficult to get a "clear shot" from the base station to every potential user of the service. Interestingly, new MMDS technology is emerging that bypasses the line-of-sight requirement and thus should expand the market potential of the technology. Because MMDS was originally designed for one-way video distribution, reverse channel capability was not designed into the systems. To be used in duplex mode (symmetric or asymmetric), a phone line and a modem were required to support upstream communication. Newer MMDS systems do indeed have two-way capability, but older systems are still dominant in the installed base.

As a high-speed Internet access technology, MMDS is promising. Because the only standards associated with MMDS deal with purely spectrum-related issues, service providers offer a wide range of channel capacities, modulation schemes, and data rates. Downstream rates of 50Mbps can be achieved with the technology. Large service providers like Sprint and Worldcom view MMDS as a way of offering not only high-speed Internet access, but also wireless telephony and other services in the residential and small business markets. As with LMDS, time will bear out whether MMDS can make substantial inroads into its target markets.

Satellite-Based Wireless Internet Access Options

Satellite systems for Internet access fall into two basic categories. The first includes systems whose satellites are parked in a geosynchronous orbit (GEO systems) about 22,500 miles from the surface of the earth. The second includes newer systems whose satellites orbit much closer to the earth. These latter systems are called "LEO" systems to indicate that the satellites are in low-earth orbit. We shall examine both types of satellite systems in the following sections.

GEO Systems

GEO satellite systems are familiar to many as providers of direct broadcast satellite (DBS) TV services. In the U.S., DirecTV and Dish Network are the largest operators of these systems. Hughes Network Systems began offering DirecPC service as an adjunct to DirecTV several years ago. In late 2000, DirecPC systems with two-way capability (the old system used the PSTN for upstream traffic) were introduced. Dish Network currently offers such a service as well under the moniker of Echo Star.

In order to subscribe to a GEO satellite service for high-speed Internet access, customers must purchase a two-way dish system. For new subscribers, this is not problematic, as they would have to purchase a dish system anyway. For households that already have satellite systems for video, the need to purchase and install a new system may be a roadblock to deployment of the service. Once connected, the user of a GEO satellite can expect two-way service with upstream data rates of 125–150Kbps and downstream rates of 400–500Kbps. In this regard, GEO satellite networks fall somewhere in between ISDN access and ADSL access. Perhaps the most important advantage of satellites for high-speed Internet access is their ubiquity of service. For households in rural areas, satellites may be the only option for high-speed connectivity.

Two important considerations regarding service quality when connected to a GEO satellite network are reliability and delay. Because GEO systems operate in the microwave band, signal quality can vary as a function of weather conditions. High winds can cause dishes to wander off their targets. Rain and fog can interfere with signals as well. These impairments might not be seen as critical problems to users who expect only video services from their provider. For users of Internet access services, however, problems are likely to surface more readily and be viewed as more critical. Moreover, because geosynchronous satellites are located approximately 22,300 miles from Earth, their use incurs substantial end-to-end delay. In fact, the end-to-end delay over a geosynchronous satellite system is about one-quarter of a second. Unfortunately, GEO satellite services are relatively new on the Internet access scene. Time and experience will indicate whether these services are robust enough for the average residential Internet user.

LEO Systems

Aside from the rather spectacular market failure of the Iridium system, the jury is still out on the potential for LEO satellite systems to support high-speed Internet access. Several companies, however, are counting on business plans that focus on that application. Services are expected to appear in the near future.

The problem with LEO satellite systems lies in the sheer numbers of satellites required to attain wide geographic coverage. Unlike their GEO brethren, satellites in low-earth orbit do not remain stationary relative to the surface of the earth. Because LEO satellites move relative to the surface of the earth, many more of them are required to serve a given user population. More satellites means more launches, and that equates to more expense. To offset that expense, LEO satellite providers will have to either sign up substantial numbers of users or offer a premium service with enough added value to justify the high service cost.

In terms of data rates, LEO services are being advertised as asymmetric and two-way. When the Teledesic network becomes available for Internet access, users can expect downstream data rates of 64Mbps and upstream data rates of 2Mbps. Such high data rates may be just the differentiator that LEO systems need to make it in the marketplace, although whether or not a large enough market for these rates exists is still in question. Nonetheless, LEO networks are currently under construction and will add to the plethora of high-speed Internet access services currently available to users.

SUMMARY

This chapter has examined a number of alternatives for providing high-speed Internet access to the residential user. Wireline (ADSL and cable modems) and wireless (LMDS/ MMDS and satellites) technologies were covered with an emphasis on technology and service offerings. The central theme that emerges from this chapter is that, whereas high-speed Internet access alternatives are in the marketplace, their widespread availability is lacking. Cable modems currently lead the way in installed base. This lead shows no signs of abating in the near term. Unless many of the problems with ADSL can be remedied, this author sees a bleak future for that technology. Interestingly, wireless technologies may eventually prove to be the wild card that trumps the rest of the high-speed Internet access players.

Once high-speed Internet access is a reality for a majority of home users, we will likely see a behavioral shift in the way that people use the Net. Applications that were not considered feasible in the past (e.g., streaming audio and video) will become popular with users who have high-speed access. This, in turn, will breed newer applications that are even more bandwidth hungry, and we will be back to the start. What new technology developments will have matured to satisfy the bandwidth cravings of the next-generation Internet user? The beat goes on.

CHAPTER 21

ISP Networks

I t's not hard to start your own business as an Internet service provider (ISP). This author has run a private ISP on an Ethernet network with one server and two phone lines. I've also seen ISPs with over 2000 customers built upon one ISDN (Integrated Services Digital Network) PRI (Primary Rate Interface) and one server hosting their DNS (Domain Name Service), RADIUS (Remote Authentication Dial-in User Service) server, Web pages, and e-mail. For a couple of hundred dollars, you can start your own ISP in less than 24 hours. By becoming a reseller of another ISP, you can provide "private label" Internet connections without having to know a thing about IP, routers, and TCP applications, while still accepting local dial-in connections from around the country.

A commercial ISP can be as simple as a server that accepts dial-in connections and a connection to the Internet. In this chapter, we will examine the design elements ISPs work with. Starting with the configuration of the ISP network, we will scale it to a large ISP and then consider what additional services customers of that ISP will expect. Much of what is discussed in this chapter has been introduced in previous chapters. Here we put it all together to see how the parts of the technology mesh to create a commercially viable business.

ACCESS

In its most basic incarnation, an ISP is a way for users to connect to the Internet. As such, the ISP needs some way for users to connect to it, and the ISP itself needs some way to connect those users to the Internet.

The average customer of the ISP will connect to the ISP via a dial-up modem link. This can be as simple as a modem attached to a customer computer dialing a modem attached to the ISP network. While convenient for the home user, modems scale very poorly in an ISP environment. Each modem that the ISP maintains needs its own attached phone line. In addition to the expense and physical real estate needed to support several dozen modems, the cost of the individual phone lines is prohibitive as well. To further discourage the use of regular plain-old telephone service (POTS) lines at the ISP, the high-speed modem options that are popular with customers (56Kbps) are not available when the call is from one modem over another on POTS.

To solve these problems, ISPs will use ISDN PRI lines or channelized T1s to their premises to support 23 or 24 simultaneous customer calls per incoming line. Access concentrators, or access servers, are used to terminate the digital facilities, providing a more cost-effective way to support a large modem pool. Access servers can scale to support over 16,000 dial-in calls per device! Today more ISPs are employing access servers that will support cable modem and DSL users as well. Business customers of the ISP generally connect to the ISP using frame relay, DSL, or private lines. Less commonly, ISDN and ATM are supported for business customers as well.

The ratio of customers to access ports is not one to one. Instead, each ISP will generally provision one dial-in connection for every 8 to 10 customers, although higher ratios such as 1:15 or even 1:20 have been known to exist. Thus, one T1 line with 24 channels could support 240 actual customers. Clearly, the higher the user-to-line connection, the more profitable the ISP can be. However, the ISP must constantly monitor the number of calls that are blocked; otherwise, they might get a reputation for poor access and lose customers.

NETWORK ACCESS POINTS (NAPS) AND PRIVATE PEERING

Once customers have dialed in to the ISP, they will generally expect some sort of access to Internet resources elsewhere. To provide this service, the ISP needs its own Internet connection. One way to provide this is to connect to another, usually larger, ISP that will be the gateway to the broader Internet. The service being provided to the small ISP is referred to as *transit*; that is, the larger ISP provides transit for the smaller ISP's traffic to and from the Internet. Larger ISPs may have multiple connections to one or more ISPs. Another approach is for the ISP to connect to a network access point (NAP).

In the early days of the Internet, when the National Science Foundation Network (NSFNET) was operational, the Internet was a collection of large academic and research networks that would connect to each other via NAPs. There were originally four of these NAPs, two on the east coast, one near Chicago, Illinois, and another on the west coast. After the decommissioning of the NSFNET and the migration of the Internet to being a commercial entity, the NAPs became the "public access" connection points to the Internet for anyone with the ability to pay a start-up fee and afford connectivity to the NAP. Connection to the NAP is only one piece of the equation. The ISP must also have agreements with other ISPs connected to the NAP in order for them to accept traffic from it. These agreements are referred to as *peering*. As with the private connections, when private peering, the smaller ISP usually pays a fee to a larger ISP for providing the transit service. However, if the ISPs are of similar size, they may have a peering agreement that does not call for the transfer of funds. Although similar in concept to a NAP, private peering arrangements are not open to all comers. This allows the peering ISPs to better control the quality of the traffic that flows between them. With the modern Internet, large ISPs are preferring private peering arrangements as the NAPs have become overloaded with IP traffic.

As the Internet exploded in popularity, the amount of traffic at the NAPs increased. Eventually they became so congested that data traveling through a NAP would be delayed. To circumvent this problem, the largest service providers established private peering points between each other. These private peering points look suspiciously like NAPs— small rooms with lots of routers, switches, and cabling—but access to them is limited to only the service providers who have contracted to peer with each other at these points. Since access is controlled, the traffic traveling through the private peering points is likely to suffer lower delays than traffic passing through the *public* NAPs.

While the small ISP can connect directly through a NAP, in most cases the cost of maintaining a connection to the NAP is prohibitive. The small ISP could peer with another ISP. Sometimes many small ISPs connect their networks together in this way in an effort to improve local connectivity. But for a global reach, the ISP needs to connect to the networks of large service providers. Unless the ISP is itself a major service provider with a great deal of traffic, national and international service providers will not establish a private peering session with it.

Instead, our ISP is most likely going to be forced to contract with a larger, or *upstream*, ISP for Internet access.

Today, there is really no "Internet" that someone can point at and say, "There is the Internet!" At best, the public NAPs are the closest to this idea that still exists. Instead, the term "Internet" refers to a collection of private networks maintained by governments, service providers, academic and research institutions, and private company networks. These networks are organized generally by size and connectivity, with the geographically largest and most well-connected networks considered Tier 1 networks. Tier 2 and Tier 3 networks are respectively smaller and less well connected.

What tier a service provider is considered to occupy has an exact definition. A service provider connected to two or more NAPs at speeds higher than 45Mbps is considered Tier 1. Since the NSFNET NAPs are geographically dispersed around the country, this assures that only the largest service providers fall into this category. A service provider that connects to only one NAP is considered to be Tier 2. This would imply that a Tier 2 ISP is a large regional ISP, but one lacking the national reach of a Tier 1. Tier 3 ISPs are those service providers that, lacking any connection to a NAP, obtain their connectivity by purchasing connections from Tier 1 and Tier 2 ISPs. Since designation as a Tier 1 ISP implies being especially large and well-connected to others, this information generally shows up prominently in marketing materials. Today, though, this definition is more fluid. Large networks are considered Tier 1 networks when other Tier 1 providers admit that they need direct peering arrangements with the new ISP because of its geographic reach and the amount of traffic it carries.

While only the financial resources and installed hardware of an ISP would prohibit it from becoming a Tier 1 or Tier 2, most start up ISPs generally will be Tier 3 service providers. Figure 21-1 outlines the relationship between the customer ISP and the upstream ISPs.

As the ISP gains customers, its initial link to its upstream service provider will generally be through a fractional T1, meaning that only a fraction of the possible 1.536 Mbps of data throughput will be used. This allows the monthly cost for Internet access to be kept low for the ISP. As new customers are added and the need for bandwidth grows, additional bandwidth on the T1 can be turned up.

As when installing dial-in circuits, ISPs desire to minimize their monthly expenses for the Internet connection, at the risk of alienating customers by providing poor service. Generally, a 1.36Mbps T1 line can support 200 56Kbps dial-in lines, allowing one upstream T1 to support 2000 paying customers, assuming a 10:1 ratio of users to dial-in lines. If the ISP supports users through cable modems, DSL, and business T1 links, then proportionally more upstream bandwidth is needed.

For purposes of redundancy, an ISP may elect to purchase two T1s, perhaps through different upstream ISPs. The addition of multiple links to different upstream service providers, while beneficial to the customer ISP, is going to add some complexity to the routing of information between the networks. This question will be addressed later in this chapter, when we discuss the Border Gateway Protocol (BGP), which is a routing protocol that

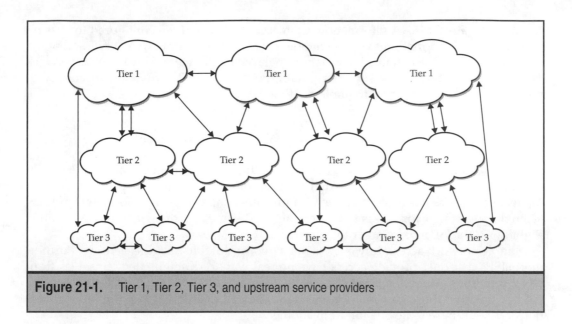

Figure 21-1.   Tier 1, Tier 2, Tier 3, and upstream service providers

differs from the distance vector and link state protocols discussed earlier in this book. Before it can consider multiple links, though, the ISP has other issues that it needs to deal with.

AUTHENTICATION, AUTHORIZATION, AND ACCOUNTING (AAA)

The first of these issues is how to make sure that only paying customers access the network. To accomplish this, some form of authentication, authorization, and accounting (commonly known as AAA) is needed. When a remote user dials into the access server, that user must first prove that she is who she says she is. This authentication is done by providing something that only she should know, generally a password along with a user name. Once the remote user has proved her identity, she is authorized to access the network and network resources, if her account payments are up to date. Finally, for legal and administrative purposes, while the user is connected, some form of accounting is performed. This can be as basic as recording the user name, the time logged on, what network parameters were assigned to the user, and how long she stays logged on. It can also keep track of the user's activities in much more detail, recording such items as which network resources she accessed and how much traffic she sent and received.

This AAA can be performed on each access server. For every user on the network, a username and password can be configured on each access server. Clearly, however, as the ISP grows and new access servers are added, this database will have to be replicated and maintained on each access server. This cannot scale. Instead, a centralized database is configured with the AAA information. Each access server is then configured to pass on authentication information from the remote client to the central AAA server. Now the user database need be maintained on only a single device no matter how large the network grows and how many access servers are configured. Chapter 23 provides a discussion of the most popular approaches to implementing AAA.

BASIC SERVICES

At its most fundamental level, we now have all the elements needed for an ISP: dial-in connections to an access server, a RADIUS server for AAA, and links to provide access to the rest of the Internet. Note that this idea presumes the presence of some sort of LAN technology, a router for access to the Internet, and some sort of switch for LAN traffic. Right now, our ISP looks similar to Figure 21-2.

Simple network access, however, is not a very profitable business venture. Furthermore, ISPs that only provide access have no real way to differentiate themselves from other providers. Since the slowest point in customers' traffic over the Internet is the 56Kbps link that those customers are using for access, even high-speed uplink connections cannot differentiate one ISP from another.

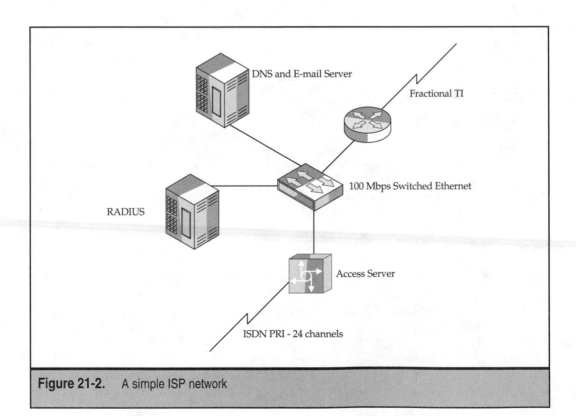

Figure 21-2. A simple ISP network

For an ISP to convince customers to purchase their Internet connections from it, that ISPs must also provide additional value to the customer. Some of these value-added services have themselves become standard. Most ISPs provide e-mail addresses for customers. Thus, an e-mail server is needed. An ISP may host its own Web site to provide information to visitors, and it may also host customers' Web sites. To this end, one or two Web servers may be needed. ISPs also provide domain name–to–IP address resolution for their customers (DNS) and will generally have one or two DNS servers on site as well. Thus, the minimum number of servers that an ISP will host on their LAN will be at three: e-mail, WWW, and DNS. As the example at the beginning of the chapter indicated, this could be done on a single computer, but that is unusual. Customers will be upset if one service is down but downright irate if all services have failed at once! In addition to any servers that may be added to the original three for sake of redundancy, ISPs may also choose to host a newsgroup server, chat servers, file servers, and game servers. In short, the more an ISP can convince its customers to go no further than its own LAN, the more money the ISP will save on upstream Internet links. Some servers, such as game servers, may even provide additional revenue for the ISP. With all of these servers comes the need to back them up and manage them. A backup server and perhaps even a backup network will be employed. To coordinate all of these devices and maintain them, a management station is then needed. To protect this increasingly valuable network, firewalls and intrusion detection systems become necessary. Finally, in order to efficiently provision new customers and update existing customers, an accounting server will need to be included as well, perhaps implementing a Web-based interface so that nontechnical employees of the ISP can add users, change passwords, and create billing statements.

In an effort to reduce the amount of traffic flowing over their upstream link, many ISPs will also implement one or more cache servers. These are servers that intercept users' Web traffic and temporarily store Web content on a local hard drive. For example, if user Alice has visited a Web site called www.hill.com, there is a chance that in the near future another user will visit the same site. Instead of sending out a new request over the upstream link to the hill.com Web server, the ISP LAN will simply send the new user a copy of the page that it has stored on its cache. This returns the Web page to the new user much more quickly than a new request to www.hill.com would, making the user happy, and saves the ISP upstream bandwidth, making good business sense.

Caching is such a good idea for the ISP and the end user that some companies are creating content distribution networks that replicate their Web pages to servers located at ISPs all over the country. This practice fulfills the ISP business model of keeping as much traffic local as possible, and it creates happy customers who are likely to visit the replicated Web sites more often because of their quick response times on the network. This in turn creates happy content providers who are generating lots of "eyeballs" (Web site hits) for their Web sites.

We see now that, although the ISP could technically get along with one server to provide expected services to end users, it will probably use a large number of servers. A large ISP may even start replicating its own servers to data centers that host dial-in lines all around the country.

These servers will be connected using current LAN technology. At one time, in order to handle the high demand and bandwidth needs of many users, ISPs were toying with ATM and FDDI as the LAN technology. The plummeting prices of Ethernet switches, faster speeds, and ease of use, however, have all but assured that 100Mbps Ethernet and 1000Mbps (Gigabit) Ethernet are the LAN technologies of choice for most ISPs.

With these things taken into account, our ISP looks more like Figure 21-3.

ROUTING CONSIDERATIONS AND BGP

One basic element that ISPs use to keep customers happy and prevent them from going to one of the many other service providers (a process known as churn) is to implement multiple upstream links to provide performance and, just as importantly, availability. This

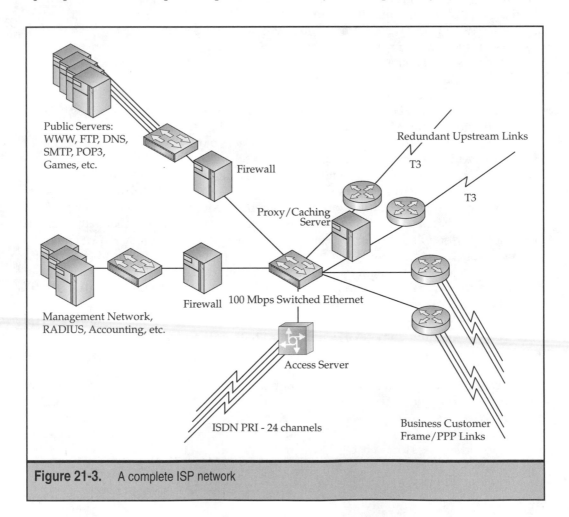

Figure 21-3. A complete ISP network

addition of links to the rest of the Internet, however, increases the complexity of the network, in particular the routing.

Consider the simple ISP or company connection to the Internet. With one upstream link, it is a fairly simple process to determine the destination of a data packet. After all, if the destination is not on the local LAN, it has to be "out there," on the other side of the upstream link.

Once more than one link becomes available, the routing decision becomes more difficult (Figure 21-4). Which link should be used to provide the quickest way to the destination network?

Traditionally, this decision can be solved with a routing protocol like those discussed in Chapter 18. But neither of the major types of routing protocol in use, distance vector (RIP, for example) or link state (OSPF), really fulfills all of needs in this scenario.

A distance vector routing protocol has several limitations. First, it does not scale very well. A network the size of the Internet is going to need to support a great many routes to all possible destinations. Simply exchanging these routes on a regular basis as RIP does will overwhelm all but the highest-bandwidth connections. Second, the information exchanged is not very detailed for a complex network. It does not even take link speed into account.

Link state routing protocols do scale well and can support networks with thousands of routers. Link state routing protocols work through sharing what is known as the *link state database* among all other routers in its area. This allows routers in the network to

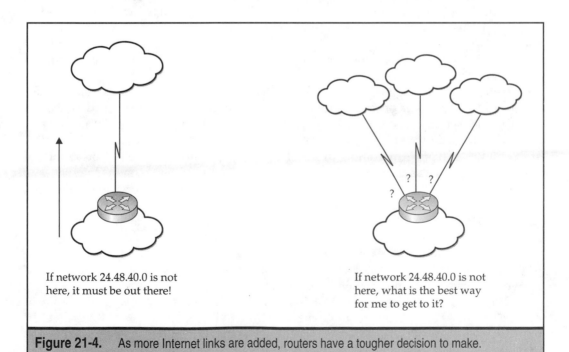

If network 24.48.40.0 is not here, it must be out there!

If network 24.48.40.0 is not here, what is the best way for me to get to it?

Figure 21-4.   As more Internet links are added, routers have a tougher decision to make.

compile a complete database of all routes in the network and incidentally to create a map of what the entire network looks like. This creates three drawbacks. First, if ISP A and ISP B are connected via OSPF, they will each have a complete topological map of the other network. This is not a popular idea among network administrators. Second, even when breaking up the link state network into areas, the summarized information from all the other areas on the Internet as a whole will overwhelm the router trying to create its topology map. Finally, political considerations on the Internet are just as important as the most efficient routing path between any two points. Link state routing protocols excel at finding the shortest path between source and destination, but they lack the capability to control traffic according to business need or regulatory restrictions.

What is needed in this case is another protocol: the Border Gateway Protocol version 4 (BGPv4, also known as BGP). While distance vector and link state protocols allow routers to find the best route within a network, BGP allows routers to find the best route when traversing multiple networks.

Discussing where BGP is used requires some clarification on terms. No doubt the reader has been astounded up to this point at the range of meaning assigned to the word "network." A network could be the physical layer between a switch and an Ethernet network interface card, an area encompassing many Ethernet links and defined by a port on a router and the subnet mask of an IP address, or a collection of IP networks connected by routers. To eliminate the confusion of the term "network," when referring to BGP the term *autonomous system* (AS) is used. An AS is the collection of networks under the administrative control of a single entity. Strictly speaking, the routed network in my office is an AS, as is the network composed of thousands of routers and tens of thousands of links spanning a multinational corporation. Each of these example ASs fulfills the definition of an AS. I control my office network, and the IT department of the multinational controls all of the parts that make their network.

When dealing with aggregated networks on the scale of the autonomous system, we use slightly different terms in describing routing. As a whole, routing protocols are described as either *interior gateway protocols* (IGPs) or *exterior gateway protocols* (EGPs). IGPs include any routing protocol designed to be used *within* an AS. RIP and OSPF, discussed in Chapter 18, are examples of IGPs. Any routing protocol designed to distribute route information between ASs is known as an EGP. BGPv4 is the only EGP currently used on the Internet.

BGP is a routing protocol that routes between ASs. For it to do this, each network needs a unique identifier. These identifiers are known as AS numbers (ASNs), and each network using BGP needs an ASN just as it needs its own set of IP addresses. Referring again to the AS examples used earlier, while each network is technically an AS, an official AS number is needed only if it is connected to the Internet and using the protocol BGP to share routing information with other networks. In this case, my home office AS does not need an ASN, but it is very likely that the multinational corporation's network will have an ASN.

While a detailed explanation of BGP is beyond the scope of this book and the protocol can be somewhat complicated to implement, in the end BGP does the same thing that RIP or OSPF does: it creates a routing table in a router so that the router can determine where to send a packet to get it closer to the destination. With that in mind, let's examine BGP at a high level to see the advantages that it offers the ISP.

In Figure 21-5, our ISP has grown from a single T1 line for upstream access to three T1 lines for performance and redundancy reasons. Each of the T1s is connected to a different upstream ISP. Note that each ISP is labeled by AS number, as BGP requires. For later discussion, additional ASs are included.

To maximize its investment in the three T1 lines, our ISP (AS42) would like each one of them to share the load of outgoing traffic. For data that has its ultimate destination in the vicinity of AS100, link A should be used, and for data that has its ultimate destination near AS200, link B should be used. Likewise for AS300 and link C. Assuming a diverse user base and diverse destinations, the traffic should more or less even itself out over the three upstream links. This will increase the performance for the ISP customer, since the quickest route is being taken and each T1 is less likely to be congested because the traffic is evenly spread out.

The ISP achieves redundancy by configuring the T1s to back each other up. Thus if link A to ISP A should fail, B and C will begin transferring data to ISP A. As long as one T1 line stays up, all destinations on the Internet are available, and as long as two or more T1 lines are up, traffic will take the shortest path available.

To assist the routers in the ISP network in making the decision as to which path is shortest, the ISP and its upstream neighbors need to exchange information about what networks each AS has knowledge of. This process is best illustrated through an example.

According to Figure 21-5, AS100 is connected to AS101. AS101 is in turn connected to AS201, which is connected to AS301. Beyond these ASs lies the rest of the "Internet."

Our ISP, AS42, will learn how to reach the IP network 200.1.1.0/24 that is part of AS101 from three directions. AS100 will tell AS42 "I can reach 200.1.1.0/24 through AS101, which is directly attached to my AS" using BGP. At the same time, AS200 will tell AS42 "I can reach 200.1.1.0/24 through AS201 and then AS101." Finally, AS300 will use BGP to tell AS42 "I can reach 200.1.1.0/24 through AS301, then AS201, and finally AS101."

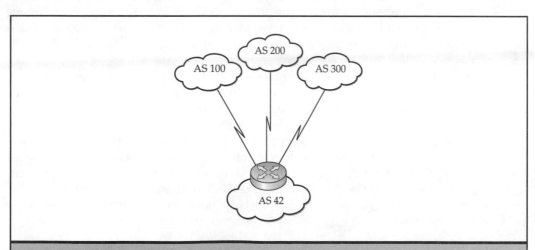

Figure 21-5. ISP with three T1 lines to upstream service providers

AS42 will compare the BGP information that each of its neighbor ASs gives it. Barring configuration options to the contrary, AS42 will conclude, "The best way to reach 200.1.1.0/24 is through AS100 because my packets will travel through the least number of ASs to reach that IP network." If the link between AS42 and AS100 should fail, routers running the BGP protocol on AS42 would then conclude, "The best way to reach 200.1.1.0/24 is through AS200 because my packets will travel through fewer ASs than if I sent them to AS300."

By reiterating this process for every network (more than 100,000 times on today's Internet), our ISP, AS42, can figure out which link makes the most sense to use for any given destination.

Notice that AS42 never really knows what is on the inside of the other ASs. AS100 could be a single room in a data center with connections at gigabit Ethernet speeds, or it could be a multinational corporation with many slower T1 links connecting remote corners of the network. BGP allows only for a rather simple counting of the number of AS "hops" between the source and the destination. While this could lead to BGP making less than optimal decisions, the simplified information allows BGP to scale very well and protects information about the internal topology of ASs that are passing traffic through them.

BGP also would allow the administrators of AS42 to take paths as they see fit rather than according to the best routing decision. For example, if the link between AS42 and AS100 were usage sensitive, with AS42 being charged by AS100 according to the amount of data sent over it, the network administrators of AS42 could decide "Always use AS200 to reach 200.1.1.0/24, and if that fails, always use AS300. If no other option exists, use AS100." The administrators of AS42 could also decide to always use some links for inbound data and some for outbound data or one group of users to use AS100 all the time and another group of users to use AS300 all of the time. BGP gives a degree of flexibility and control over routing decisions that other protocols can't match.

To summarize, once our ISP grows to the point that it needs multiple upstream links, it is going to need to implement the routing protocol BGP between itself and its upstream neighbors so that the ISP's routers can make intelligent decisions on how to use those links for all destinations on the Internet. This increases the complexity of the ISP network and makes greater demands on the technical staff.

Once the ISP has started implementing multiple links and all the servers described previously, it is highly likely that the ISP is going to have multiple IP networks on its own LAN and multiple routers. In this case, the ISP is also going to need to implement an IGP to distribute reachability information within its AS. In most cases, this IGP is going to be one of three choices: OSPF (which has been discussed in Chapter 18), Intermediate System–Intermediate System (IS-IS), which an older (but just as good) link state routing protocol very similar to OSPF, or EIGRP, which is a Cisco proprietary distance vector routing protocol specifically designed for large complex networks. Each of these IGP protocols is popular in ISP networks because they are designed for large networks and allow changes in the network to be communicated very rapidly within the AS.

ADDITIONAL SERVICES

At this point, our ISP has become a full-fledged regional ISP. All of the basic services of an ISP are being offered: connectivity, e-mail, Web pages, DNS, file servers, content caching, and redundant upstream links. At this point, however, many ISPs struggle to differentiate themselves from their competitors and keep their customers by branching out and offering additional services. Some of these options are mentioned here, with references to the chapters that discuss them in more depth.

ISPs may begin by offering dedicated access links to customers or selling them to businesses. These will generally be in the form of a frame relay connection (Chapter 16) or SDSL link (Chapter 19), and less commonly using ISDN (Chapter 12) or ATM (Chapter 17). Generally, ISPs are competing with incumbent carriers to provide the access connections, and these incumbents may have their own ISPs that are trying to resell the same connections. To compete, ISPs are starting to offer Ethernet WAN connections from local businesses to the ISP. These offer high speeds and low tariff fees.

Since access alone is not very profitable, the ISP will increase its value to the business by offering to host the business's Web pages and DNS servers. Increasingly, the ISP may host e-mail servers for the business as well. All of these services are easy enough for any business with a competent IT staff to implement, but the ISP has the advantage of dedicated staff and economies of scale to make this a worthwhile proposition for its business customers. In exchange for some loss of control and responsiveness, the business customer can offload the cost of some IT staff, the need to maintain the servers' hardware and operating systems, redundant power, physical and network security, backups, and so on. The ISP has gained another source of revenue and a customer that will not be able to leave them easily.

Once the ISP is offering access links to business customers, it is a natural progression for it to offer VoIP (Chapter 23) and VPN services (Chapter 23) to its customers. Since neither of these technologies currently work optimally between service providers, if a business customer has several remote offices and a main office, the selling ISP has strong leverage to acquire additional contracts for each remote site. Again, these are technologies that an IT staff can implement without the participation of the ISP, but economies of scale and dedicated staff with day-to-day experience in the technology prove to be a strong argument in favor of contracting with the ISP to provide these services.

Taking advantage of the shortage of IT staff, ISPs that provide VoIP networks, VPNs, and applications hosting are also trying to gain entry into the customer LAN by providing more consulting services. This could include network design, installation of routers and switches internal to the LAN, and in some cases even the physical pulling of category 5 cables for the wiring plant. It may also mean application development and Web site design. The ISP may also contract to provide firewalls, intrusion detection, and network security. In short, the competitive ISP needs to provide more than Internet connectivity and e-mail. It will become the one-stop shop for all customer networking needs.

While all of the previously listed services are sources of additional revenue and increased customer loyalty for the ISP, they all suffer from the disadvantage that they are also services that can be provided by the customer IT staff independent of the ISP. What the ISP really requires is a value-added service that the customer cannot replicate. This is where QoS and traffic engineering come into play. By implementing MPLS or Diffserv (Chapter 23 for both), the ISP can create virtual "premium" networks that offer higher classes of service than normal IP networks. These networks can be sold to business customers that have high availability or stringent delay characteristics for their networks. Using unaltered IP headers, this type of premium IP service is very difficult to achieve. MPLS allows the ISP to offer a scalable, easy-to-provide service that distinguishes them from their competitors, brings in additional revenue, and decreases the likelihood that a business customer will change providers.

As the ISP continues to grow, it will start adding points of presence (POPs) in strategic markets. Each POP could be nothing more than a remote access server that relays incoming calls to the ISP backbone via a dedicated link. It could also be a large data center that replicates the ISP's main office LAN to provide premium service in high-density markets. Many ISPs, lacking the resources to finance and build a private POP in each market, will share their access servers and the cost of maintaining the POP with other ISPs. This practice allows each ISP market penetration in an area normally not available to them, yet it allows them to minimize the financial risk by cooperating with their competitors.

Large resellers also exist that maintain POPs and access servers. These resellers enable any ISP willing to pay a membership fee to provide a "virtual" POP for their remote access clients. Some of these resellers will also maintain all of the necessary servers (e-mail, Web, DNS, file, caching) that a normal ISP would use on behalf of its ISP clients. Providers using these arrangements are known as non-facilities-based ISPs. The ISP that customers think they are connecting through does not own any hardware and is only leasing access and equipment from a reseller. In most cases, the non-facilities-based ISP can even lease technical support from the reseller! This allows the ISP to minimize its operating expenses during its initial growth period and focus its cash on marketing and building up the customer base.

SUMMARY

Becoming an ISP today is a trivial matter. With virtually no hardware or technical experience, the entire process can be outsourced. Even supplying your own networking equipment, you will find starting up a small ISP less than daunting. Despite this, or because of it, the market for the small ISP is very competitive in urban areas. To attract subscribers in this market, the ISP must increase the value that it offers its customers by providing higher speeds, reliable Internet connections, and additional services. By the time this process is completed, the small ISP is now a medium-sized ISP with more inherent complexity and greater technical demands upon its owners and network administrators.

PART VI

Convergence

CHAPTER 22

Introduction to ATM Backbones

little over a decade ago, Telcordia (then Bellcore) and the ITU-T (then CCITT) developed parallel views of the next-generation backbone infrastructure. These crystal ball projections positioned Broadband ISDN (B-ISDN) as the next-generation service and Asynchronous Transfer Mode (ATM) over Synchronous Optical Network (SONET) as the transport and switching systems. Figure 22-1 provides a look at the initial view of this universal backbone technology. While this universal backbone view of the telecommunications industry never materialized, the use of ATM as a backbone technology remains a viable choice.

Instead of being the universal backbone carrying all traffic, as predicted by the standards bodies, ATM has become a versatile backbone technology that can carry any traffic. Frame relay service providers, Internet service providers, digital subscriber line (DSL) providers, and PSTNs all are using ATM as a backbone transport technology. In this chapter, each of these applications of ATM is examined.

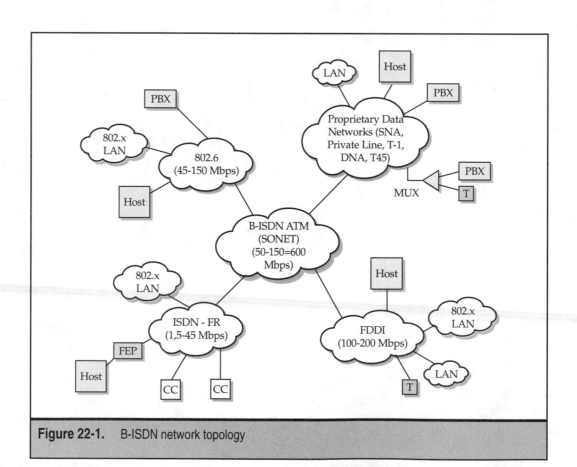

Figure 22-1.   B-ISDN network topology

ATM AND THE INTERNET

The growth of the Internet has forced service providers to try many different approaches to solving the bandwidth problem. One of these approaches is the use of ATM. ATM has shown up in two very distinct portions of the Internet. The first is as a backbone technology connecting switch/routers of an Internet service provider (ISP), and the second is as a cross-connect technology for a network access point (NAP).

As a regional or national backbone technology, ATM offers point-to-point connectivity among IP routers; high-speed interfaces (up to OC 48); class-of-service classifications; and a mature operations, administration, maintenance, and provisioning (OAM&P) support system.

As a NAP technology, ATM offers high-speed, flexible, assured connectivity among routers. These connections can be managed in real time by system administration staff or by the traffic generated by the routers.

Network Access Points

A typical network access point (NAP) has the option of using the Fiber Distributed Data Interface (FDDI), Gigabit Ethernet, or Asynchronous Transfer Mode (ATM). In some implementations, to include the Internet2 NAPs, ATM is the sole choice for connectivity. In these NAPs, routers from each of the service's ISPs and the Internet backbones are connected to an ATM switch. The switch acts as a digital patch panel of sorts, providing connectivity between the various elements. In network interconnection parlance, this is referred to as a *collapsed backbone*. Because of the scalability of ATM switches and the dedicated bandwidth of the interfaces, as the requirements of the routers increase, the port speeds and switching matrix of the switch can be increased as well. The use of ATM as a cross-connect technology for the network access points (NAPs) provides benefits. Here are some of them:

▼ Multiprotocol support is provided. ATM acting at layer 2 is independent of the upper-layer transported protocol. Thus, a NAP can handle a mixture of IPv4 and IPv6 without affecting all devices attached to the NAP.

■ ATM provides a combination of static routing and dynamic routing between interconnected routers. The NAPs act as boundary points between ISPs. The ATM component provides a barrier between these routers.

■ ATM interfaces are available in speeds up to OC-48 (2.4Gbps). This allows high-speed dedicated bandwidth between NAP components.

■ ATM systems have a lineage based in the telephone operating companies. As such, most ATM equipment has an embedded OAM&P capability providing traffic measurement, performance monitoring, service-level monitoring, and network management functions.

▲ The virtual circuit service and dedicated links between routers offer an additional level of security for traffic. The shared elements of the NAP are minimized.

ATM as a Tier 1 Transport Technology

Interexchange carriers are deploying and offering ATM services nationally and internationally. These same networks are being used to transport Internet traffic. ATM allows the creation of virtual point-to-point connections between points of presence without the need to buy fiber. The ATM network is transparent as far as the IP traffic is concerned. This capability provides a number of benefits, including these:

▼ The ATM network allows bandwidth management. As bandwidth needs grow, more bandwidth can be procured. Also, ATM allows traffic engineering: multiple IP paths can be sent on one ATM connection.

■ International connectivity allows rapid growth and expansion. The terms of the service agreement rather than construction costs limit the redesign of the network.

■ Disaster recovery options are not possible with dedicated circuit technologies. ATM's virtual circuit technology allows routing around failed nodes and links. Network management and provisioning options allow service level agreements of 99.999 percent.

▲ International interworking standards allow true multivendor interoperability. This fact enables a national ISP to select transport services according to economic terms rather than technical requirements.

The vBNS

The National Science Foundation's (NSF) very high performance Backbone Network Service (vBNS), operated by MCI, is the premier example of the integration of the Internet and ATM.

The vBNS connects a total of 140 supercomputing sites and institutions and 29 peer networks. This minimum link speed into the network is a DS-3. Each site is connected via IP over ATM. Up-to-date information about the vBNS can be found on the Internet at **http://www.vbns.net**.

USING ATM WITH ADSL

When all forms of services, such as voice, video, and data, are accessed over the same physical network, this is commonly called a *full-service network*. The international standard for full-service networks is Broadband ISDN (B-ISDN). It is possible to build a full-service network using ADSL. Here is how this can be done: On the premises, all forms of access to devices, whether TV set-top box, PC, or some other device (e.g., stereo, refrigerator), are connected to a premises ATM adapter. This ATM adapter is typically integrated with the ADSL terminal unit, the remote (modem) device. The user content (either Internet data or TV shows) is encapsulated in ATM cells, which are sent inside ADSL frames across the ADSL link.

At the DSL service provider's location, the individual DSL access lines are multiplexed onto an ATM switched network. This part of the network complies in every way with the B-ISDN specifications and standards. The ATM switching network, in turn, provides access to all services, from an Internet service provider (ISP) offering Internet access to other services such as video or graphics applications.

The advantage of this arrangement is that the equipment at the customer's location can support a common interface, such as Ethernet (10Base-T), a PPP link, a phone, or even some other option. Because of the encapsulation, this interface is of no concern to the DLS service, since all services are encapsulated in ATM cells. The only concern is the encapsulation method used within the AAL5 frame structure, since the method used on the premises must match that used on the service provider end. One of the most common implementations of Internet access over a DSL is called PPPoE (Point-to-Point Protocol over Ethernet). The name is misleading because, in most services, ATM is the underlying transport technology, with AAL5 as the framing mechanism. So the full stack should be called IPoPPPoEoAAL5oATM.

Separation of Functions

It is sometimes hard to keep all of the layers straight when dealing with complex and mixed encapsulation methods such as IP to PPP to Ethernet to AAL5 to ATM to DSL. The package (shown in Figure 22-2) always has the same sequence of headers: DSL framing carrying ATM header(s) preceding the AAL5 header, preceding the Ethernet header, preceding the PPP header, preceding the IP header.

At the lowest level, the DLS modems handle mainly the DSL superframe and the DSL frame. However, the DSLAM ATM function integrated with the DSL modems looks at the ATM header to switch and direct the ATM cell to the proper output port leading to the ATM network. Naturally, ATM switches will always look at the ATM header also, as the cell makes its way from source to destination across the ATM network.

The AAL5, Ethernet, PPP, and IP headers are looked at and processed mainly by the routers and remote access devices that are used to validate users and get them connected to the Internet. The AAL5 header allows the re-creation of the transmitted frame, the PPP

ATM Header(s)	PPP Header	Ethernet Header	IP Header	Upper Layer Data	AAL5 Trailer
Provides Switching and Multiplexing of user data streams	Provides Authentication	Provides terminal configuration	Provides Internet destination	The transported stuff	Maps the frame payload to the ATM cells and back again

Figure 22-2. PPoE header structure

protocol is used for authentication, and the Ethernet protocol is used for address configuration. Finally, the IP protocol is used by the routers to determine the destination of the packet.

It is important to keep this separation of functions in mind when trying to figure out what role each device plays in a mixed IP, ATM, and DSL network.

To put everything in perspective, it must be pointed out that the use of ATM between PPP and ADSL will continue for some time. The ATM QoS is there to improve IP's best-effort service. The ATM chipsets are plentiful and inexpensive, and PPP and ATM can work together to make each one better than when functioning on its own.

FRAME RELAY INTERNETWORKING OPTIONS

Frame relay is principally offered as a connection-oriented, permanent virtual circuit service. It provides data connectivity between geographically dispersed areas. Its distance- and usage-insensitive pricing make it ideal for the LAN interconnection. Frame relay has a problem, though: because of the transported frame's variable length and the placement of the frame check sequence field at the end of the frame, each switch handling the frame adds a random amount of queuing delay. As the networks grow larger and more traffic is added, this delay can lead to customer dissatisfaction. Enter Asynchronous Transfer Mode (ATM). As a connection-oriented, virtual circuit service, ATM is compatible with frame relay. Its fixed frame (cell) length and header error correction field minimize queuing delay. The standards organizations and the Frame Relay Forum have created interworking documents that specify the mapping of header and informational content between frame relay and ATM. Interworking function (IWF) devices are available to allow the transportation of frame relay traffic over ATM backbones.

Frame Relay Interworking Options

The Frame Relay Forum (FRF), in conjunction with the ATM Forum (ATMF), has defined two models of frame-to-ATM interworking.

The FRF.5 Implementation Agreement (IA) defines ATM–to–frame relay network interworking in a manner that allows two frame relay sites to interconnect via an intervening ATM network. The ATM network is transparent to both frame relay sites.

The "magic" of the interworking is performed by an interworking function (IWF) device that can be located in any number of places. The IWFs (FRF.5 requires two) could be in the frame relay network, the ATM network, or even the end system itself. In essence, FRF.5 involves the first IWF, encapsulating frame relay into AAL5 frames, which are then mapped into ATM calls; and the second IWF, removing this ATM/AAL5 encapsulation. Some of the frame relay fields are also mapped into corresponding fields at both the ATM and AAL layers.

FRF.8 defines frame relay–to–ATM service interworking; here, frame relay is converted into ATM; this differs from FRF.5's encapsulation approach, discussed earlier. The standards use the term "mapping" as opposed to "conversion," most likely due to the protocol conversion restrictions placed on the Bell operating companies after divestiture in 1984.

When we considers the popularity of frame relay among users and the penetration of ATM into the carrier backbones, these frame relay–to–ATM interworking options become quite significant.

LAPF to AAL5 Encapsulation/Mapping (FRF.5)

When a frame relay frame is transported over an ATM network, both the encapsulation and mapping functions must take place. The FRF.5 agreement defines the following sequence for this transport.

The LAPF frame has its Flag and Frame Check Sequence (FCS) fields removed, as they are not needed (that is, they are redundant with the AAL5 functions). Once these fields are removed, the resulting data unit is placed into the payload of an AAL5 frame.

Depending on the mapping options in effect, various LAPF header fields can be mapped into the equivalent functions at the ATM and AAL layers. For example, the Discard Eligibility (DE) setting of the LAPF frame can be mapped into the ATM layer's Cell Loss Priority (CLP) bit. Other options include mapping the Forward Explicit Congestion Notification (FECN) function and the frame relay data link connection identifier (DLCI).

Frame Relay to ATM Service Interworking (FRF.8)

FRF.8 defines the conversion mechanisms between frame relay and ATM/AAL5, which are more complex than those defined in FRF.5. A typical conversion mechanism follows.

First, the information field of a frame relay frame must be mapped into the payload field of an AAL5 frame. Unlike the encapsulation approach employed by FRF.5, in FRF.8 the LAPF frame header and trailer are stripped off. The next issue involves the optional translation of the multiprotocol encapsulation techniques used in the respective technologies (i.e., RFC 2427 [formerly RFC 1490] versus RFC 1483). The last mapping option involves the need to translate between the permanent virtual circuit (PVC) management protocols used by frame relay into the operations, administration, and maintenance (OAM) cells used for similar purposes in ATM networks. The mapping of the payload and management protocol fields was not required in FRF.5, as both ends of the link spoke frame relay. Now, with one end operating as a pure ATM node, these issues become significant because they can affect both the functionality and the interoperability of the service interworking.

Frame to ATM Traffic Management Issues

In either implementation, FRF.5 or FRF.8, the frame relay QoS parameters must be matched to those provided by an ATM network. Version 2.0 of the ATM Forum's Broadband Intercarrier Interface (B-ICI) specification provides numerous guidelines for matching the service characteristics of various technologies to ATM transport. Several different formulas and techniques are provided to assist the designer of frame relay–to–ATM interworking solutions.

The formulas map the frame relay port speed to the ATM connection's peak cell rate. The frame relay committed information rate (CIR) parameter is mapped to the ATM connection's sustained cell rate (SCR). These formulas allow for a similar service level agreement from the ATM network as delivered from the frame relay network.

ATM AS THE PSTN BACKBONE

The PSTNs, like other high-technology companies, are in a cash crunch, yet to maintain their market they must modernize. ATM allows this modernization while providing a natural migration path from the existing time division network to the next-generation packet infrastructure. ATM allows service aggregation, providing a higher level of utilization of existing facilities. Add to this the Quality of Service (QoS) attributes of ATM, and multiple service-level agreements can be meet on these same facilities. Few PSTNs have employed the full capabilities of ATM, but all use ATM in one fashion or another. ATM's circuit emulation service allows PSTNs to transport leased-line services interleaved with traditional packet services (Internet). Recently ATM has been integrated into switching platforms (both real time and digital cross-connects) to allow easier integration of ATM and traditional PSTN services.

ATM as the Universal Transport

The market has shown that ATM, as a service offering, does not compare to frame relay, yet as a backbone technology ATM is experiencing major growth. ATM's adaptation layers and traffic classes support the transport of any data types (interactive packet data, real-time applications, or guaranteed services). When compared to individual transport systems or the bandwidth necessary for non-QoS technologies (such as IP), ATM's aggregated service is a very cost-effective solution. Because of the mature state of ATM and the interworking facilities that are available, ATM is considered an incumbent technology, which can be used to support the high-margin services (including voice) that are offered by the PSTNs. Figure 22-3 depicts ATM as a universal backbone supporting the existing services offered by a PTSN. As newer technologies are introduced, ATM will find a role to fill, for instance in the development of Multiprotocol Label Switching (MPLS). Within the MPLS architecture, ATM is a transport option for the MPLS data. For the near future, ATM will remain the convergence-based technology that minimizes cost while maximizing flexibility at the lowest risk.

ATM QoS

ATM has become the preferred access technology where bandwidth management is a major issue. At line rates below OC-48 (2.4Gbps) conservation of the bandwidth becomes critical. ATM allows allocation of bandwidth to the users according to their administrative requirements. As long as IP QoS efforts are viewed with skepticism, ATM is the technology of choice where multiple types of service delivery, service aggregation, and service transport are required.

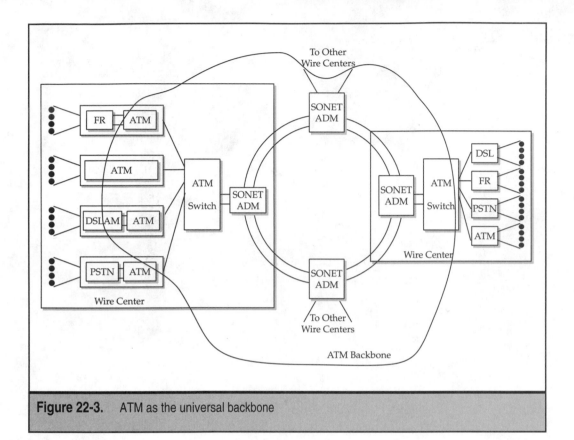

Figure 22-3. ATM as the universal backbone

ATM BACKBONE SUMMARY

In the decade since B-ISDN and ATM were defined as the next-generation networking technology, much has changed due to the influence of the Internet and the Internet Protocol (IP). Yet ATM has remained a viable, stable, and flexible backbone technology for frame relay networks and as a switching technology for DLS access lines (as can be seen in the levels of ATM equipment sales worldwide). ATM's popularity in the Internet community is waning due to other, newer technologies, but it remains a building block for the PTSN community.

CHAPTER 23

Internet Protocol
as the Core

We live in the age of convergence, which promises to provide traditionally separate services such as voice and data over a single network. Internet Protocol (IP) networks, including the Internet, are getting most of the focus today. The Internet is much more than e-mail and the World Wide Web; companies are creating intranets, extranets, and virtual private networks on top of this infrastructure. An IP-based backbone network offers an inexpensive medium for all forms of communications, but such convergence goes well beyond cost savings. It turns the mere transport of bits into a commodity, allowing for the development of new and exciting value-added services, such as unified messaging, desktop video conferencing, video- and voice-enabled e-commerce, and other applications yet to be imagined.

To deliver such a diverse set of services, we have developed mechanisms that deliver Quality of Service (QoS) so that each application can perform optimally in this merged environment. This chapter provides an overview of current QoS approaches that potentially allow IP to match the performance of such services as Frame Relay. Since many organizations are using the Internet to implement new applications such as virtual private networks (VPNs) and voice over IP (VoIP), security has become paramount. Cryptography and firewalls are not the only tools used today, but they are the basis of Internet security and will be introduced in this chapter. The chapter then covers VPNs and VoIP. Finally, a brief discussion covers the carrier IP-based networks that deliver these converged services.

QUALITY OF SERVICE

As IP is used more and more to support multiple traffic types, provisions for the support of QoS mechanisms are required. QoS describes the performance parameters that a network needs to deliver to an application. For example, voice is delay sensitive, so a network used for voice requires low latency. Other QoS parameters include jitter (delay variation) and bandwidth. Generally, the QoS mechanisms under consideration for use in IP environments fall into one of three categories: one approach operates *end-to-end*, another operates *hop-by-hop*, and a third *bypasses* IP altogether. Here, we'll discusses the latter two QoS approaches, focusing on the IP bypass technique called Multiprotocol Label Switching (MPLS).

Differentiated Services

Hop-by-hop IP QoS approaches do not require that calls be set up prior to transmitting session data. These approaches use a set of predefined rules that result in packets following one another through paths in the network. The packets merely carry some form of identifier that associate them with one flow or another.

As a result of the Type of Service (TOS) limitations, the Differentiated Services (DiffServ) working group of the Internet Engineering Task Force (IETF) has created a new meaning for this field in the IP header. This new meaning is compatible with proposed evolution to IPv6. The objective of DiffServ's redefinition of the TOS field is to provide service level agreements (SLAs) between routers. These SLAs enable the creation of unidirectional classes of service (COS) across the network. Three types of COS are defined: best effort, assured forwarding, and expedited forwarding.

DiffServ is a simple mechanism that changes the name of the TOS field to DS. The least significant 6 bits of the DS field contain a Differentiated Services Codepoint (DSCP), which indicates the manner in which the particular packet should be treated by routers handling the packet. Depending on the value in the DS field, routers will treat the packet according to a set of predefined rules, called per hop behavior.

Two popular strategies service providers use when implementing DiffServ to deliver QoS are weighted random early discard (WRED) and weighted round robin (WRR). When congestion occurs, WRED throttles transmission back, sending increasingly fewer packets over time. In WRR, packets are sent using a round-robin procedure, but each packet queue is given a weight—bandwidth can then be divided among the queues based on the weighting.

DiffServ's primary advantage is that it's a simple method of classifying various application services. It is a flexible way to prioritize traffic, allowing for efficient resource sharing and some approximation of guaranteed service. DiffServ can be used alone, but it can also be combined with other protocols. It is commonly integrated with current implementations of MPLS, which interoperates with DiffServ by using the DS field to assign priority to MPLS packets to create different COS. DiffServ can easily be used at the edges of an MPLS network with MPLS traffic engineering in the core. MPLS is discussed in the next section.

NOTE: Documentation on DiffServ can be found at the IETF's Web site: **http://www.ietf.org/html.charters/diffserv-charter.html**.

Multiprotocol Label Switching

MPLS is a standards-based IETF variant of several successful proprietary schemes. MPLS is said to *bypass* IP because no part of the IP packet is processed for QoS purposes. The objective of MPLS is to improve the performance and scalability of IP. As an added benefit, this "virtual circuit" mechanism can also be used for routing in VPNs.

The MPLS tag is four octets long. It specifies criteria for handling the packet to provide a given QoS level. The label points to a predefined network path that offers the QoS level the packet requires. If a better path exists (one that better meets the user's QoS demands), a node can relabel the packet upon egress. Nodes that can perform this function are known as *label swapping routers*.

Once the packet is accepted by an MPLS network, processing the IP packet header is no longer required. The packet makes its way through the network based solely on the contents of the MPLS label. Thus, an MPLS network can appear as a single hop to IP.

The MPLS label supports the uncoupling of the routing function from the forwarding function in MPLS networks. Traditional IP routers concern themselves primarily with the routing function (which outbound link to use to move a packet most efficiently from source to destination). Traditional routers rarely concern themselves with forwarding details, which some would argue is a Data Link layer function. An MPLS-capable router uses the contents of the MPLS label to indicate which route to use based on the application's QoS demands.

MPLS is a Layer 2.5 protocol that sits between IP and whatever Data Link protocol is carrying the IP packets outside of the MPLS environment. Once inside the MPLS network, the contents of the IP header no longer need to be examined to determine the route. In fact, the route is now determined at least partially by the user's QoS requirements. The MPLS ingress node can use one of several criteria to decide what the specific value of the label should be. QoS requirements drive the selection of a particular label in an MPLS network.

The terminology used in MPLS is as distinctive as that of any other major networking architecture. As usual, a group of new acronyms need to be defined and used. Below are some of the most important MPLS terms:

▼ **MPLS Domain**　A more correct term than "MPLS network."

■ **Label Switching Router (LSR)**　The "engine" of the MPLS domain. The LSR is defined as any device capable of supporting MPLS. The LSR could be an IP router, a frame relay switch, an Asynchronous Transfer Mode (ATM) switch, or something else entirely. All that is important is the MPLS functionality.

■ **Edge LSR (or LER)**　These devices stand at the borders of the MPLS domain. The original MPLS documents make no distinction between LSRs and edge LSRs except for their position in the MPLS domain. The confusion from having both called LSRs has resulted in some documents inventing the acronym *label edge router* (LER) to distinguish edge LSRs from core LSRs. The distinction is an important one: LSRs in the MPLS domain only have to switch MPLS labels and understand MPLS protocols, while LERs must also support non-MPLS functions such as traditional routing on at least one port.

▲ **Label Distribution Protocol (LDP)**　This protocol is used to share label information among all LSRs, edge and core. Generally, an LDP Request is used to ask another LSR on a link for a label based on the destination IP address. The LDP Mapping reply supplies the label to be used on the link for this stream of IP traffic.

Figure 23-1 shows an MPLS domain composed of two LERs and two LSRs with a client attached to the LER on the left of the diagram and a server on the right. Neither client nor server need be MPLS aware and remain typical IP devices. Once the LER receives an IP packet with a Layer 2 header, the edge LSR consults the IP routing table and inserts a label between the IP packet and proper Layer 2 header for transport across the MPLS cloud. The contents of the label can be determined by a number of criteria, including the QoS requirements of the user. Then the whole unit is sent across the MPLS cloud. During user traffic operation, no routing table lookups take place in the core LSR devices. Instead, the unit is switched based on label value, which changes hop-by-hop. At the far end of the MPLS cloud, the LER removes the label and routes the packet to the server using the proper Layer 2 header.

MPLS supports differential handling of multiple traffic types by selecting the most appropriate path through the network. Such path selection can be based on information contained in the IP packet header (such as the encoding of the DiffServ field), or on the outcome of a signaling exchange between the user and the network (such as Resource Reservation Protocol [RSVP] messages).

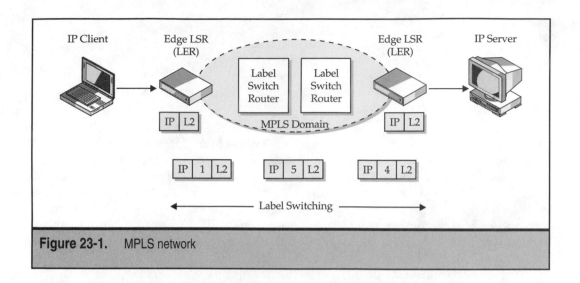

Figure 23-1. MPLS network

Implementing MPLS in an existing IP router internetwork requires changes to the network. MPLS is a "shim" protocol that fits between IP and the Data Link layer. As such, MPLS adds a layer of complexity and overhead to an existing network. On the administrative side, MPLS requires that label selection criteria be defined and that paths through the network be available to match the desired result (such as QoS delivery).

MPLS is thought to be suitable for use in large public and private internets. MPLS is sometimes heralded as a mechanism for integrating IP and ATM networks because labels can be selected based on information mapped from either or both of these protocols. Interestingly, some have suggested that to eliminate the overhead associated with MPLS, we replace the Virtual Path Identifier/Virtual Channel Identifier (VPI/VCI) bits in the ATM header with the MPLS label. One could argue that if the appropriate application programming interfaces (APIs) are available, why bother with MPLS at all? IP QoS (such as DiffServ) could be mapped directly to ATM QoS (variable bit rate—non-real-time, VBR-NRT service) without the need for a shim protocol.

The network architecture at the heart of the MPLS domain is of no concern to MPLS itself. The links between LSRs in the domain could be provided by frame relay, ATM, SONET/SDH, or something even more exotic. All that is important is that all LSRs in an MPLS domain understand the type of shim label used and how to process the shim label. The Layer 2 header within the MPLS domain can be seen as just a "wrapper" for the shim label, although when frame relay and ATM are used to construct an MPLS domain, it is more likely that the frame relay header or ATM cell header acts as the shim label.

While we tend to focus on MPLS in an IP/ATM environment, it is truly protocol independent. That is a great advantage and means that MPLS can provide a migration path from our existing architectures to more streamlined, wholly-optical networks. Some vendors see a day in the near future when the MPLS label will represent a frequency of light to be used over an optical network using DWDM and optical switches.

> **NOTE:** Documentation on MPLS can be found at the IETF's Web site: **http://www.ietf.org/ html.charters/mpls-charter.html**.

SECURITY

The Internet is a collection of the users and networks that connect via the various ISPs. When a person or organization connects a LAN or host to the Internet, the local network becomes a part of the Internet. Just as in the real world, the Internet population is unfortunately not always friendly and respectful. Although most ISPs define user guidelines and require users to sign them, once the links are installed, there is little control over who uses or accesses the network at any particular time. What's more, virtually every major university and many secondary schools are online, opening the door to a sea of novice users, students, and hackers. This makes security a particularly important issue for those organizations that are connected to the Internet.

Two important aspects of Internet security impact e-commerce, VPNs, and even VoIP: *cryptography* and *firewalls*. Each has different functions. Cryptography's main use is to protect data traversing a network, although it can be used for other purposes, such as for authenticating data and users. Firewalls chiefly are concerned with protecting internal networks from external threats, but they often have VPN and other capabilities as well. Security is a large topic, involving much more than these two aspects. One must also consider viruses and worms, development of security policies, internal threats, and a myriad of other issues. However, understanding cryptography and firewalls is an important first step to understanding security overall.

Introduction to Cryptography

Cryptography is the science of writing in secret code. In data and telecommunications, cryptography is necessary when communicating over any untrusted medium, such as the Internet. Regardless of the goodness or evil of the computer system or network, the best way to protect data from theft or alteration is to use cryptography techniques. Cryptography is also the basis for authentication mechanisms, which is as crucial as making data unreadable by intruders.

The general purpose of cryptography is to transmit or store information so that only the intended parties can read it. Here are a few salient terms:

▼ **Plaintext** The initial raw unencrypted data.

■ **Ciphertext** The unreadable result of encryption.

■ **Algorithm** A set of mathematical rules used to transform plaintext into ciphertext and *vice versa*. Algorithms are generally well-known, which allows them to be rigorously tested by the global community.

▲   **Key**   A number used in the calculations defined by the algorithm; the same key or another key can be used to decrypt the ciphertext back into usable plaintext. The larger the key, the harder it is to break a code. Computers have made it easier to attack ciphertext today using brute force methods rather than by attacking the mathematics. With a brute force attack, the attacker merely generates every possible key and applies it to the ciphertext. Any resulting plaintext that makes sense offers a candidate for a bona fide key.

Three types of cryptographic schemes are in use: hash (or one way) functions, secret key (or symmetric) cryptography, and public key (or asymmetric) cryptography.

Hash functions, also called message digests, are algorithms that effectively use no encryption key. Instead, hash functions apply an irreversible mathematical transformation to the original data so that the plaintext is not recoverable from the ciphertext. Another feature of hash functions is a low probability that two different plaintext messages will result in the same hash ciphertext (an event known as a *collision*). In addition, the length of the hash ciphertext will always be the same regardless of the length of the input, thus making even the size of the plaintext message unrecoverable.

Hashing mechanisms are typically used to provide a sort of digital fingerprint of a file, which is often used to verify that the file has not been altered while in storage or transmission. They are also commonly employed by many operating systems for storing passwords (such as UNIX's publicly viewable password file), so that passwords are never kept on a system in plaintext.

A common example of a hash function is MD5. Described in RFC 1321, MD5 was developed by Ron Rivest after potential weaknesses were reported in MD4, an earlier algorithm. It is similar to MD4, except that it processes the data multiple times, sacrificing processing speed for additional security. Another popular hash function is the Secure Hash Algorithm. Both are used to verify data integrity in VPN environments and as part of digital signatures used in public key certificates.

From secret decoder rings found in cereal boxes, to the Enigma machine used by the Germans during World War II, secret key cryptography (SKC) is probably the best known form of cryptography. In SKC, a single key is used for both encryption and decryption. The key must be known to both the sender and receiver. In this scenario, for example, Alice applies the key and algorithm to the plaintext to create the ciphertext and sends the encrypted message to Bob, who applies the same key value to decode the message. Because a single key is used for both functions, secret key cryptography is also called *symmetric* encryption.

The most widely used secret key cryptography scheme is the Data Encryption Standard (DES), designed by IBM in the 1970s and adopted by the National Bureau of Standards in 1977 for commercial and unclassified government applications. DES is a block cipher; it accepts 64 bits of input and, after a series of transformations, creates a new 64-bit block. DES uses a 56-bit key and was designed to yield fast hardware implementations and slow software implementations. Triple-DES (3DES) is a variant of DES, with a number of modes that use either two or three different keys, coupled with three encryption steps. DES/3DES are widely used to encrypt data in VPNs, particularly IPSec implementations. Another scheme,

RC4 (Rivest Cipher version 4) is popular in secure e-commerce/ Web environments. Rijndael (pronounced "rhine-dahl", it is named for the inventers, Dr. Joan Daemen, and Dr. Vincent Rijmen), the new Advanced Encryption Standard (AES), was adopted in 2000 by the U. S. government.

Public Key Cryptography

The largest problem with SKC is ensuring that both Alice and Bob have the secret key. Just as it would be when trying to share copies of a physical key when Alice and Bob are in different locations, trying to distribute secret key values between them across an untrusted medium such as the Internet is difficult.

Public key cryptography (PKC) was invented to address the key distribution issue. While the British and American governments had secretly created PKC schemes before anybody else, public credit is given to Martin Hellman and Whitfield Diffie of Stanford University. They created the Diffie-Hellman algorithm in 1976. Often called *asymmetric* cryptography, this scheme requires two keys: applying one key encrypts the plaintext while applying the other key decrypts the ciphertext. It does not matter which key is applied first and which is applied second, though one is necessarily chosen to be the public key (a number distributed publicly and unsecurely), and the other is designated as the private key (a number kept private by the owner and not given to anybody else). The conceptual operation of PKC is shown in Figure 23-2.

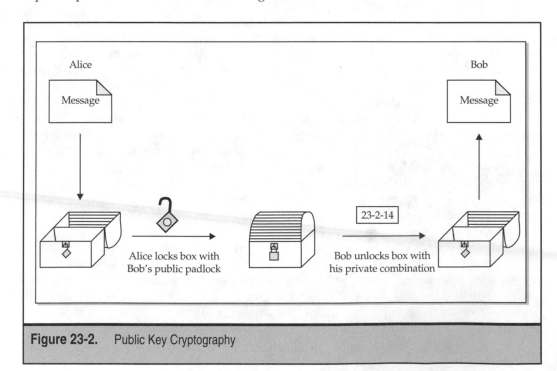

Figure 23-2.   Public Key Cryptography

The most important element to understand is that Alice knows (or has) something that Bob has made publicly available while he has kept a related item private.

For our analogy, we'll use combination padlocks. Imagine that Bob has manufactured many combination locks that he has placed in various public areas, such as post offices. He is the only person who knows the combination (assume nobody can reverse engineer these super strong padlocks to determine the combination). When Alice wishes to send a message to Bob, she can put the message in a lock box, obtain one of his public padlocks, and use it to lock the box. Now Alice can send the message via an untrustworthy courier service to Bob without fear because only Bob can unlock the box. Bob uses his memorized combination to open the padlock and retrieve the message. This exchange method works for any message, which could in fact be a secret key. Note that this analogy breaks down when discussing the use of PKC in digital signatures, but it works when trying to understand the basics of the approach.

PKC employs large numbers that make calculation relatively slow when compared to SKC, so messages generally are not encrypted with PKC. Instead, public keys are used to encrypt short secret keys, which are used for the actual encryption of data. Protocols such as Secure Sockets Layer (SSL) rely on a key exchange similar to the following generic example:

1. A client sends a connection request to a server.

2. The server responds with its public key.

3. The client encrypts a secret key (sometimes these keys are only valid for a short time, or session, and are thus called session keys) with the server's public key.

4. The server decrypts the message with its private key to extract the secret key.

5. The secret key has been securely distributed, so both parties have the same secret key and can now encrypt and decrypt data.

The most common PKC scheme used today is called RSA, named for its inventors: Ronald Rivest, Adi Shamir, and Leonard Adleman. The RSA scheme uses a variable size encryption block and a variable size key. The public and private keys are derived from a very large number, n, that is the product of two prime numbers chosen according to special rules; these primes will typically have 100 or more digits each. An RSA key public key includes a number n and a number e, which is derived from the factors of n; its corresponding private key comprises n and d, another derived number. Encryption is accomplished by solving the following equation: $ciphertext=plaintext^{e} \bmod n$. Decryption uses a similar equation: $plaintext=ciphertext^{d} \bmod n$. The public key components (e, n) can be distributed without fear that an attacker can determine the private key values (d, n) from this information alone. The ability for computers to factor large numbers is rapidly improving, however. The protection of RSA, of course, is that users can increase the key size to always stay ahead of the computers.

Many security methods rely on PKC to secure data traversing the Internet. To fully take advantage of the power of this method, some sort of infrastructure must be created to support it. A nascent public key infrastructure (PKI) is beginning to take shape, involving an array of protocols, services, and standards. While there is no single accepted definition, it appears that digital certificates and certificate authorities (covered in the next section) are a part of the mix. PKI usually addresses issues such as the following:

▼ **Registering keys**   Once a key pair has been created, the public key portion must be registered with a trusted third party who issues a certificate. Certificate authorities vouch for the identity of the user or site who has applied for the certificate.

■ **Obtaining keys**   Distribution of keys and certificates has been the greatest challenge for those involved with PKI. Some protocols exchange keys at the beginning of a session, as in the case of SSL and some virtual private network environments.

■ **Recovering keys**   Potentially one of the most contentious debates is how to gain emergency access to encrypted data. Usually this means some method for legitimate users to read information encrypted with a lost key. The U.S. government still champions the idea of "key escrow" to allow investigating bodies the ability to read encoded messages and files. Civil libertarians and a host of other interested parties continue to voice strong opposition.

■ **Revoking keys**   If a private key has been somehow compromised, the associated certificate must be revoked. This is generally done by maintaining a CRL in a repository at a certificate authority. CRLs can be consulted when verifying digital signatures.

▲ **Evaluting trust**   The whole point of creating digital certificates is to provide some degree of trust. Digital signatures are used to determine whether a certificate is valid. Certificates often authorize specific operations.

PKI continues to evolve, and a lot of work remains to be done. To facilitate the development of open standards, the IETF established the PKIX Working Group (**http://www.ietf .org/html.charters/pkix-charter.html**) in 1995. Its mandate is to create the necessary framework to support a PKI based on the X.509 standard.

Digital Certificates

Despite the power of PKC, some major non-technical issues need to be resolved. How and where can a party's public key be found? On a local server, directly from the party, or from a trusted third party? How does a recipient determine whether a public key really belongs to the sender? How does the recipient know that the sender is using a public key for an authorized, legitimate purpose? When does a public key expire? How can a key be revoked in case of compromise or loss?

The answers to these questions hinge on the trust relationship that must exist between the communicating parties. Alice and Bob cannot have a comfortable electronic relationship unless they both have confidence that they know to whom they are talking, what each is authorized to do, what is being said, when it is being said, and that the conversation is private. The PKI approach to resolve these issues involves the use of *certificates*.

All of us are familiar with the basic concept of a certificate in the real world. A certificate identifies us to others, indicates what we are allowed to do, provides specific authorization for the holder, indicates how long the certificate is valid, and identifies the authority that granted the certificate. For example, a driver's license typically identifies the holder's name and date of birth, address, class of vehicle(s) that the holder can operate, expiration date, serial number, and identification of the authority granting the license. In addition, most states include a photograph and some states include organ donor and/or blood type information. In the U.S., individual states issue the licenses, but national agreement dictates that one state's license is honored nationwide. Furthermore, each state takes appropriate precautions to verify a person's identity before granting the license, so that a driver's license is considered a bona fide identification document with which to obtain employment, get on an airplane, or prove one's age.

In the PKC environment, certificates are digital documents that bind a public key to an individual, organization, corporate position, or another entity. They allow Bob, for example, to verify Alice's claim that a given public key does, in fact, belong to her. In their simplest form, certificates contain a public key and a name; they can also contain an expiration date, the name of the authority that issued the certificate (and that is vouching for the identity of the user), a serial number, any pertinent policies describing how the certificate was issued and/or how the certificate may be used, the digital signature of the certificate issuer, and perhaps other information.

Certificates are necessary in the cyber world because paper obviously does not exist. Digital certificates will, in fact, be the basis for the future of paperless electronic commerce and will provide the mechanism for a wide range of business decisions. The specific functions of the certificate are listed below.

▼ **Establishment of Identity**   Verifying who the certificate holder is.

■ **Assignment of Authority**   Establishing what actions the holder may or may not take, based upon this certificate.

▲ **Securing Confidential Information**   For example, encrypting the session's symmetric (secret) key for data confidentiality.

Certificates are used in many applications, perhaps most commonly for secure communication using a Web browser or in VPNs. They are obtained from a certificate authority (CA), which is simply an entity or organization that issues certificates. The CA can be any trusted central administration willing to vouch for the identities of those to whom it issues certificates and their association with a given key.

CAs have policies governing issuance of certificates and identification of requesting parties, and operational procedures for renewing and revoking certificates. All certificates have an expiration date of usually one, two, or five years after issuance. The CA must be able to renew a certificate prior to expiration upon request of the user. Revocation is harder and more crucial; certificates could have to be revoked in case any information in the certificate changes (such as job function, authorization level, account number, and so on), the user violates some certificate usage policy, or the private key is lost or compromised. When a certificate is revoked, this information must be made available as soon and as widely as possible. When certificates are revoked prior to their expiration period, they are added to a certificate revocation list (CRL). When Bob checks the certificate associated with a message sent by Alice, Bob should also check the CRL to be sure that Alice is not on it!

Many commercial CAs are in the global market place, including GTE Cybertrust (formerly Baltimore Technologies), Digitrust, Entrust, and Verisign (which bought Thawte, another popular CA). No government entity gave these companies the right to be CAs; rather, the market decides who is to be trusted. It doesn't matter from which company you obtain a certificate; the goal remains the same: to provide a public key mechanism for identification and key exchange. This is an absolute requirement in a world that relies so heavily on the Internet for commerce and exchanging corporate information.

Introduction to Firewalls

A *firewall* is a mechanism for protecting a local network from external, untrusted networks. Like the physical firewall used in some buildings to limit the spread of fire, a network firewall protects a user's site from purposeful and inadvertent attacks from the outside. While firewalls are important aspects of securing a system from outside intrusion, system managers should be aware that most attacks are launched from the inside.

Firewalls enforce the security policy that each site adopts. The sophistication of the firewall depends on the site's security philosophy. Firewalls can be classified into three generic types, which might be employed in combination:

▼ **Packet filters**   Block packets based on a set of rules derived from the direction the packet is traveling, the protocol employed, host address(es), protocol port number (such as the TCP/IP application), the physical interface, and/or other factors. Packet filtering is generally employed in routers. The quintessential example of a type of packet filter is the configurable access control lists (ACLs) found in Cisco routers.

■ **Higher-layer inspection firewalls**   Similar to packet filters, these firewalls are more sophisticated due to their awareness of all layers in the protocol stack. They also typically maintain state tables to enable the scanning of packets within the context of the overall data stream. Higher-layer firewalls are sometimes called "stateful" firewalls because they typically store state information aout connections. The most popular example of a higher-layer inspection firewall is CheckPoint's Firewall-1 product. CheckPoint actually patented the first "stateful inspection" technique.

▲ **Proxy Servers**   One or more systems that appear to provide a set of services to the outside, but actually act as a proxy for the real server. An external client does not connect directly to an internal server, but instead attaches to the proxy, which, in turn, attaches to the internal server. Proxy servers can come in circuit-level or the application-level forms. The Raptor product from Symantec (who bought Axent, who bought Raptor, who made the original Eagle firewall) is the best known application proxy. While not often seen outside the academic environment, SOCKS (which is not an acronym, but an abbreviation for Socket Security) is a good example of an generic circuit-level proxy.

Firewalls, like most network devices, come in a wide variety of shapes and sizes. Thus even the basics of configuring a firewall can vary drastically from product to product. If a firewall does not have a particular capability, configuration issues do not apply to that function at all. Low-end firewalls are essentially limited to packet filtering.

However, almost all high-end firewalls can perform a basic set of functions, including the ability to filter and inspect traffic, perform or incorporate some type of traffic/user/device authentication, and translate between an IP address space used internally on an intranet and externally on the Internet (a proxy server characteristic). These firewalls can also scan both e-mail and Java or ActiveX applets for potential harmful content, and they can encrypt sensitive information sent between the firewall and some remote device.

In addition to this basic set of functions, many sophisticated firewalls can load balance logical collections of servers to reduce device congestion and user delays, as well as manage other firewalls and routers with regard to ACLs and related concepts. Top-of-the-line firewalls can even provide encryption key management and certificate authority functions for various e-commerce and VPN applications.

This last set of functions begins to position the firewall as a device that overlaps with VPN devices in the network. In fact, some firewall vendors position their products as VPN devices, bundling as much functionality into the firewall as possible.

Relying on a single security point is not a good idea. Like the varied defenses of a medieval castle, network security should be implemented in a layered fashion. Figure 23-3 shows an example network that employs a packet filtering router as a first line of defense,

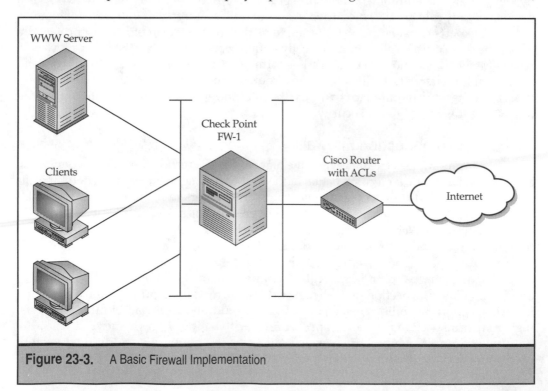

Figure 23-3.   A Basic Firewall Implementation

with a more sophisticated firewall providing additional protection. More layers can be added as necessary to provide greater protection to more valuable network and information assets. To best secure a network, a security policy should be developed first, usually after conducting a security audit. That is the only way to be sure an organization is spending the appropriate amount of money on the appropriate level of security.

Packet Filters

Packet filters block or permit packets, based on a set of rules that specify characteristics such as those listed here:

▼ State of the connection (new or already established)

■ Direction the packet is traveling (inbound or outbound)

■ Communications protocol used (such as IP, IPX, ICMP, TCP, UDP)

■ Source and/or destination host addresses

■ Protocol port number (the higher layer application)

▲ Physical interface

A network firewall administrator configures these rules ahead of time. The rules are applied consecutively and all packets are subject to a pass/fail decision. If the packet header passes all requirements, the packet is forwarded onto the public network. If a packet fails any rule test, it is discarded. An audit trail is maintained for future reference.

Some experts consider packet filtering to be the weakest form of firewall protection, in part because of the difficulty in correctly defining and specifying the rules. However, packet filters are the most easily accessible form of protection in most environments and provide an excellent start. Given that most routers today have packet filtering capabilities, and routers are required to connect to the Internet, every organization should at least implement this basic type of security.

Higher Layer Inspection Firewalls

A step up from simple packet filtering is the higher layer inspection firewall. The device can be software, hardware, or a combination of the two. Many higher layer inspection firewalls are just software that runs on a server, although some routers have this capability as well.

As with packet filter rules, higher layer inspection rules vary. A possible outbound rule is, "Allow no Web connections to go to the hacker website *www.hackersrus.org*"; a typical inbound rule is, "Allow no inbound Telnet sessions at all." Note the high degree of granularity that higher layer inspection allows.

While ordinary packet filters can also operate on some fields of the TCP/UDP headers, a higher layer inspection firewall makes this feature its main purpose and adds many more inspection capabilities to simple TCP/IP "pseudo-header" packet filtering. So the higher layer inspection firewall is a packet filter with a bit more awareness.

The awareness of a higher layer firewall gives the network an additional level of protection. Because these firewalls maintain state information and understand the typical behavior of protocols and applications, they are more effective at stopping some kinds of attacks. For example, a simple packet filter won't stop a Denial of Service attack such as a SYN Flood because each packet is considered in a vacuum. A higher-layer firewall can recognize the aberrant behavior of such attacks and block them.

Proxy Servers

A *proxy* is nothing more than some intermediate device that acts in place of one of the end-communicating devices. Thus, a proxy host will pretend to be one of the network's local hosts. The reason to employ a proxy host is that the network manager needs to focus only on that one system; if security measures are properly implemented at that host, all the internal hosts are protected. Proxies come in two flavors: both application- and circuit-level proxy servers employ proxy agents, which often makes the distinction between the two a little fuzzy.

Application-level proxies are hardware or software specific to a given application. They operate at the Application Services layer of TCP/IP. When the internal client system attempts to establish a connection to a WWW server on the Internet, the proxy intervenes—it acts like the server from the perspective of the client and acts like a client from the perspective of the server.

A circuit-level proxy differs in that it is application-independent. It operates at the Transport layer of TCP/IP. When the WWW client attempts to make a connection to the WWW server, the proxy again intervenes, but this time at the TCP level. The circuit-level proxy has no knowledge of the higher layer protocol. Furthermore, if this client were to attempt to establish an FTP connection to this same server, the same proxy host would intervene to create a new TCP connection; in the case of application-level hosts, a different proxy agent would be needed for the WWW and FTP processes.

VIRTUAL PRIVATE NETWORKS

The ubiquity and low cost of the Internet makes it an attractive medium for transporting corporate data, essentially making the Internet a corporate WAN or remote access mechanism. A VPN is a logical network that provides user privacy over a public network. This definition includes networks such as frame relay, but it is especially associated with the Internet today. To accomplish the trick of treating a public network as though it were private requires some form of tunneling and often some form of remote access authentication. QoS and security are also great concerns when using the public Internet to deliver important data.

Tunneling

Tunneling is just a fancy form of encapsulation. The difference between the two terms is that encapsulation follows the usual "top-down" approach of enveloping a higher layer protocol in a lower protocol's overhead, while tunneling generally involves lower layer protocols being placed inside higher layer protocols. For example, an IP packet (Layer 3) is usually encapsulated in a Point-to-Point Protocol (PPP) frame (Layer 2), which is then transmitted. In a tunneled environment, an Internetwork Packet Exchange (IPX) packet (Layer 3) might be placed inside a PPP frame (Layer 2), which is in turn put into an IP packet (Layer 3)—the entire packet would then be placed into yet another frame before transmission.

Tunneling has traditionally been necessary because the data to be transported is not understood by the network it is to traverse. IPX cannot be transported natively over an IP network such as the Internet. Think of IPX as a car and the Internet as a lake—you can't drive a car across a lake, but if you put your car on a ferry boat (the tunneling protocol), it can get across the lake. Today, as legacy environments such as NetWare and SNA have moved to IP as a Network layer, VPNs are primarily concerned with moving IP data over an IP backbone. This means that transporting non-native protocols is a non-issue, and thus privacy (the P in VPN) becomes the only real problem to address.

Tunneling itself deals only with the concept of transport, and provides nothing in terms of privacy. Despite claims to the contrary, encrypting tunneled data is the only way to truly secure transmissions. Tunneling protocols are the agreements between tunneling devices that often offer some form of security in addition to assuring transport of information.

The concept of tunneling protocols is not new for data communications. Tunneling methods have been used for many years, standardized as Data Link Switching (DLSw) and Remote Source Route Bridging (RSRB). Tunneling protocols address a variety of issues, including the following:

▼ **Network Services** Corporations are multinetwork operating system environments. No single vendor can provide a solution for all possible scenarios; consequently, multivendor multiprotocol traffic is a fact of life for most corporations. Another fact is the growing use of a single protocol's backbone (TCP/IP). By providing protocols identification capability, the tunneling protocols allow multiple LAN protocols to be transported on a single backbone network.

■ **Interoperability** Sharing critical information requires interoperability and connectivity. A common backbone protocol suite and a tunneling system that can handle multiple protocols achieve both.

■ **Performance** Shared information is useful only when the information is presented without much delay. The performance is the responsibility of the network. Tunneling protocols provide a means to monitor the performance (round-trip delay) and notify the users of problems.

■ **Reliability** Shared information is also useful only when the information is error-free. The reliability of communications is also the responsibility of the network. The tunneling protocols provide a means to monitor the error performance of the network.

▲ **Security** This is a hot topic in corporate communications. Tunneling protocols provide a means to invoke encryption, authentication, and accounting schemes— the building blocks of most security systems.

Five tunneling protocols are briefly described below:

▼ **Generic Routing Encapsulation (GRE)**   The IETF refers to this tunneling technique as *encapsulation*. A number of RFCs have been published that deal with encapsulating an alien protocol over an IP-based Internet. RFC 1701 defines GRE, but it does not define the payload (alien protocol) or the native protocol. It defines only the capabilities of a tunneling system. RFC 1702 defines the mechanism for generic tunneling over an IP network for Mobile IP, RFC 2003 defines IP tunneling over IP, and RFC 1234 defines tunneling IPX over IP.

■ **Layer 2 Forwarding (L2F)**   Created by Cisco, L2F tunnels either PPP or Serial Line Internet Protocol (SLIP) over an IP network. L2F has largely been relegated to legacy status, with most providers migrating their existing customers to L2TP or other forms of tunneling. Since the basic unit used to transport tunneled data is a Layer 2 (such as PPP) frame, L2F is referred to as a Layer 2 protocol.

■ **Point-to-Point Tunneling Protocol (PPTP)**   A tunneling protocol proposed by Microsoft and supported by a group of hardware vendors, it encapsulates PPP inside GRE version 2 protocol. PPTP follows the PPP connection establishment procedures and can employ the PAP or Challenge Handshake Authentication Protocol (CHAP) for authentication. PPTP is also an example of a Layer 2 protocol.

■ **Layer 2 Tunneling Protocol (L2TP)**   A blend of L2F and PPTP tunneling protocols, L2TP is a compromise between Cisco (L2F) and Microsoft (PPTP) to standardize the tunneling marketplace. As its name suggests, L2TP is just like its predecessors and is considered a Layer 2 protocol.

▲ **IP Security (IPSec)**   A security protocol described in IETFs RFCs 2401–2412, it has become the industry's VPN protocol of choice. IPSec is a Layer 3 tunneling protocol and can only transport IP data. Its focus is therefore not on multiprotocol transparency, but on securing IP packets traversing the IP VPN. Unlike most tunneling protocols, IPSec inherently provides data integrity and privacy mechanisms in two forms called Authentication Header (AH) and Encapsulating Security Payload (ESP). Originally IPSec lacked user authentication capabilities, but proprietary and standards-based approaches are now being implemented. The most common implementation of the protocol uses "tunnel mode" ESP, which encapsulates an encrypted IP packet into unencrypted outer IP packet for transport. A transport mode is also defined but is not generally used. Refer to the "IP VPN" blueprint for an illustration of a generic IPSec VPN.

Remote Access Authentication

When deploying a remote access solution in either a dial-up or VPN environment, proper administration of that service requires careful examination of authentication, authorization, and accounting (AAA). Authentication addresses how each user is to be validated.

Authorization deals with the rights of the individual users after successful validation. Accounting methodologies relate to usage issues.

As networking technologies have improved, so have security options. Network administrators typically use more than one of these systems in concert with one another to provide a multi-tier security system. Some authentication options include call back, password authentication protocols such as CHAP, dynamic password systems such as SecurID, and Network Operation System (NOS) authentication such as that provided by Windows NT.

After a user is authenticated, an authorization takes place to determine what the user is allowed to do. It can be configured by restricting the exposure of the internal network to remote users in addition to simplifying the view of the internal network for the less technical remote user. A restriction example would be a user having access to a specific subset of network resources (such as file servers, print servers, or fax servers). One approach to controlling these access rights is to direct the user login request to a specific network operating system (such as an NT domain) where the respective accounts contain the access privileges for each user. Other approaches include setting up a predefined tunnel from the remote access server to a remote VPN device, allowing only specific users (perhaps the network support personnel) to use higher speed ISDN connections, or automatically creating a Telnet session to a specific UNIX host.

Accounting systems allow the tracking of authentication and authorization transactions from the beginning to the end of a connection. Statistics about each connection are captured and available to network administrators. The information might be used for billing purposes, based on connect time or bytes transferred, or to enhance security—to prevent multiple user logins occurring from different locations due to sharing of username and password or theft.

Most of the remote access management schemes used today fit the client/server model. The remote access server to which the user dials in is typically the client (in spite of the remote access *server* name), which issues requests to the server containing the AAA database. In some cases, the AAA server might, in turn, issue a second request to another AAA server. The first server is called the *proxy* AAA server.

Server-based access management is a client/server architecture that allows all the security information to be located in a single, centralized database versus databases in all the remote access servers. In some cases, the security database is a distributed database located in several systems. The key is that changes to the database are done in a few systems, rather than the large number of remote access servers an ISP or large company might have. Figure 23-4 depicts a remote user login sequence using an AAA server.

1. Following is an explanation of the figure: An access request is sent from the user to the remote access server. This login request typically contains the username and password and is transferred from the user to the remote access server via one of several protocols, such as PAP or CHAP. Access to the remote access server can be accomplished via the switched dial-up network.

2. The remote access server builds an access request message, typically a TCP/IP or UDP/IP message, which gets forwarded through the intranet to the

authentication/authorization/accounting (AAA) server. The remote access server is previously configured with the Network layer address of one or more AAA servers.

3. If the AAA server is able to authenticate the user, it forwards a user profile to the remote access server. If the AAA server is unable to authenticate the user (e.g., the user is trying to use a restricted ISDN link), an Access Denied message would be sent back to the remote access server.

4. The remote access server parses the user access profile information and completes the login sequence to the user (e.g., the remaining steps of the PPP negotiation are completed). An example of the user access profile information might be information that directs the remote access server to establish a Telnet connection to a specific IP host.

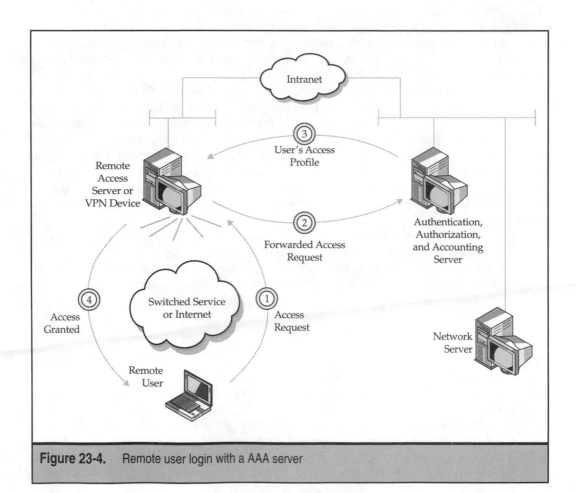

Figure 23-4.   Remote user login with a AAA server

The two most popular client/server protocols that authorization and accounting systems use are Remote Authentication Dial-In User Service (RADIUS) and Terminal Access Controller Access Control System+ (TACACS+).

RADIUS is a server-based AAA scheme developed by Livingston Enterprises, a router vendor in Pleasanton, California (now Lucent Technologies Remote Access Business Unit). RADIUS is an open protocol running over UDP, described in RFCs 2138 and 2139 (which make older specifications in RFCs 2058 and 2059 obsolete).

RADIUS has received significant support within the router and security communities, with products available from many companies, including ADC Kentrox, Ascend, Bay Networks (now Northern Telecom), Check Point, Cisco, IBM, Motorola, Raptor Systems, Secure Computing, Shiva, and 3Com. It is a scalable protocol, meaning that it can support a network comprising a large set of users. It has, therefore, also received wide support from many Internet service providers.

TACACS is another AAA system for remote network access. Developed by Cisco Systems, TACACS refers to a family of protocols. The original TACACS supported only authentication to a central server. A later, more robust version called Extended TACACS (XTACACS) added authorization and accounting features. RFC 1492 describes both TACACS and XTACACS.

The current flavor of the protocol is called TACACS Plus (TACACS+) and it is being promoted by Cisco in favor of the previous two versions. Originally a Cisco proprietary solution, TACACS+ has been submitted for approval as standard by the IETF. TACACS+ is conceptually similar to RADIUS, in that it provides AAA capability and supports an environment where a single or few TACACS+ servers can support many remote dial-in servers. Like RADIUS, TACACS+ is highly scalable, so that it can work with networks with very few or very many users. TACACS+, is arguably more mature than RADIUS, but it is not widely deployed nor supported; indeed, few vendors besides Cisco and Novell support TACACS+.

In a Cisco-only environment, however, TACACS+ is the recommended access management solution because it has a richer set of optional features than Cisco's implementation of RADIUS. Current implementations use TCP as a transport to ensure reliable delivery of the AAA messages, ostensibly providing a more secure and complete accounting log.

IP TELEPHONY

Initially, the attraction of IP telephony was reduced toll-call costs. Today, with long-distance calls relatively inexpensive, the savings driver is almost nonexistent. However, using Voice over IP (VoIP) is still extremely attractive as we continue our trend toward a totally converged network.

If all we did was duplicate the old phone services in this new world of IP, we'd actually be taking a step backward. Fortunately, our thinking about VoIP has matured since its debut in 1995. The industry has begun to realize that the true power of IP telephony isn't that it can replace existing PSTN services; instead, it can bring us new applications and services.

Traditional and IP Telephony Compared

In many ways, IP telephony is similar to traditional telephony. The goal is still to provide a connection across a network so people can communicate, and while the details might differ, both use digital encoding schemes, and both require some form of signaling. The big difference, of course, is whether you deliver the voice over a packet or a circuit network.

An IP voice call starts out in the same manner as a traditional phone call (as described in Chapter 10), with User A picking up the phone and being connected to their local exchange carrier (LEC). However, User A calls a VoIP gateway, which is a local call. The gateway answers and requests a user ID, which is entered via the telephone key pad. If authorized, a second dial tone is returned and the called party's number can now be entered. Buffering the phone number, the originating gateway must now find the appropriate terminating gateway. In many circumstances, gateway operators inform each other of their presence and keep a database of gateway IP addresses associated with local access and transport areas (LATAs) and exchange codes. Alternatively, the originating gateway could issue a multicast IP packet to locate a terminating gateway. Once located, the originating gateway sends its counterpart a series of IP packets to request and set up a connection at the destination. If accepted, the terminating gateway dials the called party and sends the DTMF signals to the LEC just as if it were the originating telephone.

Once User B answers the phone, active circuit switched connections exist between User A and the originating gateway and between the terminating gateway and User B. The digitized voice signal is sent through the network as a stream of IP packets routed to the destination gateway. The voice transmission between the line card on the LEC's switch and the respective gateway uses traditional PCM digitization, however the IP packet transport will likely employ one of the new low bandwidth compression schemes, so the gateways will have to perform any necessary conversions.

Components

Figure 23-5 shows a number of the components that can make up a VoIP network. Rather than depict a single solution, it shows the most common components and their possible utility in the network.

The central component of a VoIP network, or any other IP network for that matter, is a routed backbone. Many makes and models of routers can be purchased, with each manufacturer claiming that its approach to router design has some fundamental advantage over its competitors. Leaving comparative politics aside for the moment, let us observe that the faster the router the better, and that the routers in the core of the network must support any additional services that the IP services require. That might entail the integration of routing with lower layer switching, the support of special routing or reservation protocols, the ability to communicate and use QoS information of various types, or anything else that the client devices or gateways require.

At the edges of the routed "cloud" are the devices that connect the protocol-independent IP network to the telephony users who wish to use it: the VoIP gateways. The

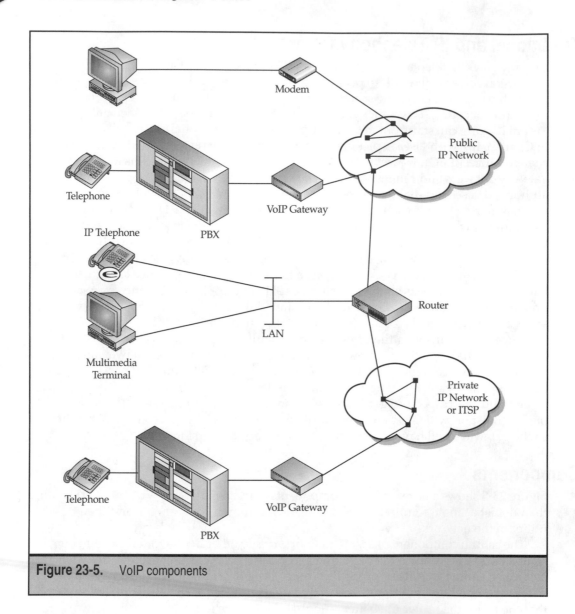

Figure 23-5. VoIP components

fundamental purpose of the gateway is to digitally encode and packetize the voice signal and send it to the desired destination IP address, where another gateway does the reverse, making the voice signal suitable for the phone system or private branch exchange (PBX) at the destination.

Finally, the user devices in most cases are the phone systems and PBXs that make up the customer premises equipment (CPE) of the traditional telephony network. The important exception to this is the presence of the multimedia-equipped personal computer, which can fill the gap between the roles of gateway and telephone system. In fact, with the addition of a relatively simple software package, most modern PCs have no trouble speaking to a VoIP gateway via the IP side of things, which can greatly extend the reach of IP telephony services.

Standards

Early implementations of IP voice were largely proprietary. Equipment from various vendors could not interoperate, meaning that all VoIP hardware and software had to be purchased from a single manufacturer. Today, companies are forming alliances to ensure that their equipment is interoperable, and the result is the emergence of standards. As IP telephony evolves, so will the standards and increasing adherence to standards. As we have learned from the ISDN experience, no standards, no acceptance.

Standards are needed for all parts of the IP telephony infrastructure. Today, we expect to be able to purchase a phone from any manufacturer, take it home, plug it in, and have it work. For IP telephony, standards are required inside the IP network, at the interface to customer premises equipment, and at the interface to the PSTN. Gateways must interoperate with Signaling System 7 (SS7); IP phones must support standard compression techniques; QoS implementations, as they emerge, must be based on standards.

H.323

Chief among the major standards that relate to IP telephony is the International Telecommunication Union-Telecommunication Standardization Sector (ITU-T) H.323 specification entitled Packet-Based Multimedia Communications Systems.

In addition to voice, H.323 addresses video and data conferencing functionality. In fact, many in the industry view H.323 as a videoconferencing specification and not as an IP telephony standard. This is ironic given that the only media an H.323 terminal must support is voice. H.323, then, is the internationally standardized way of doing VoIP.

The H.323 document is an "umbrella" specification as it calls upon and integrates numerous other standards. For example, H.323 relies on the H.225 protocol for ISDN-like (Integrated Services Digital Network) call signaling while the H.245 protocol controls the actual media streams flowing between the terminals. H.323 also calls upon the T.120 data conferencing protocols to support applications such as shared white boards. H.323 also references a rather comprehensive suite of audio and video codecs.

One other note about H.323 and IP telephony: The H.323 specification attempts to avoid naming a specific set of transport protocols. H.323 gives IP and Internetwork Packet Exchange (IPX) nearly identical treatment. In theory, any packet-based protocol that supports both reliable and unreliable modes of transmission could support H.323. In practical terms, however, the TCP/IP protocol suite seems to be getting all the attention when it comes to H.323 compliant products.

The IETF is also significant in the realm of VoIP as it is responsible for standardizing all things related to IP. In the area of real-time services over IP, the most significant IETF contributions are the Real-Time Transport Protocol (RTP) and the Resource Reservation Protocol (RSVP). The RTP Request for Comments (RFC) defines how real-time information can be transported over an IP network, while the RSVP specification provides the ability to reserve the bandwidth necessary for a given RTP session. The RTP protocol is a required part of H.225, and therefore, of the H.323 protocol suite.

H.323 also encompasses a variety of voice encoding standards, including G.711 (pulse code modulation, or PCM, 64Kbps), G.726 (adaptive differential pulse code modulation, or ADPCM, 32Kbps), and G.723.1 (algebraic code-excited linear prediction, or ACELP, 5.3Kbps). Video encoding standards include H.261 and H.263, both of which operate in the 28.8Kbps range.

Yet another significant VoIP related "standard" is the Voice Over IP Service Interoperability specification as put forth by the VoIP Forum. This specification enhances and clarifies the relevant VoIP standards (e.g., H.323 and RTP) to promote interoperability and value-added services. It should not surprise you that the VoIP Forum's Implementation Agreement (IA) mandates the use of TCP/IP as a transport protocol. Eliminating options that were left open in the ITU-T specifications is one way that these forum agreements help promote interoperability.

The gatekeeper and gateway are optional elements in H.323. The H.323 gatekeeper provides address translation, resolution, and admissions control services for the registered endpoints within its zone, which encompasses all terminals, gateways, and other devices that fall under its control. The H.323 gateway is the means to connect between packet-based VoIP devices and the circuit-switching network, using either ISDN or plain old telephone service (POTS). This includes the voice conversion to and from digital IP packets as well as signaling conversion between the two networks. If a call is set up between two VoIP devices (devices that adhere to the H.323 standard), no gateway services are required.

Finally, the multipoint control unit (MCU) is an optional H.323 endpoint device that facilitates communications between three or more terminals engaged in a multipoint conference. An MCU will always contain at least one multipoint controller (MC)—which facilitates the multipoint connections—and can contain one or more multipoint processors. While the MCU could be a stand-alone device, it can also be contained within an H.323 gateway or gatekeeper. In fact, a single physical device can house the MCU, gateway, and gatekeeper functions. A terminal device can have an MC that facilitates conference calls, but a terminal with an MC is not considered an MCU, since the MC within the terminal is not "callable."

Megaco/SIP

While not dead by any means, H.323 is becoming a legacy VoIP standard. Its complexity and ITU-centric approach to telephony has caused its star to fade over the past few years. A new breed of emerging standards has competed quite well with H.323, and one called MEGACO appears to be the leading contender today.

MEGACO is a work in progress by the IETF and is currently specified in RFC 2805. MEGACO is the result of integrating the earlier Simple Gateway Control Protocol (SGCP) and IP Device Control Protocol (IPDC) specifications into a single standard.

MEGACO describes a distributed set of gateways under control of one or more call agents (CA). At a minimum, a functional MGCP application will include at least one CA and a single media gateway. A signaling gateway is also required when MEGACO environment interconnects with an SS7-based network. This device conveys the SS7/Digital Subscriber Signaling System 1 (SS7/DSS1) signaling received from the circuit mode network to and from the CA. The signaling gateway and the CA are often co-located.

The CA's role is to provide call control intelligence that is physically removed from the packet to circuit mode translation functionality provided by the gateways. In other words, the CA is responsible for conducting signaling exchanges with gateways (MEGACO) and end-systems for either establishing or tearing down packet-to-packet or packet to circuit media connections.

The media gateway's role is to convert audio formats and transmission techniques between the conventional circuit switched network (CSN) and a packet-based VoIP network. Typical media gateway functions include the conversion of G.711 64Kbps Mu/A law speech, delivered as a Time Division Multiplexing (TDM) channel (DS0), to and from a low bit rate voice compression technique, which is then transported as a packet stream. The media gateway functionality may be distributed across multiple pieces of equipment, all of which are controlled by a single CA entity (which may also be distributed across multiple computing platforms). The CA communicates with the gateways using MEGACO. MEGACO is asymmetric in that the CA always issues commands while the gateways always issue responses.

It is important to note that to the outside world, the distributed set of gateways and the CA entity appear to be a single VoIP gateway. The CA can interface with the end user using several different VoIP signaling protocols. Interestingly enough, the H.323 standard can be used, though the industry has shifted its focus to the IETF's Session Initiation Protocol (SIP).

The IETF has produced an alternative multimedia terminal signaling protocol known as the Session Initiation Protocol (SIP). This protocol is defined in the IETF RFC 2543 and is seen by some as a competitor to the significantly more complex suite of H.323 protocols. SIP and MEGACO, in contrast, complement each other and interwork easily due to the use of the Session Description Protocol (SDP) in both specifications. In effect, the CA simply converts incoming connection requests (either SS7/ISUP or H.323) into the appropriate SIP verbs (invitation and ACK).

Some debate exists as to the relative merits of using the IETF's SIP as opposed to the ITU's H.323 for multimedia terminal signaling. On the one hand, H.323 is an international standard, is more mature, and is somewhat transport agnostic (i.e., one can find mention of IPX in the H.323 series of specifications). On the other hand, if it is to be IP-based telephony, then who better to standardize it than the IETF? Regardless of which terminal signaling approach the industry embraces, MEGACO enjoys a relative safe berth in that it can function in both environments.

In addition to being simpler and IP-centric, the SIP approach currently enjoys far better support for mobile users due to its client registration and server location functionality. SIP operates in a client/server fashion similar to other Internet protocols such as SMTP or HTTP. SIP consists of two parts: the user agent, which resides in the end system and supports multimedia sessions, and the network server part, which provides addresses resolution and user location functions. It should be noted that the user part typically contains both a client and server function as this will allow the terminal to both place and receive calls.

Applications

The idea of a single-cable, media-independent data/phone system has had some resurgence with the increasing acceptance of VoIP. In fact, at its simplest, a LAN-based PBX system is nothing more than a reversed VoIP gateway attached to the office Ethernet on one side, and the local loop or T-1 to the telephone company's CO on the other. Sound card–equipped PCs, either with or without attached handsets, can serve as "intelligent phones." Those without PCs or not wishing such a high degree of integration can use purpose-built "IP phones" that look like regular POTS handsets, with the exception of the Ethernet connector in the place of the regular phone plug. The network manager can remotely administer such phones, and the users might not even be aware that they are using a VoIP system.

Basic telephony is relatively simple to obtain with such a system; in fact, it is easier than over the wide area because of the greater degree of control over traffic and usage that an administrator has over the LAN as compared to the wide area network (WAN), or certainly, over the Internet. The fact that LAN bandwidth is almost always much more than that required by data traffic helps ensure good sound quality. Users simply pick up their phones or click the "make a phone call" button on their PC desktops, and the connection between the client and the gateway is established. The connection allows dialed phone numbers to be sent to the gateway, which then places the call for the caller, using either traditional or IP-based telephony.

The real payoff comes not simply with the convergence of voice and data on the LAN, but with the impending integration of them—that is, the intelligent routing of customer data to the user who is on the phone with that customer, the ease with which statistics can be kept on phone usage, the linking of automatic call distribution (ACD) functions with the data operations of a call center, and other such "next generation" applications. Much emphasis has been placed on trying to achieve a high level of integration between telephony and data systems, but the essential differences between the two have always posed a great deal of difficulty. The advent of IP telephony might allow this idea to live up to its promise.

This convergence of voice and data onto a single network has great, exciting ramifications for e-commerce. A major obstacle to Internet commerce sprang up around the issue of payment when the first retail Web sites appeared, because people were uncomfortable providing credit card information via an HTML form. Many early businesses asked shoppers to place their orders via the Web and then call in the credit card information via the telephone to alleviate those concerns. Similarly, while a well-designed Web site can provide an immense amount of information about a supplier's product or service, sometimes the sale can only be made by the intercession of a salesperson or customer service agent who can answer questions, provide advice, or just give the shopper that "taken care of" feeling—something that is difficult to do from a browser.

So the next generation of Web sites are those that integrate VoIP technology to allow the shopper to go beyond the graphics and text of HTML and click on a "talk to someone" button at the bottom of the page. Integrating voice services into the stream of IP data with the PC as communications device (rather than making the user call the company herself on the conventional telephone) means that the sales transaction can now move along as a multimedia experience, with the salesperson perhaps leading the customer through the site to areas of interest, making suggestions, and generally helping the customer through the rest of the order. Marketing professionals can barely contain their excitement at the thought of such a sales channel, as it combines the ability to provide large amounts of information to the customer with the warmth of personal assistance.

As a caveat, some issues still remain when replacing or augmenting traditional circuit-switched voice with IP telephony applications. For example, E-911 requires the phone system to report a caller's location, but that capability doesn't yet exist in most VoIP vendor implementations. VoIP also can have problems when traversing multiple tandem switches due to conversion from 64Kbps PCM to compressed forms used in packetized voice. While not insurmountable problems by any means, these two examples illustrate that moving to IP telephony is not a magic fix and does entail some difficulties. However, the potential boon of new applications has motivated a great many people to resolve these issues.

CARRIER NETWORKS

As the industry continues to work on QoS and security issues to make the Internet a more reliable medium for business, it has recognized that all components are interrelated. Service providers no longer can simply be "pipes" into the network but must move up the value chain and become one-stop shops for all communications needs. Technology convergence has led to service convergence, and carriers have deployed sophisticated networks to support it all.

One of the most important aspects of carriers is the high bandwidth provided, typically over fiber, and sometimes combined with QoS mechanisms. Backbones are primarily in the OCN (greater than DS-3) range of speeds, with some delivering OC-48 or even OC-192. Beware that providers can claim to have an OC-192 network but have only a small segment of the network actually operating at that speed.

Individually, providers have millions of miles of fiber in the ground in the United States and globally. ATM still finds a home in these backbones, but many carrier networks have already eliminated ATM and are running Packet-over-SONET, with or without MPLS. The next logical evolutionary step is some sort of Packet-over-Wave or Packet-over-Lamda technology.

With the need for more bandwidth, carriers have aggressively deployed Dense Wave Division Multiplexing (DWDM, covered in Chapter 10) technology. To better deliver services in the converged environment, a building trend away from multilayered networks to more streamlined, pure optical switching architectures has occurred. Current discussions about Generalized MPLS (GMPLS), which is sometimes called multiprotocol Lamda switching, revolve around wavelength management in optical switches. Within the next couple of years, we will see providers collapse the network from the traditional IP/ATM/SONET/DWDM architecture down to simply IP+GMPLS/DWDM.

An important aspect of QoS has nothing to do with bandwidth characteristics. Reliability is a critical QoS parameter, and carriers have deployed highly redundant networks to ensure that service will not be interrupted. Some minor differentiation between networks exist, but all offer redundancy in addition to latency and bandwidth guarantees as part of provider Service Level Agreements.

In a carrier network nodes typically contain a few high-speed backbone routers that connect to the backbone and several lower-speed routers that facilitate customer access to the network. In addition, at least one high-speed router for larger customers with DS-3 or greater connections usually is present. Just as in the backbone itself, these nodes have a high degree of redundancy, with multiple links between access routers and backbone routers and between the backbone routers. Traditionally this has involved mesh connectivity, though today technologies such as Cisco's Spatial Reuse Protocol (SRP) have been deployed to achieve the same level of redundancy while offering more efficient growth.

Refer to the "IP Carrier Network" blueprint for an example carrier network.

SUMMARY

Convergence of multiple applications onto a single IP-based network has finally arrived after many years of promises. It offers service providers a way to streamline their networks and deliver services in a more cost-efficient manner. While there is still much debate on the best approaches to offer QoS on these networks, and Internet security concerns have become paramount, the convergence trend continues as new applications are developed. VPNs and VoIP are only the beginning.

Our economy today requires all aspects of business to be automated and networked and demands efficient, cost-effective use of bandwidth. Originally, the advantage of convergence was somewhat mundane, limited to cost savings enabled by running voice and data over a common network, flexible billing, unified service ordering and provision, and so on. Today the vision is much more expansive as we can now provide a rich set of communications services and multimedia applications. From integrating TV and the Internet, to establishing virtual communities, the possibilities are constantly expanding in new and unimagined ways.

While the protocols behind the Internet have been in existence for more than two decades, the Internet has only recently become a common tool for businesses and consumers. Things we believed were radical and exciting a few years ago are now viewed as routine. We are at the beginning of a new era of communications that will have a profound impact on our lives. Just as the industrial revolution changed the world forever, the communications revolution has similarly changed the world. It is exciting to be at the dawn of this new age.

INDEX

B

Q

R

U

V

INTERNATIONAL CONTACT INFORMATION

AUSTRALIA
McGraw-Hill Book Company Australia Pty. Ltd.
TEL +61-2-9417-9899
FAX +61-2-9417-5687
http://www.mcgraw-hill.com.au
books-it_sydney@mcgraw-hill.com

CANADA
McGraw-Hill Ryerson Ltd.
TEL +905-430-5000
FAX +905-430-5020
http://www.mcgrawhill.ca

**GREECE, MIDDLE EAST,
NORTHERN AFRICA**
McGraw-Hill Hellas
TEL +30-1-656-0990-3-4
FAX +30-1-654-5525

MEXICO (Also serving Latin America)
McGraw-Hill Interamericana Editores S.A. de C.V.
TEL +525-117-1583
FAX +525-117-1589
http://www.mcgraw-hill.com.mx
fernando_castellanos@mcgraw-hill.com

SINGAPORE (Serving Asia)
McGraw-Hill Book Company
TEL +65-863-1580
FAX +65-862-3354
http://www.mcgraw-hill.com.sg
mghasia@mcgraw-hill.com

SOUTH AFRICA
McGraw-Hill South Africa
TEL +27-11-622-7512
FAX +27-11-622-9045
robyn_swanepoel@mcgraw-hill.com

**UNITED KINGDOM & EUROPE
(Excluding Southern Europe)**
McGraw-Hill Education Europe
TEL +44-1-628-502500
FAX +44-1-628-770224
http://www.mcgraw-hill.co.uk
computing_neurope@mcgraw-hill.com

ALL OTHER INQUIRIES Contact:
Osborne/McGraw-Hill
TEL +1-510-549-6600
FAX +1-510-883-7600
http://www.osborne.com
omg_international@mcgraw-hill.com

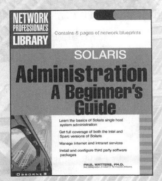
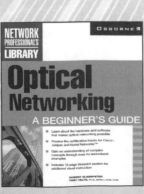

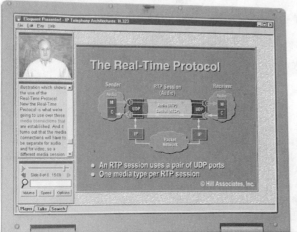